METHODS IN MOLECULAR BIOLOGY

Series Editor
John M. Walker
School of Life and Medical Sciences
University of Hertfordshire
Hatfield, Hertfordshire, AL10 9AB, UK

For further volumes:
http://www.springer.com/series/7651

Functional Genomics in *Medicago truncatula*

Methods and Protocols

Edited by

Luis A. Cañas and José Pío Beltrán

CSIC-UPV, Institute for Plant Cell and Molecular Biology (IBMCP), Valencia, Spain

Humana Press

Editors
Luis A. Cañas
CSIC-UPV
Institute for Plant Cell and Molecular Biology
(IBMCP)
Valencia, Spain

José Pío Beltrán
CSIC-UPV
Institute for Plant Cell and Molecular Biology
(IBMCP)
Valencia, Spain

ISSN 1064-3745 ISSN 1940-6029 (electronic)
Methods in Molecular Biology
ISBN 978-1-4939-8632-3 ISBN 978-1-4939-8633-0 (eBook)
https://doi.org/10.1007/978-1-4939-8633-0

Library of Congress Control Number: 2018946782

Printed on acid-free paper

This Humana Press imprint is published by the registered company Springer Science+Business Media, LLC part of Springer Nature.
The registered company address is: 233 Spring Street, New York, NY 10013, U.S.A.

Preface

Functional genomics in *Medicago truncatula* has been proved to be useful for genetic studies to resolve specific developmental or metabolic processes that are legume-specific or shown important differences with those described in *Arabidopsis thaliana*, the most studied model plant. Recent advances in the understanding of nitrogen fixation, nodulation, compound leaf, inflorescence and plant architecture, floral development, pod formation, secondary metabolites biosynthesis, and other developmental and metabolic aspects have been achieved based on the analysis of different *M. truncatula* mutants. *M. truncatula* was proposed in the 1990s as a model system due to their key attributes: diploid and small genome size, autogamous nature, short generation time, and can be readily transformed. Currently, the genome sequence of this legume species is essentially complete and a series of functional genomics tools have been successfully developed and applied. Mutagens that cause insertions or deletions are being used in this model system because these kinds of DNA rearrangements are expected to assist in the isolation of the corresponding genes by Target Induced Local Lesions IN Genomes (TILLING) approaches. Different *M. truncatula* mutants have been obtained following γ-irradiation or fast neutron bombardment (FNB), ethyl-nitrosourea (ENU) or ethyl methanesulfonate (EMS) treatments, T-DNA and activation tagging, use of the tobacco retrotransposon *Tnt1* to produce insertional mutants, stable gene silencing by RNAi, and transient post-transcriptional gene silencing by virus-induced gene silencing (VIGS). Emerging technologies of targeted mutagenesis and gene editing, such as the CRISPR-*Cas9* system, could open a new era in this field because crops created through genome engineering might prove to be more acceptable to the consumers than plants that carry foreign DNA in their genomes. This book contains valuable information on the most used methods and protocols to achieve different types of mutagenesis and reverse/forward genetics technologies in *M. truncatula*. Several reviews on the specific use of these techniques in functional studies on the genetic control of important developmental and metabolic processes in this model legume are also included.

Valencia, Spain

Luis A. Cañas
José Pío Beltrán

Contents

Contributors

GREGOIRE AUBERT • *INRA, University of Bourgogne Franche-Comté, Agroécologie, AgroSup Dijon, Dijon, France*

RAJEEV AZAD • *Department of Biological Sciences and BioDiscovery Institute, University of North Texas, Denton, TX, USA; Department of Mathematics, University of North Texas, Denton, TX, USA*

JÉRÉMIE BAZIN • *Institute of Plant Sciences Paris-Saclay, IPS2, Univ. Paris-Diderot, CNRS, INRA, Univ. Paris-Sud, Univ. Evry, Sorbonne Paris-Cité, University of Paris-Saclay, Orsay, France*

JOSÉ PÍO BELTRÁN • *CSIC-UPV, Institute for Plant Cell and Molecular Biology (IBMCP), Valencia, Spain*

FATHI BERRABAH • *Department of Biology, Faculty of Nature and Life Science, University of Ziane Achour, Djelfa, Algeria*

TON BISSELING • *The Laboratory of Molecular Biology, Department of Plant Sciences, Wageningen University, Wageningen, The Netherlands*

DAVID BURKS • *Department of Biological Sciences and BioDiscovery Institute, University of North Texas, Denton, TX, USA*

LUIS A. CAÑAS • *CSIC-UPV, Institute for Plant Cell and Molecular Biology (IBMCP), Valencia, Spain*

ORNELLA CALDERINI • *CNR, National Council of Researches, Institute of Biosciences and Bioresources, Perugia, Italy*

YUHUI CHEN • *School of Life Sciences, Lanzhou University, Lanzhou, China*

RUJIN CHEN • *School of Life Sciences, Lanzhou University, Lanzhou, China*

MARTIN CRESPI • *Institute of Plant Sciences Paris-Saclay, IPS2, Univ. Paris-Diderot, CNRS, INRA, Univ. Paris-Sud, Univ. Evry, Sorbonne Paris-Cité, University of Paris-Saclay, Orsay, France*

SHAUN J. CURTIN • *United States Department of Agriculture, Plant Science Research Unit, St Paul, MN, USA*

BRIGITTE DARCHY • *INRA, University of Bourgogne Franche-Comté, Agroécologie, AgroSup Dijon, Dijon, France*

REBECCA DICKSTEIN • *Department of Biological Sciences and BioDiscovery Institute, University of North Texas, Denton, TX, USA*

RICHARD A. DIXON • *BioDiscovery Institute and Department of Biological Sciences, University of North Texas, Denton, TX, USA*

ADEL M. ELMAGHRABI • *Biotechnology Research Center (BTRC), Tripoli, Libya; School of Biosciences, Cardiff University, Cardiff, UK*

CRISTINA FERRÁNDIZ • *Instituto de Biología Molecular y Celular de Plantas (CSIC-UPV), Ciudad Politécnica de la Innovación, Valencia, Spain*

DENNIS FRANCIS • *School of Biosciences, Cardiff University, Cardiff, UK*

CAROLIEN FRANKEN • *The Laboratory of Molecular Biology, Department of Plant Sciences, Wageningen University, Wageningen, The Netherlands*

CONCEPCIÓN GÓMEZ-MENA • *Instituto de Biología Molecular y Celular de Plantas (CSIC-UPV), Ciudad Politécnica de la Innovación Edf. 8E, Valencia, Spain*
KARINE GALLARDO • *INRA, University of Bourgogne Franche-Comté, Agroécologie, AgroSup Dijon, Dijon, France*
MARIE GARMIER • *Institute of Plant Sciences Paris-Saclay IPS2, CNRS, INRA, Université Paris-Sud, Université Evry, Université Paris-Saclay, Orsay, France; Institute of Plant Sciences Paris-Saclay IPS2, Paris Diderot, Sorbonne Paris-Cité, Orsay, France*
CHAN MAN HA • *BioDiscovery Institute and Department of Biological Sciences, University of North Texas, Denton, TX, USA*
CAROLINE HARTMANN • *Institute of Plant Sciences Paris-Saclay, IPS2, Univ. Paris-Diderot, CNRS, INRA, Univ. Paris-Sud, Univ. Evry, Sorbonne Paris-Cité, University of Paris-Saclay, Orsay, France*
IDA ELISABETH JOHANSEN • *Plant and Environmental Sciences, University of Copenhagen, Copenhagen, Denmark*
OLGA KULIKOVA • *The Laboratory of Molecular Biology, Department of Plant Sciences, Wageningen University, Wageningen, The Netherlands*
HEE-KYUNG LEE • *Noble Research Institute, Ardmore, OK, USA*
CHRISTINE LELANDAIS-BRIÈRE • *Institute of Plant Sciences Paris-Saclay, IPS2, Univ. Paris-Diderot, CNRS, INRA, Univ. Paris-Sud, Univ. Evry, Sorbonne Paris-Cité, University of Paris-Saclay, Orsay, France*
CHENGGANG LIU • *BioDiscovery Institute and Department of Biological Sciences, University of North Texas, Denton, TX, USA*
OLE SØGAARD LUND • *Plant and Environmental Sciences, University of Copenhagen, Copenhagen, Denmark*
RICHARD C. MACKNIGHT • *Department of Biochemistry, University of Otago, Dunedin, New Zealand*
KIRANKUMAR S. MYSORE • *Noble Research Institute, Ardmore, OK, USA*
MÉLANIE NOGUERO • *INRA, University of Bourgogne Franche-Comté, Agroécologie, AgroSup Dijon, Dijon, France*
SERGIO OCHATT • *Agroécologie, AgroSup Dijon, INRA, University of Bourgogne Franche-Comté, Dijon, France*
FRANCESCO PANARA • *ENEA, Italian National Agency for New Technologies, Energy and Sustainable Economic Development, Trisaia Research Center, Matera, Italy*
ANDREA PORCEDDU • *Department of Agriculture, University of Sassari, Sassari, Italy*
HÉLÈNE PROUST • *Institute of Plant Sciences Paris-Saclay, IPS2, Univ. Paris-Diderot, CNRS, INRA, Univ. Paris-Sud, Univ. Evry, Sorbonne Paris-Cité, University of Paris-Saclay, Orsay, France*
PASCAL RATET • *Institute of Plant Sciences Paris-Saclay IPS2, CNRS, INRA, Université Paris-Sud, Université Evry, Université Paris-Saclay, Orsay, France; Institute of Plant Sciences Paris-Saclay IPS2, Paris Diderot, Sorbonne Paris-Cité, Orsay, France*
HILARY J. ROGERS • *School of Biosciences, Cardiff University, Cardiff, UK*
EDELÍN M. ROQUE • *Instituto de Biología Molecular y Celular de Plantas (CSIC-UPV), Ciudad Politécnica de la Innovación Edf. 8E, Valencia, Spain*
MYRIAM SÁNCHEZ • *INRA, University of Bourgogne Franche-Comté, Agroécologie, AgroSup Dijon, Dijon, France*

EL HOSSEYN AIT SALEM • *Institute of Plant Sciences Paris-Saclay IPS2, CNRS, INRA, Université Paris-Sud, Université Evry, Université Paris-Saclay, Orsay, France; Institute of Plant Sciences Paris-Saclay IPS2, Paris Diderot, Sorbonne Paris-Cité, Orsay, France*
JOANNA SERWATOWSKA • *Unidad de Genómica Avanzada (LANGEBIO), CINVESTAV-IPN, Guanajuato, Mexico*
CHRISTINE LE SIGNOR • *INRA, University of Bourgogne Franche-Comté, Agroécologie, AgroSup Dijon, Dijon, France*
CÉLINE SORIN • *Institute of Plant Sciences Paris-Saclay, IPS2, Univ. Paris-Diderot, CNRS, INRA, Univ. Paris-Sud, Univ. Evry, Sorbonne Paris-Cité, University of Paris-Saclay, Orsay, France*
RICHARD D. THOMPSON • *INRA, University of Bourgogne Franche-Comté, Agroécologie, AgroSup Dijon, Dijon, France*
VANESSA VERNOUD • *Agroécologie, AgroSup Dijon, INRA, University of Bourgogne Franche-Comté, Dijon, France*
JAMES L. WELLER • *School of Natural Sciences, University of Tasmania, Hobart, TAS, Australia*
JIANGQI WEN • *Noble Research Institute, Ardmore, OK, USA*

Chapter 1

Grain and Forage Legumes: Nutritional Value and Agriculture Sustainability

José Pío Beltrán and Luis A. Cañas

Abstract

Humanity faces great challenges with respect to the use of energy, the production of food and feed, and the management of the Earth through sustainable practices. Agriculture can play a key role to give appropriate responses to these challenges. By the end of this century, human population will grow up to around 10,000 million people, meaning we must be able to produce food and feed for more than an additional number of 3300 million people. Legumes together with cereals have been combined to produce healthy food along the history of agriculture in all geographical areas of the planet. However, recently, the use of legumes, mainly in the developed countries, has been neglected therefore compromising human health and sustainable production of food and feed. Agronomy has always been driven by technology and innovation. The development of genomic tools in legume model systems such as *Medicago truncatula* will allow to make progress into the knowledge of critical processes of legumes biology such as nitrogen fixation, including the mechanisms controlling nodulation through soil nitrogen sensing, drought and flooding tolerances or the understanding of key factors governing the vegetative development of legumes, the control of inflorescences architecture or floral transition, and fruit set and seed development and composition. Traditional breeding combined with genome editing techniques will drive the production of grain and forage legume varieties for the future.

Key words Legumes, Nutritional value, Agriculture sustainability, *M. truncatula*, Genomic tools

1 Feeding the World in the Twenty-First Century: In the Search of Food Security

Human population is continuously increasing, thanks to its ability to obtain food and feed. Although its speed of growth is already declining, it is expected that global human population will increase from the current 7500 to 9000 million people by 2050 and over 10,000 million by the end of the century [1]. Today, there are still more than 800 million people suffering from hunger or malnutrition; therefore, at the end of the century, we should be able to procure additional food to more than 3300 million people.

According to the Food and Agriculture Organization of the United Nations (FAO), food security exists when all people, at all

Luis A. Cañas and José Pío Beltrán (eds.), *Functional genomics in* Medicago truncatula*: Methods and Protocols*, Methods in Molecular Biology, vol. 1822, https://doi.org/10.1007/978-1-4939-8633-0_1,

times, have physical and economic access to sufficient, safe, and nutritious food to meet their dietary needs and food preferences for an active and healthy life. This constitutes a great challenge that requires to consider food and feed production, energy use, and management of natural resources at the same time when we demand our agri-food sector to react in a competitive and sustainable manner. José Graziano da Silva, General Director of FAO, calculates that to reach food security by 2050, we should be able to increase food production by 60% and that with the available technologies, an increase of energy consumption of 50% and of 40% of water will be required. Despite legume crops might offer lower yields than cereals, farmers frequently include legumes in crop rotations due to their capacity to fix atmospheric nitrogen. We foresight that a sustainable agriculture to feed the world will not be possible unless both pulse and forage legumes play a key role. Consistently, FAO supported the International Year of Pulses in 2016 due to their potential to contribute to hunger and malnutrition eradication by means of healthy and diverse diets taking advantage at the same time of the ability of legumes to the maintenance of soil structure and to replenish soil fertility when used in crop rotations [2].

2 Agriculture Sustainability

Climate change is already affecting agriculture. It is expected that there will be a drop in crop yields due to climatic stress factors such as fluctuations in temperature, drought, flooding, and salinity; therefore, the main recommendations of the European Plant Science Organization (EPSO, *see* www.epsoweb.org/) include among the most innovative areas in the plant sector in the coming decades the improvement of resource use efficiency and stewardship, the enhancement of yield and yield stability for increased resilience in dynamic environments, the improvement of plant health for resilient production, and the development of plants with improved composition for human and animal nutrition and health, and to accomplish these aims, sustainable strategies have to be applied. Legume crops could greatly contribute in all these aspects; therefore, actions are necessary to increase genomic resources such as functional genomics and use innovative plant breeding techniques to improve the yield and nutritional quality of crops as well as to obtain new varieties of legumes with enhanced resilience to climate change.

The regulation of floral transition and reproductive development in legumes including processes such as pollination, early fruit set, and seed development are key factors for fruit and seed yields as well as forage production. Inflorescence architecture conditions flower and fruit formation and therefore fruit yield (Fig. 1a–k). The molecular and genetic analysis of floral development of model

Fig. 1 Morphological diversity of inflorescences, flowers, and fruits in legumes. Subfamily Papilionaceae: *Pisum sativum* flower (**a**) and *Medicago truncatula* flower (**b**). Subfamily Mimosoideae: *Acacia dealbata* inflorescence (**c**) with a great number of stamens in their flowers. Subfamily Caesalpinioideae: *Ceratonia siliqua* (carob tree) inflorescence, male flowers with five stamens (**d**), female flowers with carpels but without developed stamens (**e**), hermaphrodite flowers with both stamens and carpels (**f**). (**g**) Open pea pod showing the seeds. (**h**) Helical and prickly pod of *M. truncatula.* (**i**) Pod and seeds of acacia. (**j**) Immature pod of carob tree. (**k**) Mature pods of carob tree showing the seeds

plants such as *Antirrhinum majus* or *Arabidopsis thaliana* uncovered the MADS-box family of transcription factors controlling key aspects of floral transition and floral organ identity. The understanding of these processes in legumes could allow us to combine biotechnological strategies to plant breeding to obtain, for example, a forage legume such as alfalfa remaining vegetative producing only shoots and leaves therefore avoiding the use of energy to produce undesirable flowers and fruits. However the inflorescences of *Arabidopsis* and *Antirrhinum* are simple, and flowers are formed directly from the primary inflorescence meristem. Legumes have compound inflorescence architecture with flowers developing from secondary or even higher-order inflorescence meristems. The formation of secondary inflorescences suggested the existence of a specific genetic function in legumes. We showed that this function in pea is under the control of the *vegetative1* gen [3]. Plants with loss of function of *veg1* display a phenotype in which secondary inflorescences are substituted by vegetative branches with the consequence that plants are not able to produce any flower, and they remain vegetative. *Veg1* and its orthologous genes in other legume species can be used to manipulate floral transition a developmental key factor in forage legumes (Fig. 2a, b).

Nitrogen and water irrigation are limiting factors to crop production. Plants need soils containing enough nitrogen that once it has been used, can be replenished by means of both organic and inorganic fertilizers. Current agricultural practices are contributing to climate change by releasing greenhouse gases such as nitrous oxide derived mainly from nitrogen fertilizers or methane produced by intensive livestock farming. To avoid putative environmental

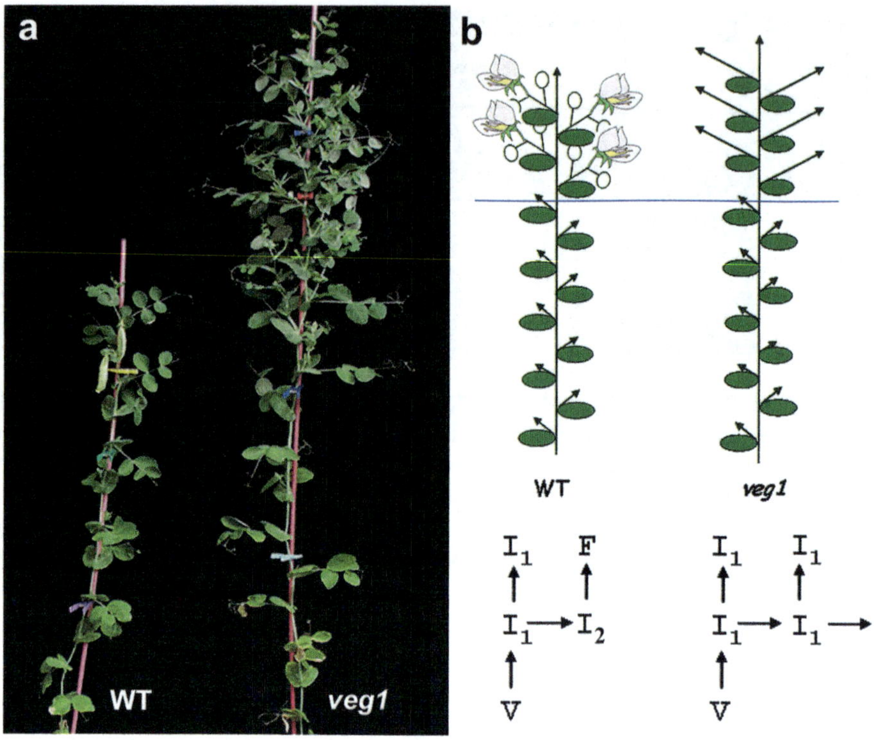

Fig. 2 Wild-type pea plant compared with a *vegetative1* mutant. (**a**) Phenotype of a wild-type (WT) pea plant compared with a *vegetative1* (*veg1*) mutant which shows an intense vegetative growth and absence of flowers. (**b**) Schematic representation of the architecture of a WT pea plant compared with a *veg1* mutant showing the transition from the vegetative meristem (V) to the floral meristem (blue line). In the mutant, a series of primary inflorescence meristems (I_1) are formed but no secondary inflorescence meristems (I_2) and flowers (F)

nondesired effects derived from energy consumption or release of gases with greenhouse effect, it is advantageous to cultivate legumes that in symbiotic association to soil nitrogen-fixing bacteria such as rhizobium have the ability to incorporate atmospheric nitrogen. One limitation for the use of legumes as a source of nitrogen input into agricultural systems is the suppression of nitrogen-fixing nodules when soils are satiated with nitrogen. The molecular characterization of the regulatory controls allowing to break the link between the amount of nitrogen in soils and nodule formation needs to be addressed [4]. The legume-rhizobia interaction leading to nodule formation is mediated by a complex molecular signal exchange that needs to be untangled. To invade plant roots, bacteria might enter via root hairs or by means of intercellular invasion that has been observed in many legumes. Major advances have been produced in the understanding of the intracellular pathway and nodule formation studying the interaction between rhizobia and the model legumes *Lotus japonicus* and *Medicago truncatula*. The characterization of the mechanism of the intercellular root infection leading to nodule development could provide new insights to modify key steps of nitrogen-fixing nodule formation to eventually engineer this ability into major nonlegume crops [5]. Climate change is also producing significant modifications in environmental temperatures, precipitation patterns, and occurrence of drought in diverse agricultural areas. Droughts, floods, and extreme temperatures have caused important decreases in crop production over the last decades [6]. Screening for natural genetic variation in drought and flooding tolerance traits including root architecture, water use, nitrogen fixation, and yields is helping to identify the best resources for genetic improvement in legumes such as soybean [7]. Pulses and forage legumes are fundamental crops to develop a sustainable agriculture able to reach global healthy food security. Actions to be taken include obtaining new varieties with the capacity to adapt to changes in environmental conditions. To this purpose we need to better understand critical biological processes such as efficient nitrogen fixation, legume vegetative development, control of inflorescence architecture and floral transition, and fruit set and seed development.

Time is running out if we consider that till our days to obtain a new variety with an interesting trait and to put it in the market takes us at least a decade. According to the great diversity of legumes and considering that materials in seed banks available were developed in many cases without considering the changes in climate change of today, it is necessary to combine individual legume studies including both the most used in agriculture and the so-called orphan legumes and the development of legume model systems such as that of *Medicago truncatula*, especially to design tools for functional genomics with the aim of understanding basic biological processes and to develop new strategies for legume improvement.

3 Nutritional Value

The use of legumes can be observed associated with cereal crops since the origins of agriculture. Appropriated diets require proteins, carbohydrates, and lipids besides different minerals and vitamins. Pulse legumes produce seeds of high nutritional value (www.ndb.nal.usda.gov/ndb). It is interesting to note that protein contents in legumes are double or triple than those of cereals. For example, 100 g of seeds contain between 9 and 29 g of proteins in chickpeas, between 14 and 36 g in peas, and from 19 to 39 g in beans, lentils, or faba beans, while lupines and soybean vary between 26 and 55 g [8]. Specifically, humans need food supplying all essential amino acids. Those can be obtained from animal-derived foods. However, cereal proteins are deficient in lysine and threonine, while proteins from legumes do not supply the required amounts of sulfur-containing amino acids (methionine and cysteine) and tryptophan. On the other hand, legume seeds are rich in lysine and other valuable amino acids. Therefore, along the history of agronomy, cereals and legumes were cultivated together in very remote geographical areas in order to combine their crops to obtain nutritious food: wheat, barley, oat, and rye combined with pea, faba bean, or chickpea at the Mediterranean basin, corn with beans in Central America, yam with beans or peanuts in South America, sorghum and millet with cowpea in Central Africa, or rice together with several *Vigna* species in Asia. Combined legume and cereal proteins have similar biological value to animal proteins. Besides, legume proteins do not contain gluten although the presence of proteinase inhibitors might affect in some cases protein digestibility. Legumes used as crops have low lipid contents between 1 and 7% per dry weight of seeds with the exception of peanuts, lupins, and soybeans that have a high content of unsaturated fatty acids, mainly oleic, linoleic, and linolenic acids. Legumes are also an important source of carbohydrates mainly starch and raffinose oligosaccharides. Up to 36% of the seed dry weight may be starch with variable composition in amylose and amylopectin depending on the legume. These complex carbohydrates are slowly absorbed when digested therefore are useful in the control of the blood levels of glucose and insulin. Starch together with fiber levels in diets helps to control lipidic metabolism. Fiber content in pulses varies between 3 and 36 g per 100 g of seed dry weight. Other nutritional values of grain legumes are related to their relative high content in minerals as Fe, Mg, and Zn or in vitamins as thiamin, niacin, and folic acid (www.ndb.nal.usda.gov/ndb).

According to the World Health Organization (WHO), the projected deaths by major causes considering all World Bank income groups and all ages can be classified as derived from communicable diseases, maternal and perinatal conditions, and nutritional deficiencies or from chronic diseases or from injuries. Except in the low-income countries where the major death cause will continue to be the group of communicable diseases, in all other countries all over the world, the projected causes of death will be derived from chronic diseases such as obesity, diabetes, cardiovascular diseases, or cancer [9]. Major causes originating chronic diseases are smoking, unhealthy eating, obesity, and physical inactivity. Here we have an opportunity to contribute to prevent chronic diseases by means of healthy diets. Regular consumption of pulse seeds may provide some health benefits, particularly to prevent hypercholesterolemia and hypertension. The hypocholesterolemic effect of soybean protein has been supported by the US Food and Drug Administration in 1999. Data available from diverse studies using animal models and human trials suggest that this beneficial activity can be associated with diets including grain legumes, although it is necessary to elucidate both the identity and the mode of action of the active components in the seeds [10]. Bowman-Birk inhibitors (BBI) from legumes, such as soybean, pea, lentil, and chickpea, have potential health-promoting properties within the gastrointestinal tract. These potential pharmacological benefits have been linked to their intrinsic ability to inhibit serine proteases and both trypsin- and chymotrypsin-like proteases that could be involved in carcinogenesis. The elucidation of the mode of action of BBI would provide insights into their potential as colorectal chemopreventive agents, providing a strong base for the development of legume crops and their products as health-promoting food [11]. Since the 1970s, the proportion of overweight and obese people has grown at an alarming rate in many countries. According to the WHO in 2016, more than 1900 million people were overweight, while 650 million were obese as a consequence of eating too much or bad eating habits. Obesity consists in a high risk of suffering diabetes and other chronic diseases. Prevention of excess weight gain can be accomplished with relatively small changes in lifestyle behaviors to control body weight. Legumes can be a valuable food by which the needs of the undernourished or populations suffering hunger could be met. They can be incorporated into meat products, such as sausages and burgers, to lower the energy density of these foods while providing important nutrients. Replacing energy-dense foods with pulse-derived substances has been shown to have beneficial effects on the prevention and management of obesity and related disorders, such as cardiovascular disease, diabetes, and metabolic syndrome [12].

Finally, forage legumes are of capital importance in feeding animals. Concentration and chemical constituents of lignin are factors that may influence cell wall digestibility of forages. Proper animal nutrition greatly depends on adequate combinations of fiber and proteins. The apparent inhibition of CW digestion by lignin was much greater in grasses when alfalfa, bird's-foot trefoil, smooth bromegrass orchard grass, and red clover feedstuffs are compared [13]. This is especially relevant to feed ruminants where the pregastric fermentation in the rumen by the microbiota facilitates the digestion of the cell wall components. However, besides the positive high proteic content in forage legumes, the presence of other substances like proteinase inhibitors, phytic acid, tannins, or lectins has to be considered. Much research is needed to better understand the risk-benefit factors that represent these compounds. For example, it has been proposed that proper amounts of tannins in ruminant diets could diminish the risk of pasture bloat, and efforts to engineer proanthocyanidin biosynthesis in plants are currently being carried out [14, 15]. On the contrary, an excess of tannins could inhibit the digestion and decrease the palatability of the feedstuff.

4 Genetic Improvement of Legumes: Molecular Breeding and Functional Genomic Tools

Grain and forage legumes are known to be an important source of proteins and other nutrients both for human food and animal feed. The variability in seed protein content among different varieties of legumes is mainly due to their genomes and to differences in agronomical growth conditions. In addition, several anti-nutritional substances have been found in some legume crops. Their importance varies according to their amount, their effects (factors with toxic effects or reducing protein digestibility), and with the possibility to be reduced or eliminated in the different stages of food processing. To improve the content of limiting amino acids in legumes as well as other important agronomical and nutritional traits by conventional breeding programs seems very difficult. We have to maximize the efforts toward developing high-yielding and nutritious legume cultivars possessing resistance/tolerance to the major stresses especially in the context of climate change. Although conventional breeding approaches have been successful to address the issue of low productivity in several legume crops, this is not happening at the desired success rate. Therefore, it is essential to intensify legume genetic enhancement programs using advanced breeding approaches wherein the potential of genomics needs to be exploited for accelerated development of improved cultivars possessing high yield, genetic resilience against stresses, and enhanced nutritional quality [16].

To design an effective breeding program for a legume species based on biotechnological approaches, it is required a good knowledge of the legume genome and the availability of functional genomic tools to facilitate the generation of mutant phenotypes. Despite numerous genome-wide association studies that have already been performed, the challenge to determine the function of genes, gene products, and their interaction is still open. Functional genomics aims to discover the biological function of particular genes and to uncover how sets of genes and their products work together. Transgenic plants are proving to be powerful tools to study various aspects of plant sciences. The emerging scientific revolution sparked by genomic-based technologies is producing enormous amounts of DNA sequence information that, together with plant transformation methodologies, is opening up new experimental opportunities for functional genomics analysis. The most conventional approach to the analysis of gene function is loss-of-function mutagenesis by chemicals or fast neutrons that introduce random mutations or deletions in the genome. Transferred DNA (T-DNA) tagging or transposon tagging methods were developed to generate loss-of-function mutations because these tag sequences can be used to identify the genes disrupted by these elements. Gain-of-function approaches have been used as an alternative or complementary method to loss-of-function approaches as well as to confer new functions to plants. Gain-of-function is achieved by increasing gene expression levels through the random activation of endogenous genes by transcriptional enhancers or by the expression of individual transgenes under the control of a strong promoter.

Different functional genetic tools, such as chemical and physical mutagenesis, reverse genetics approaches using T-DNA or transposon insertional mutagenesis, transient and stable posttranscriptional gene silencing by VIGS or RNAi, the development of TILLING platforms, etc., have been successfully applied to gene function studies related to different developmental or metabolic processes in legumes. Emerging technologies of targeted mutagenesis and genome editing, such as the CRISPR/Cas9 system, are opening a new era in this field. Recently, the applications of CRISPR/Cas9 have been focused toward the study of SNF in legumes, which should greatly advance our understanding of the basic mechanisms underpinning the legume-rhizobia interactions and guide the engineering of the SNF pathway into nonlegume crops to reduce the dependence on the use of nitrogen fertilizers [17]. An optimized agrobacterium-delivering CRISPR/Cas9 system has been developed in forage legumes which could successfully induce targeted genome modifications in *Medicago truncatula*. By using this system, monoallelic and biallelic homozygous mutants for *MtPDS* were obtained in the T0 generation, which suggested that such strategy for vector construction is amenable, and this CRISPR/Cas9 system could serve as an effective tool to facilitate studies of gene function in *M. truncatula*. Further study will extend the range of applications of CRISPR/Cas9 to cultivated legume forages with more complicated and larger genomes such as alfalfa and clover, which will benefit and speed up the forage legume improvement [18]. In conclusion, future advances in legume crops to produce more food and feed contributing to a sustainable agriculture will require synergy among several fields of research, including traditional breeding, genetics, physiology, biotechnology, and crop management.

Acknowledgments

We gratefully acknowledge funding from the Spanish Ministry of Economy and Competitiveness (http://www.idi.mineco.gob.es/portal/site/MICINN) grant BIO2016-75485-R.

References

1. Gerland P, Raftery AE, Ševcíková H, Li N, Gu D, Spoorenberg T et al (2014) World population stabilization unlikely this century. Science 346:234–237
2. Foyer CH, Lam HM, Nguyen HT, Siddique KH, Varshney RK, Colmer TD, Cowling W, Bramley H, Mori TA, Hodgson JM et al (2016) Neglecting legumes has compromised human health and sustainable food production. Nat Plants 2:16112
3. Berbel A, Ferrandiz C, Hecht V, Dalmais M, Lund OS, Sussmilch FC, Taylor SA, Bendahmane A, Ellis TH, Beltran JP et al (2012) *VEGETATIVE1* is essential for development of the compound inflorescence in pea. Nat Commun 3:797–804
4. Murray JD, Liu CW, Chen Y, Miller AJ (2017) Nitrogen sensing in legumes. J Exp Bot 68:1919–1926
5. Ibanez F, Wall L, Fabra A (2017) Starting points in plant-bacteria nitrogen-fixing symbioses: intercellular invasion of the roots. J Exp Bot 68:1905–1918
6. Lesk C, Rowhani P, Ramankutty N (2016) Influence of extreme weather disasters on global crop production. Nature 529:84–87
7. Valliyodan B, Ye H, Song L, Murphy M, Shannon JG, Nguyen HT (2017) Genetic diversity and genomic strategies for improving drought and waterlogging tolerance in soybeans. J Exp Bot 68:1835–1849
8. Iqbal A, Khalil IA, Ateeq N, Sayyar Khan M (2006) Nutritional quality of important food legumes. Food Chem 97:331–335
9. Daar AS, Singer PA, Persad DL, Pramming SK, Matthews DR, Beaglehole R, Bernstein A, Borysiewicz LK, Colagiuri S, Ganguly N et al (2007) Grand challenges in chronic non-communicable diseases. Nature 450: 494–496
10. Arnoldi A, Zanoni C, Lammi C, Boschin G (2015) The role of grain legumes in the prevention of hypercholesterolemia and hypertension. Crit Rev Plant Sci 34:144–168
11. Clemente A, Sonnante G, Domoney C (2011) Bowman-Birk inhibitors from legumes and human gastrointestinal health: current status and perspectives. Curr Protein Pept Sci 12:358–373
12. Rebello CJ, Greenway FL, Finley JW (2014) A review of the nutritional value of legumes and their effects on obesity and its related co-morbidities. Obes Rev 15:392–407
13. Buxton DR, Russell JR (1988) Lignin constituents and Cell-Wall digestibility of grass and legume stems. Crop Sci 28:553–558
14. Cañas LA, Fresquet S, Roque E, Rochina MC, Beltrán JP (2017) Forage legumes with improved nutritional value: condensed tannins to avoid pasture bloat. In: Clemente A, Jiménez-López JC (eds) Legumes for global food security, Chapter 8. Nova Science Publisher Inc, NY, pp 183–222
15. Fresquet-Corrales S, Roque E, Sarrion-Perdigones A, Rochina M, Lopez-Gresa MP, Diaz-Mula HM, Belles JM, Tomas-Barberan F, Beltran JP, Cañas LA (2017) Metabolic engineering to simultaneously activate anthocyanin and proanthocyanidin biosynthetic pathways in *Nicotiana* spp. PLoS One 12:e0184839
16. Pandey MK, Roorkiwal M, Singh VK, Ramalingam A, Kudapa H, Thudi M, Chitikineni A, Rathore A, Varshney RK (2016) Emerging genomic tools for legume breeding: current status and future prospects. Front Plant Sci 7:455
17. Wang L, Wang L, Zhou Y, Duanmu D (2017) Use of CRISPR/Cas9 for symbiotic nitrogen fixation research in legumes. Prog Mol Biol Transl Sci 149:187–213
18. Meng Y, Hou Y, Wang H, Ji R, Liu B, Wen J, Niu L, Lin H (2017) Targeted mutagenesis by CRISPR/Cas9 system in the model legume *Medicago truncatula*. Plant Cell Rep 36(2):371–374

Chapter 2

Model Legumes: Functional Genomics Tools in *Medicago truncatula*

Luis A. Cañas and José Pío Beltrán

Abstract

Many researchers have sought along the last two decades a legume species that could serve as a model system for genetic studies to resolve specific developmental or metabolic processes that cannot be studied in other model plants. Nitrogen fixation, nodulation, compound leaf, inflorescence and plant architecture, floral development, pod formation, secondary metabolite biosynthesis, and other developmental and metabolic aspects are legume-specific or show important differences with those described in *Arabidopsis thaliana*, the most studied model plant. Mainly *Medicago truncatula* and *Lotus japonicus* were proposed in the 1990s as model systems due to their key attributes, diploid genome, autogamous nature, short generation times, small genome sizes, and both species can be readily transformed. After more than decade-long, the genome sequences of both species are essentially complete, and a series of functional genomics tools have been successfully developed and applied. Mutagens that cause insertions or deletions are being used in these model systems because these kinds of DNA rearrangements are expected to assist in the isolation of the corresponding genes by Target-Induced Local Lesions IN Genomes (TILLING) approaches. Different *M. truncatula* mutants have been obtained following γ-irradiation or fast neutron bombardment (FNB), ethyl-nitrosourea (ENU) or ethyl-methanesulfonate (EMS) treatments, T-DNA and activation tagging, use of the tobacco retrotransposon *Tnt1* to produce insertional mutants, gene silencing by RNAi, and transient post-transcriptional gene silencing by virus-induced gene silencing (VIGS). Emerging technologies of targeted mutagenesis and gene editing, such as the CRISPR-*Cas9* system, could open a new era in this field. Functional genomics tools and phenotypic analyses of several mutants generated in *M. truncatula* have been essential to better understand differential aspects of legumes development and metabolism.

Key words Model legumes, *M. truncatula*, Functional genomics, Targeted mutagenesis, Reverse genetics, Gene editing, Legume development, Secondary metabolites

1 Introduction

Legumes are a large group of plants widely appreciated not only by their nutritional properties but also by their important roles in food security and sustainable agriculture. The Leguminosae or Fabaceae family is currently classified into six subfamilies: Cercidoideae, Detarioideae, Duparquetioideae, Dialioideae, Papilionoideae, and Caesalpinioideae, which includes the mimosoid clade [1].

Luis A. Cañas and José Pío Beltrán (eds.), *Functional genomics in* Medicago truncatula*: Methods and Protocols*, Methods in Molecular Biology, vol. 1822, https://doi.org/10.1007/978-1-4939-8633-0_2,

Leguminosae is the third largest angiosperm family in terms of species number after Asteraceae and Orchidaceae, with close to 770 genera and over 19,500 species. The model and crop legumes are most heavily studied cluster in one large clade within the Papilionoideae subfamily, which includes less than half of the species diversity of legumes (445/503 genera and 14,000 species).

Since Gregor Mendel (1822–1884), the Austrian monk, discovered the basic principles of heredity through experiments in his garden using the garden pea (*Pisum sativum* L.) as model system, most researchers have sought a legume species that could serve as a model system for genetic studies to resolve specific developmental or metabolic processes that cannot be studied in other model plants. Nitrogen fixation, compound leaves, inflorescence and plant architecture, floral development, pod formation, secondary metabolite production, and other developmental and metabolic topics are legume-specific or show important differences with those described in *Arabidopsis thaliana*, the most studied model plant.

2 The Need of Model Legumes: Developmental and Metabolic Differences with Other Model Plants

The very low levels of microsynteny (8%) between *A. thaliana* and *Medicago truncatula*, one of the most studied model legumes, suggest that the structure of the *Arabidopsis* genome may not be a good indicator of how legume genomes are arranged. Elucidating the sequences and functions of the genomes of three legume species, *M. truncatula*, *Lotus japonicus*, and *Glycine max* (soybean), has been a more than decade-long effort, and currently the genome sequences of these species are essentially complete [2, 3].

A good example on the need of model legumes to study certain developmental processes that cannot be studied in other model plants is the genetic control of inflorescence and floral development, which is similar in many respects in different plant species. Inflorescence in legumes is indeterminate, and many Leguminosae species, such as pea (*P. sativum*), *M. truncatula*, or *L. japonicus*, show compound inflorescences of double raceme, while the inflorescences of *Arabidopsis thaliana* and *Antirrhinum majus* are simple racemes. In addition, the ontogeny of most legume flowers is drastically different from the floral ontogeny of other model plants like *A. thaliana* or *A. majus*. In these model species, during the transition from vegetative to reproductive development, a unique inflorescence meristem is produced before the floral meristem, while in legume species, a primary (I_1) and secondary (I_2) inflorescence meristems are produced before the floral meristem (Fig. 1 upper right and e). In *A. thaliana* and *A. majus*, the initiation of floral organ primordia is a centripetal and sequential process [4, 5]. That is, first the sepal primordia are formed; second the primordia that will form petals appear, immediately followed by

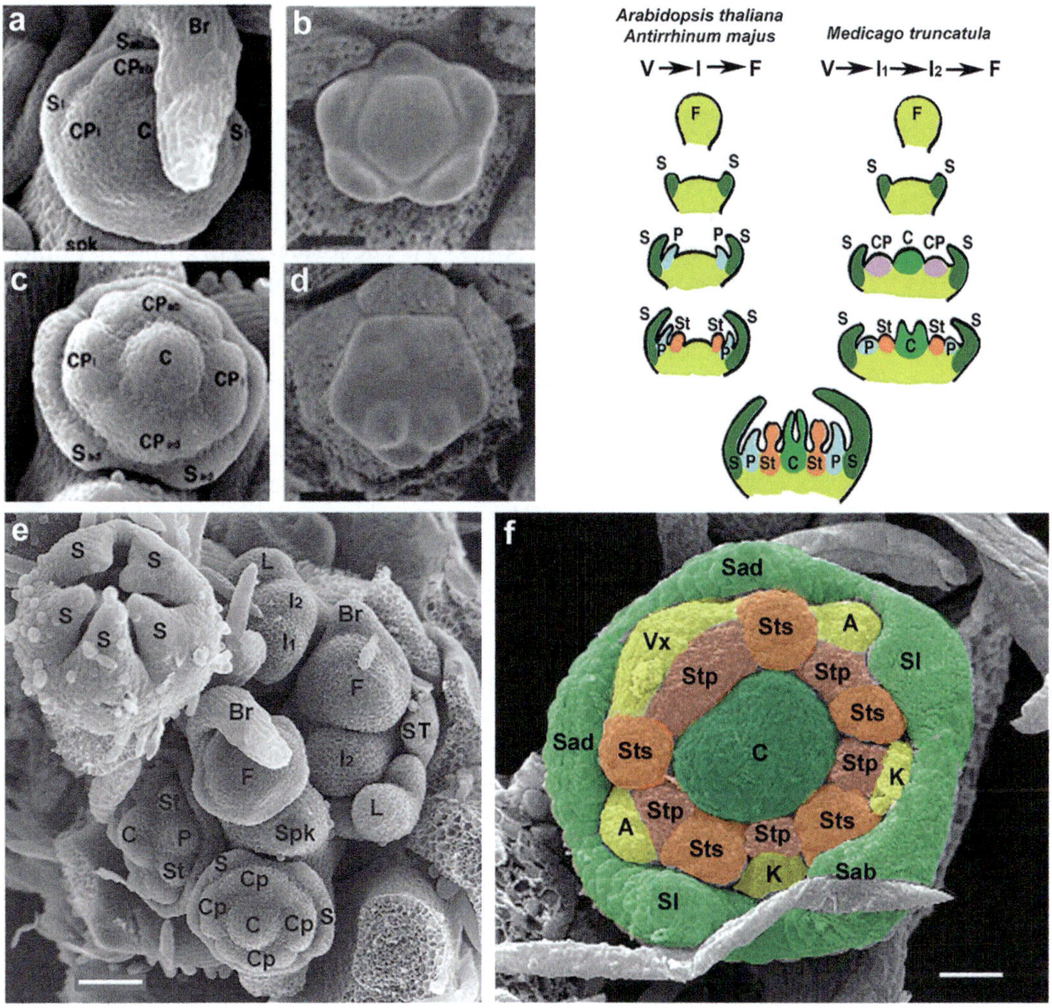

Fig. 1 Comparative floral ontogeny among *Medicago truncatula* and other model plants. (**a**) Abaxial-adaxial initiation of organ primordia in *M. truncatula*. (**b**) Simultaneous organ primordia initiation in *Antirrhinum majus*. (**c**) Early carpel differentiation in *M. truncatula*. (**d**) Late carpel differentiation in *A. majus*. (**e**) Different stages of floral development in *M. truncatula*. (**f**) Common primordia compartmentation in *M. truncatula* to establish petal and stamen territories. Upper right: schematic representation of floral ontogeny in *M. truncatula* compared with two model species, *Arabidopsis thaliana* and *A. majus*. In legume species a primary (I_1) and secondary (I_2) inflorescence meristems are produced before the floral meristem. The initiation of floral organ primordia is a centripetal and sequential process in *A. thaliana* and *A. majus*. In contrast, the order of floral organ initiation in *Medicago* is unidirectional in all whorls starting from the abaxial position of the flower with a high degree of overlap. *L* leaf, *Br* bract, *Spk* spike, *ST* stipule, *F* floral meristem, *I1* primary inflorescence meristem, *I2* secondary inflorescence meristem, *Sad* adaxial sepals, *Sl* lateral sepals, *Sab* abaxial sepal, *P* petal, *CP* common primordia, *Stp* antepetal stamen, *Sts* antesepal stamen, *C* carpel, *Vx* vexillum petal, *A* allae petals, *K* keel petals. Adapted from Benlloch et al. [6]

the stamen primordia; and finally the residual central part of the floral meristem forms the carpel primordium (Fig. 1 upper right). Using scanning electron microscopy (SEM) analysis, we have characterized the early developmental events of the wild-type *Medicago truncatula* flower and selected morphological characters as markers to break it down into eight different developmental stages [6]. The order of floral organ initiation in *Medicago*, in contrast to *Arabidopsis* and *Antirrhinum*, is unidirectional in all whorls starting from the abaxial position of the flower with a high degree of overlap (Fig. 1a, b). Other important differences with these model species are the early carpel differentiation (Fig. 1a, b) and the existence of four common primordia from which petals and stamens differentiate (Fig. 1c, d) [7–9]. The formation of common primordia, as opposed to discrete petal and stamen primordia, has been described in many legume and non-legume plants [8–13]. The common primordia in *Medicago* are established at the adaxial and abaxial flower positions, and at the two sides of the adaxial-abaxial floral axis, limiting externally with the sepal primordia and internally with the carpel primordium. The proper organ differentiation within each of the four common primordia must play a central role in the ontogeny of *M. truncatula* flowers (Fig. 1f). This raises the question of whether the known A, B, C, and E functions are sufficient to control the formation of petals and stamens from common primordia or whether new functions are needed.

Other interesting example on the need of model legumes to study certain developmental processes that cannot be studied in other model species is the presence of compound leaves in their shoot and nitrogen-fixing nodules in their roots. A simple leaf consists of a single undivided blade with an entire, serrated, or lobed margin, and a compound or dissected leaf consists of multiple blades (leaflets), organized as pinnate, palmate, or higher-ordered structures (Fig. 2a–e). Different targeted mutagenesis and reverse genetic approaches have been used to better understand how compound leaves are generated.

Legumes have traditionally been model systems to investigate interactions of roots with microbes. Legume nitrogen fixation starts with the formation of a nodule. *Rhizobium*, a common soil bacterium, invades the root and multiplies within the cortex cells. The plant supplies all the necessary nutrients and energy for the bacteria. Root nodules can be seen 2–3 weeks after planting, depending on legume species and germination conditions. When nodules are young and not yet fixing nitrogen, they are usually white or gray inside. As nodules grow in size, they gradually turn pink or reddish in color, indicating nitrogen fixation has started. The pink or red color is caused by leghemoglobin that controls oxygen flow to the bacteria. Nodules on many perennial legumes such as alfalfa and clover are finger-like in shape. Two main types of nodule have been described: determinate and indeterminate. Determinate nodules are found on certain tribes of tropical legumes (*Glycine*, *Phaseolus*, and *Vigna*) and on some temperate legumes such as *Lotus*. Determinate nodules lose meristematic activity shortly after initiation; thus, growth is due to cell expansion resulting in mature nodules which are spherical in shape. Indeterminate

Fig. 2 Compound leaves, root nodules, and flowers of different model legumes. (**a**) *Pisum sativum* compound leaf. (**b**) *Medicago truncatula* cv. Jemalong 2HA leaf. (**c**) *Medicago sativa* cv. Regen SY27 leaf. (**d**) *Glycine max* leaf. (**e**) *M. sativa palmate1* mutant showing five leaflets instead of three. (**f**) Close view of 4-week-old *M. truncatula* A20 root inoculated with *Sinorhizobium meliloti*, the pink color is due to the presence of leghemoglobin and indicates that the indeterminate nodules are fixing nitrogen. Adapted from Ninjatacoshell (Wikimedia Commons, distributed under a CC BY-SA 3.0 license). (**g**) *Pisum sativum* flower. (**h**) *Medicago truncatula* flower. (**i**) *Lotus japonicus* flower. (**j**) *Glycine max* flower

nodules are found in the majority of legumes whether in temperate regions or in the tropics (*Pisum*, *Medicago*, *Trifolium*, *Vicia*, etc.), and they maintain an active apical meristem that produces new cells for growth over the life of the nodule. This class of nodules have a generally cylindrical shape, which may be extensively branched (Fig. 2f). Mature nodules may resemble a hand (palm) with protruding portions (fingers), although the entire nodule is generally less than 1/2 inch in diameter. When legume plants encounter low soil nitrogen concentrations, they release flavonoids to form a symbiotic relationship with rhizobia, and rhizobia respond by releasing a nodulation factor (*nod*), which stimulates nodule formation in plant roots. Exposure to *nod* factor triggers the formation of deformed root hairs, which permit rhizobia to enter the plant and form an infection thread, which is an intercellular tube that penetrates the cells of the host plant. Infection triggers rapid cell division in the root cells, forming a nodule of tissue. The relationship between a host legume and the rhizobia is symbiotic, providing benefits to both participants. Once the rhizobia have established themselves into the root nodule, the plant provides carbohydrates in the form of malate and succinate, and the rhizobia provide ammonia for the formation of amino acids. Many legumes are popular agricultural crops specifically because they require very little fertilizer. Used properly some legumes can even serve as fertilizer for later crops, binding nitrogen in the plant remains in the soil.

The best-studied crop legumes like alfalfa (*Medicago sativa* L.), pea, or soybean (*Glycine max* L.) are either tetraploid, have large genomes with abundant repetitive DNA, or are difficult to transform and regenerate. To avoid these problems, two legume species, *Medicago truncatula* and *Lotus japonicus*, were proposed in the 1990s as model systems (Fig. 2g–j). Key attributes of these species include their diploid genome, autogamous nature, short generation time, small genome size, and feasibility to be readily transformed using *Agrobacterium tumefaciens* and then regenerated to yield fertile transgenic plants [14–16]. Despite their similar attributes as experimental systems, an important distinction between *L. japonicus* and *M. truncatula* is that they represent the two major classes of nodule ontogeny: determinate and indeterminate development, respectively [17, 18].

The garden pea is in some senses the founding model for genetics, through Mendel's work in 1866. Pea has been further developed as a model for development traits such as leaf form [19], tendril formation [20], transposon evolution [21], defensive responses [22], starch and sugar synthesis [23], hormonal control of shoot branching [24, 25], inflorescence architecture [26], and floral development [27–34]. However its large genome size and the difficulty to produce transgenic plants have been the main drawbacks to use the garden pea as model system. Soybean has been utilized as model for studies of seed development [35], root hair development and early nodulation responses, mineral uptake, and protein and oil biosynthesis. Common bean has been used to characterize the molecular basis of photoperiod sensitivity and determinacy [36].

Medicago and *Lotus* will serve as effective models for the cool-season legumes [37] and soybean for the many crop species in the Phaseoleae [38]. However, these are not one-way streets, since the crops also inform the genomic models. For example, *Medicago* and *Lotus* have been mostly used in studies of nodulation, mycorrhization, and plant-symbiont signaling [39]. *M. truncatula* has been used for the study of phenylpropanoid and isoflavonoid pathways and secondary metabolites [40–43], various defense responses, [44] and root architecture [45]. Information from these two model systems may be useful for forage legumes (e.g., alfalfa and clovers) even for the study of agronomic traits (yield and growth habits), because *L. japonicus* and *M. truncatula* are related to the forage legume species *Lotus corniculatus* (bird's-foot trefoil) and *Medicago sativa* (alfalfa), respectively.

3 The *Medicago truncatula* Genome

To develop and apply functional genetics tools in a certain plant species, it is essential to have the most complete information on its genome. The genus *Medicago* is mainly composed of forage species with high symbiotic nitrogen fixation potential and high protein content. The best known *Medicago* species is alfalfa (*M. sativa*), which is a perennial, allogamous tetraploid. In contrast to alfalfa, *M. truncatula* is one of the 32 described *Medicago* species that is an annual, autogamous diploid. *M. truncatula*, also called barrel medic, is a native of the Mediterranean basin. Endemic populations exhibit drought and salinity tolerance and are able to grow in a wide range of soil and environmental conditions. In addition, *M. truncatula* possesses high genetic and phenotypic variation both within and between populations, although individual plants are highly homozygous [46]. *M. truncatula* has been extensively used as model system in the last years for molecular and functional genetics research in legumes. It is generally accepted that knowledge about certain shared characteristics of legumes, such as the pathways involved in symbiosis with rhizobia and synthesis of flavonoids and glycosides, is considerably transferable from models to crops. The biochemistry of legumes is distinct from that of other plant groups, and many unique molecules with biomedical application, such as isoflavones and plant sterols, are found among legume secondary metabolites [47]. The use of this model legume over the last two decades has improved our understanding of genomic structure and gene function for legumes.

Functional genomics is greatly aided by knowledge on genome sequence and transcriptome of the target species, and legumes are the plant family with the greatest amount of genomic data currently available [48, 49]. *M. truncatula* has a diploid genome ($2n = 16$), which is relatively small (~375 Mbp); its improved sequence is already available and annotated. A re-sequencing approach is ongoing which proved efficient in gene discovery via GWAS analysis. The *M. truncatula* consortium is founded by the National Science Foundation and the Samuel Robert Noble Foundation in the USA and by the European Union in Europe. This consortium has made significant developments in express sequence tags (EST) and genome sequencing. A collection of 226,923 ESTs that represent 36,878 unigenes and > 87 Mb genomic sequences of the 200 Mb euchromatic regions have been made publicly available. The gene space of eight chromosomes has been almost fully sequenced (http://www.medicago.org).

The draft sequence of the genome of *M. truncatula* cultivar A17 was published in the journal *Nature* in 2011 [2]. The sequencing was carried out by an international partnership of research laboratories involving researchers from the University of Oklahoma (USA),

J. Craig Venter Institute (USA), Genoscope (France), and Sanger Centre (UK). Partner institutions included the University of Minnesota (USA), University of California-Davis (USA), the National Center for Genomic Resources (USA), John Innes Centre (UK), Institute National de Recherche Agronomique (France), Munich Information Center for Protein Sequences (Germany), Wageningen University (the Netherlands), and Ghent University (Belgium).

The *M. truncatula* Genome Sequencing Consortium began in 2001 funded by a grant from the Samuel Roberts Noble Foundation. In 2003, the National Science Foundation and the European Union 6th Framework Programme provided most of the funding. In 2009, the 84% of the genome sequence assembly, based on bacterial artificial chromosomes (BACs), was essentially completed (http://www.medicago.org/genome/genome_stats.php). In July 2013, the version 4.0 of the *M. truncatula* genome was released (http://www.jcvi.org/cgi-bin/medicago/overview.cgi). This version combined sequences gained from shotgun sequencing with the BAC-based sequence assemblies, which has helped to fill in the gaps in the previously mapped sequences. A parallel group known as the International Medicago Gene Annotation Group (IMGAG) is responsible for identifying and describing putative gene sequences within the genome sequence.

The development of new molecular and genetic tools is essential in order to optimize the exploitation of this model plant. To support functional genomics in *M. truncatula*, the Noble Research Institute has developed a compendium or atlas of gene expression profiles for the majority of *M. truncatula* genes covering all its major organ systems (roots, nodules, stems, petioles, leaves, vegetative buds, flowers, seeds, and seed pods) with detailed developmental time series for nodules and seeds, using the Affymetrix Medicago Gene Chip®. These data have been supplemented with transcriptome data from plants subjected to various kinds of abiotic and biotic stresses and data from specific cell and tissue types that will aid gene function determination, biological discovery, and molecular breeding efforts. The *M. truncatula* Gene Expression Atlas (MtGEA) web server (https://mtgea.noble.org/v3/) was developed to maximize the use of publicly available Affymetrix Gene Chip data and aid efforts to "interpret" the *Medicago* genome through functional genomics. This web server aims to archive all publically available *M. truncatula* gene expression data derived from the use of the Affymetrix Gene Chip. The web service allows the user to identify and analyze the expression of one or more genes of interest, using the Affymetrix Medicago probe-set identification number, gene sequence, and gene annotation information (such as gene ID, keyword in description, GO term, KEGG bincode or protein domain name) as starting point. The web server also seamlessly interconnects with the *M. truncatula* genome browser, which allows the user to query using various genomics features (e.g., a particular chromosome coordinate or an EST ID mapped to the genome) and links back to the corresponding expression profiles in MtGEA. The user may further study gene expression profiles based on co-expression analysis and differential expression analysis. Direct download of the whole dataset organized by experiment is also reliable. Data output can be selected and is provided as a tabular form compatible with common analytical and visualization software. In Chapter 3 of this book, there is a review on key current genomic databases that are specific to *M. truncatula* as well as those that contain *M. truncatula* data in addition to data from other plants. Strategies to obtain archived *M. truncatula* genomic data originally deposited into custom databases that are no longer maintained, but are now accessible in general databases, are discussed.

4 Functional Genomics Tools Developed in *Medicago truncatula* to Study Developmental Processes in Legumes

Different functional genomics techniques have been developed and successfully applied in recent years to *M. truncatula*. Many mutations characterized in this legume genome are the product of chemical mutagenesis by means of ethyl-nitrosourea (ENU) [50] or ethyl-methanesulfonate (EMS) treatments [48], in which the mutagenic mechanism of single base transition is expected to yield alleles with a wide range of phenotypes. Mutagens that cause insertions or deletions have been widely used because these kinds of DNA rearrangements are expected to assist in the isolation of the corresponding genes by target-induced local lesions in genomes (TILLING) approaches.

Some chemically induced mutants were obtained long time ago in the garden pea, like the *Stamina pistilloida* (*Stp*) mutant induced by diethyl sulfate treatment. This floral homeotic mutant shows a loss-of-B-function phenotype with sepals instead of petals in the second whorl and carpels instead of stamens in the third whorl. *Stp* is the pea homolog of the *Antirrhinum* gene *Fimbriata* (*Fim*) and of *UNUSAL FLORAL ORGANS* (*UFO*) from *Arabidopsis*. Differences between *Fim/UFO* and *Stp* mutant phenotypes and expression patterns suggest that expansion of *Stp* activity into the leaf was an important step during evolution of the compound leaf in the garden pea [9, 51].

In *M. truncatula* the ethyl-methanesulfonate (EMS) has been extensively used as alkylating agent to develop different mutant collections. EMS induces single base pair C/G to A/T substitution in nucleotides. The mutagenized seeds are germinated, and the resulting plants are self-pollinated to produce M1 progenies.

The M1 plants are then grown, and a TILLING M2 collection is established by growing few seeds from each M1 plant. Total genomic DNA is purified from each M1 plant and pooled. The mutant collections are usually screened with reverse genetic approaches. TILLING involves the identification of mismatches in heteroduplexes formed by single-stranded DNA from the wild-type and mutant alleles of the target locus. The target sequences are generated by PCR amplification from bulked DNA isolated from single M1 plants using labeled primers appropriate for the detection strategy employed. The amplicons are then heated, causing strand separation, reannealed in order to form heteroduplexes, and cleaved by an endonuclease active on single-stranded DNA at the mismatch point and the products separated by electrophoresis [48]. In Chapter 5 of this book, there is a complete description of the TILLING methodology. The protocol described is straightforward to set up, and screenings are routine. The advantages of TILLING include the fact that it is non-GMO that a saturating population is easily attained, and that point mutations are relatively well-tolerated, allowing reasonable population sizes to be used and an allelic series of mutations at a given locus to be isolated.

Several EMS mutant collections of *M. truncatula* are currently available. Within the framework of the European Grain Legume Integrated Project (GLIP, http://www.pcgin.org/GLIP/GLIP.htm), two mutant collections were established. The two collections showed the same number of M2 lines that were obtained from M1 populations with different sizes. The number of mutations detected in the two EMS populations was 1 every 485 kb. A pilot reverse genetic experiment with 56 target genes revealed an efficiency of 13 independent alleles per exon screened, 67% of which were missense and 5% nonsense mutations. An Italian functional genomics initiative produced a small collection of TILLING mutants with about 2500 M2 lines and a reported efficiency of about four independent alleles for target sequence [48]. A catalogue of mutant phenotypes was developed, and services for reverse screening with target sequence are available (http://inra.fr/legumbase).

In the last decades, several mutant collections have been obtained in *M. truncatula* by seed irradiation using appropriate doses of γ-rays [52] or fast neutron bombardment (FNB) [53–63]. Most of these mutants were obtained by FNB of the genotype A17, although the γ-irradiation was used in the Jemalong J5. Reverse screening of FNB populations has been carried out by the DeTILLING strategy [60]. This approach allows detection of mutants by PCR on bulks of DNAs of FNB mutants. The wild-type target amplification is avoided by a strategy that combines restriction enzyme digestion of the template and the use of poison primers. With this strategy a mutant recovery rate of 29% has been obtained from a population of 156,000 M2 plants (4 genes out of 14 screened). In Chapter 4 of this book, there is a detailed procedure that has been used for screening of mutants derived from fast neutron bombardment mutagenesis in *Medicago truncatula*. These mutant resources have been used in screening and characterization of different categories of mutants including symbiotic nitrogen fixation, nodule development, and growth and patterning of leaf, stem, and root system architecture in legumes.

Map-based cloning required to identify the mutation of interest is helped by strategies such as transcriptional cloning [64], which has allowed the identification of FNB-induced mutations. This approach relies on the identification of mutated genes through detailed genome-wide transcriptomic analyses. Also genome-wide analyses of FNB mutant are expected to benefit of the recent development of a *M. truncatula* genome-wide TILLING array by NimbleGen (Roche), an innovator in genomics focused on developing target enrichment solutions for next-generation sequencing. NimbleGen technologies allow researchers to selectively target DNA sequences of interest in their genome of interest (http://sequencing.roche.com/products/nimblegen-seqcap-target-enrichment.html).

Reverse genetic studies in *M. truncatula* did not only take advantage of the many mutant populations available but also of techniques based on post-transcriptional gene silencing (PTGS). In this case plants are transformed with a construct that produces double-stranded RNAs that will guide sequence-specific mRNA degradation of the target gene. The phenotype of the transformed plants can gradually vary from wild type to knockout; thus, many transformants are needed to obtain the desired effect. Mild effects can be beneficial in case of essential genes whose complete loss-of-function may cause lethal phenotypes.

Transient post-transcriptional gene silencing (PTGS) by virus-induced gene silencing (VIGS) has been efficiently used in pea (Fig. 3a–d) and most recently in *M. truncatula* (Fig. 3e–h). Virus-induced gene silencing is a PTGS technique that can be used transiently by scrubbing leaves or introducing the viral vector in the plant by agro-infiltration. VIGS is being used for large-scale forward genetics screening by inoculation of cDNA library and subsequent identification of the gene involved in the process of interest [65]. Viral vectors working on *M. truncatula* have been recently described. Grønlund et al. [66] successfully used a pea early browning virus (PEBV)-based vector for both transient expression of reporter genes and for silencing of the *phytoene desaturase* (*PDS*) gene that causes a bleaching phenotype. Serwatowska et al. [33] used the same viral vector to simultaneously silence the two pea and *M. truncatula AGAMOUS* genes (*PsAGa/PsAGb* and *MtAGa/MtAGb*, respectively) with redundant functions, obtaining

Fig. 3 Virus-induced gene silencing (VIGS) in two model legumes: *Pisum sativum* and *Medicago truncatula*. (**a**) Wild-type *P. sativum* flower. (**b**) Dissected wild-type pea flower showing five fused sepals in W1, five petals in W2, ten stamens in W3, and a single carpel in W4. The corolla contains three types of white petals: a single

loss-of-C-function floral phenotypes (Fig. 3a–h). Várallyay et al. [67] constructed two VIGS vectors based on the sunn-hemp mosaic virus (SHMV) that can systemically infect *M. truncatula* without causing severe symptoms and reported a successful silencing of the *Chlorata 42* gene. However, large-scale screenings based on VIGS analysis have not been reported to date for *M. truncatula*. In Chapter 8 of this book, there is a complete description of the VIGS technology applied to *M. truncatula*. Agro-inoculation of the pea early browning virus vectors pCAPE1 and pCAPE2 can establish infection in several genotypes of *M. truncatula* and can reduce target gene RNA levels to an extent that allow investigation of gene function.

Gene silencing by means of RNAi in *M. truncatula* has been extensively used to study gene function, but it has not been a matter of a functional genomics approach as for *Arabidopsis* and the AGRIKOLA collection [68]. Nevertheless, many gene functions have been characterized exploiting RNAi gene silencing [31, 32, 34, 45, 69–105]. A good example on the use of RNAi approaches for gene silencing is the generation of loss-of-function mutants with silenced floral organ identity MADS-box genes, producing homeotic alterations. For instance, Benlloch et al. [31] obtained a loss-of-B-function mutant in *M. truncatula* generated by RNAi gene silencing the MADS-box *MtPISTILLATA* (*MtPI*) gene. The mutant showed homeotic transformations with sepals instead of petals in the second whorl and carpels instead of stamens in the third whorl.

MicroRNAs are key regulators in development processes or stress responses in plants. In the last decade, several conserved or non-conserved microRNAs have been identified in *Medicago truncatula*, and different strategies leading to the inactivation of microRNAs in plants have been described. In Chapter 9 of this book, a protocol for an effective inactivation of microRNA using a STTM strategy in *M. truncatula* transgenic roots is described.

Fig. 3 (continued) standard petal (or vexillum), two wing petals, and two fused keel petals surrounding the reproductive organs, ten stamens and a central carpel. Nine stamen filaments are joined in a staminal tube, and the tenth, the vexillary stamen filament at the adaxial region, is free standing. (**c**) Loss-of-C-function in a pea flower by VIGS of the two *PsAGAMOUS* (*PsAGa* and *PsAGb*) paralogous genes. (**d**) Dissected flower of a double *PsAGab* mutant showing a great number of petals instead of stamens in W3 and a new flower in W4. (**e**) Wild-type *M. truncatula* flower. (**f**) Dissected *M. truncatula* wild-type flower showing five fused sepals in W1, five yellow petals in W2, ten stamens in W3, and a single carpel in W4. (**g**) Loss-of-C-function in a *M. truncatula* flower by VIGS of the two *MtAGAMOUS* (*MtAGa* and *MtAGb*) paralogous genes showing a great number of petals instead of stamens in W3 and a new flower in W4. (**h**) Dissected flower of a *MtAGab* double mutant showing multiple petals instead of stamens in W3 and a new flower in W4. Adapted from Serwatowska et al. [33], under a Creative Commons Attribution 4.0 International License (CC BY 4.0)

Insertional mutagenesis is an efficient reverse genetic tool to produce large-scale mutant collections to identify both key symbiotic, metabolic, and developmental genes, as well as others, of agronomical interest. T-DNA tagging has been the strategy of choice for many mutant collections in the model plant *Arabidopsis thaliana*, and it has allowed fundamental discoveries in gene functions and advances in both basic and applied plant research [106]. Unfortunately, only *A. thaliana* can be transformed easily by the floral-dip method which allows the generation of large numbers of mutant populations. Genetic transformation of *M. truncatula* can only be achieved by in vitro tissue culture-based protocols, mainly via somatic embryogenesis, which require great efforts to produce the number of mutants that would allow significant genome coverage. Activation tagging is an application of T-DNA mutagenesis aimed at obtaining gain-of-function mutations. This can be achieved in *M. truncatula* introducing enhancer sequences randomly in the target genome via a T-DNA shuttle and then analyzing the genomic regions flanking the insertion sites in individuals showing phenotypic alterations. Laffont et al. [107] reported the isolation of a compact roots *M. truncatula* mutant via T-DNA mutagenesis with a mutation in the *cra1* gene (not tagged). Carelli et al. [108] reported a mutant on the *Lha* gene obtained in *M. truncatula* via activation tagging with lack of hemolytic saponins. In Chapter 6 of this book, the detailed procedure to obtain and screen an activation-tagged population in *Medicago truncatula* is described.

Although large-scale insertional mutagenesis using T-DNA is not, up to date, feasible in legumes with probed efficiency, large-scale collections of mutants have been developed using the *Tnt1* retrotransposon of tobacco in *M. truncatula*. d'Erfurth et al. [109] demonstrated that, in the *M. truncatula* R108 genotype, this element has the ability to transpose during the early steps of in vitro regeneration with a high rate of insertion in transcribed genomic regions. Sequence analyses of insertion sites showed the virtual absence of insertion site preference. The average amount of new insertions per regenerated line was calculated in the order of ~25. Based on these data, it was shown that a collection of 14–16,000 Tnt1 lines will store tagging events for about 90% of *M. truncatula* genes [110–113]. The collection maintained at the Noble Foundation (https://medicago-mutant.noble.org/), which also includes the first mutants generated by the CNRS in France, is based on the genotype R108-c3. The mutant population was generated through callus induction, subculture, and regeneration from a starting transgenic line harboring three homozygous copies of *Tnt1* insertion. The population consists of 21,700 regenerated lines that encompass more than 500,000 *Tnt1* insertions. Based on the genome size, average gene length, and random insertion nature of *Tnt1*, this mutant population covers about 90% of genes in the *M. truncatula* genome. In Chapter 7 of this book, there is a complete description to produce *Tnt1* mutant populations in *M. truncatula*.

Another collection of about 1000 lines from the same R-108 line was produced by the CNR-IGV in Italy [114]. In the framework of the GLIP, 8000 *Tnt1* mutants were produced from the Jemalong 2HA (2HA3-9-10-3) line. The GLIP collection is maintained by various labs that participated in the project, and a subset of new *Tnt1* plants were merged with the collection at the Noble Foundation few years ago. Iantcheva et al. [47] reported that *Tnt1* transposition efficiency in Jemalong 2HA has a lower efficiency with only 10–15 new insertions per line and a variable percentage of regenerated plants without transposition. The adoption of 2HA line for mutagenesis instead of R108 was motivated by the highest DNA homology to the line used for genome sequencing (Jemalong A17) and for the presence of active and characterized endogenous retroelements [115].

Tnt1 mutant collections have been screened with both forward and reverse genetic approaches [29, 32–34, 47, 60–62, 91, 102, 103, 109–114, 116–125]. Forward approaches have been based on cloning of host sequence flanking the insertion sites and subsequent identification of events linked to the studied mutation. Based on the duplicated *Tnt1* long terminal repeat (LTR) sequences, several molecular approaches including thermal asymmetric interlaced (TAIL)-PCR and inverse-PCR have been used to recover the host sequences flanking the insertion sites [126]. Segregation analysis of each cloned insertion site can then be used to select the event linked to the mutation. Alternatively, the insertion sites associated with the mutations can be selected by segregation analysis prior to host sequence cloning by employing a sequence-specific amplification polymorphism (S-SAP)-based protocol. Confirmation of the identity of the mutation can be obtained by means of complementation tests based on the reintroduction of the wild-type gene sequence in the mutated background. Alternatively one could obtain independent alleles of the target gene and compare their similarity to the original mutant phenotype. This can be done using TILLING and *Tnt1* mutant populations as demonstrated by many publications that report successful recovery of alleles by reverse screening [112]. The power of the *Tnt1* mutagenesis approach is also witnessed by the prevalence of publications reporting successful gene cloning based on such strategy compared to the others.

A good example on the use of the *Tnt1* reverse screening applied to the identification of tagged mutants in a particular gene is the *single leaflet* mutation, which is the result of a *Tnt1* insertion in the gene *MtSGL1*, the *FLORICAULA/LEAFY* ortholog in *M. truncatula*, which is involved in the control of compound leaf development [117]. Other nice example is the publication of

Roque et al. [32] reporting the *mttm6-1* mutant generated by a *Tnt1* insertion in the *AP3*-like gene *MtTM6*, which shows partial loss-of-B-function phenotype in its flowers. The second whorl contains petals with green tissues distributed in small areas, and occasionally a carpel-like structure is fused to a petal, whereas the third whorl is occupied by carpel-like structures. Other floral homeotic mutants have been generated by *Tnt1* insertion of different MADS-box genes, like *MtPIM*, *MtPI*, *MtNGL9*, *MtNMH7* (eu*AP3*), *MtTM6* (paleo*AP3*), *MtAGa*, and *MtAGb* [29, 31–34].

The use of a well-standardized in situ hybridization protocol is vital for the systematic analysis of gene function in *M. truncatula*. Different strategies have been developed and implemented during the last decades aiming to decipher the function of particular genes. Among the different techniques, mRNA in situ hybridization remains essential to fully understand gene function. In Chapter 10 of this book, a complete protocol for the in situ localization of gene transcripts in *M. truncatula* is described. It is optimized for use on sections of paraffin-embedded tissues with digoxigenin-labeled probes. By means of this protocol, the differential expression patterns of duplicated floral MADS-box transcription factors have been elucidated in *M. truncatula*. When duplicated genes are analyzed, the tissue and cellular location of the transcripts is the only technique that account for small variations in the pattern of gene expression that occurred after duplication and diversification [32–34]. In Chapter 11, an in situ hybridization method using Invitrogen™ ViewRNA™ ISH Tissue Assay (Thermo Fisher Scientific) optimized for Medicago root and nodule sections is described. The method is based on branched (b)DNA signal amplification technology originally developed for use in microplate format and further adapted for detection of (m)RNAs in mammalian tissue sections.

Emerging technologies of targeted mutagenesis and gene editing, such as the CRISPR-*Cas9* system [127, 128], could open a new era in this field because crops created through genome engineering might prove to be more acceptable to the consumers than plants that carry foreign DNA in their genomes. The CRISPR-*Cas9* reagent is the most recent mutagenesis platform and is highly effective at generating site-directed double-stranded breaks (DSB) in *M. truncatula*. In Chapter 12 of this book, a complete protocol for the construction of reagents using two genome engineering platforms that have generated mutant plants in *M. truncatula*, *M. sativa*, and soybean systems is described. The reagents are easy to assemble, can be quickly retrofitted to test novel regulatory sequences for improved efficiency, and can be used for more advanced genome engineering strategies such as gene insertion or gene replacement.

In other chapters of this book, reviews on the use of different functional genomics tools to developmental studies in *M. truncatula* such as seed, compound leaf, root, nodule, flower, and fruit development, or salt and osmotic stress tolerance, have been included.

5 *Medicago truncatula* and Metabolic Engineering in Legumes

M. truncatula is a good model to study a series of secondary metabolites that are almost exclusively produced in legumes. This is the case of isoflavonoids, which are mainly produced in legumes but not in the model plant *A. thaliana*. *M. truncatula* has also being used as a model to study the lignin biosynthesis pathway in legumes, which also shows differences in *Medicago* when compared with *Arabidopsis*. Other interesting example is the production of flavan-3-ol polymers, also called proanthocyanidins (PAs) or condensed tannins, which showed differences in their biosynthetic pathway in *M. truncatula* from that in *Arabidopsis* [40–43].

Medicago species possess a wide range of flavonoid and terpenoid natural products, many of which, for example, the isoflavonoids and triterpene saponins, have important biological activities impacting both plant and animal (including human) health. Second, *M. truncatula* serves as an excellent model for alfalfa, the world's major forage legume, and forage quality is determined in large part by the concentrations of products of secondary metabolism, particularly lignin and condensed tannins. In Chapter 20 of this book, there is a review on the recent progress in understanding the pathways leading to flavonoids, lignin, and triterpene saponins through utilization of genetic resources in *M. truncatula*.

Forage legumes have a higher nutritional quality compared to grasses and provide an important input of protein in animal nutrition. More recently public and scientific debate has reassessed forage legume importance for the quality of livestock nutrition, and welfare has having relevant consequences on the quality of final products (meat, milk, etc.) and ultimately on human health. This happened because of the occurrence of bovine spongiform encephalopathy (BSE) related to the traditional use of offal in animal feedlots as a source of protein [48]. In addition, legumes contain secondary metabolites that can prove useful for their antioxidant, antiviral, antimicrobial, antidiabetic, anti-allergenic, and anti-inflammatory properties [129, 130]. These properties are related to secondary molecules such as flavonoids and saponins. In forage legumes saponins can be toxic to monogastric animals and reduce forage palatability for ruminants. Mutant analysis in *M. truncatula* has unveiled the genetic control of key biosynthetic steps for saponins related to oxidation and glycosylation [102, 108], opening possibilities of biotechnological manipulation of saponins in alfalfa.

Modest levels of proanthocyanidins or condensed tannins (PAs) in forages reduce the occurrence of pasture bloat and at the same time promote increased dietary protein nitrogen utilization in ruminant animals [130]. The lack of PAs in the leaves of the major forage legume such as alfalfa (*Medicago sativa*) has prompted studies for the understanding of the molecular and cellular biology of PA polymerization, transport, and storage helped by the functional genomics tools available for *M. truncatula*. Recent positive achievements were obtained by biotechnological strategies based on the overexpression of different MYB transcription factors that induced PAs accumulation in both alfalfa and clover leaves [125] using iterative processes, such as successive rounds of crosses between different transgenic lines or the sequential transformation of transgenic plants with additional transgenes. Both approaches have two major drawbacks: the long and labor-intensive processes involving several breeding generations and the fact that the different transgenes are unlinked, leading to segregation in subsequent generations.

New and improved alfalfa cultivars have been obtained in the last decades by conventional breeding programs. However, the alfalfa plant has high protein content but lacks PAs in their vegetative parts. The presence of PAs in the alfalfa plants could prevent pasture bloat, improve the efficiency of conversion of plant protein into animal protein (ruminal protein bypass), reduce greenhouse gases, reduce gastrointestinal parasites, and inhibit insect feeding. Pasture bloat usually occurs in animals grazing wheat legume or legume-dominant pastures, particularly alfalfa, ladino, and red or white clovers. Prevention of pasture bloat is difficult, and death rates as high as 20% are recorded in cattle-grazing bloat-prone pasture, and in pastoral areas, the annual mortality rate from bloat in dairy cows may approach 1%. There is also economic loss from depressed milk production in nonfatal cases and from suboptimal use of bloat-prone pastures. The introduction by metabolic engineering of several genes involved in the anthocyanin and PA biosynthesis into these new cultivars would be particularly important to ruminant livestock producers [131].

Preliminary results obtained by our group to de novo produce PAs in the alfalfa forage are promising because we have developed a multigenic construct able to simultaneously introduce four transgenes into the plant genome. To engineer the anthocyanin and proanthocyanidin biosynthetic pathways in two *Nicotiana* species, we incorporated four transgenes to the plant chassis (Fig. 4a–c). We opted to perform a simultaneous transformation of the genes

Fig. 4 (continued) construct performed by the GB2.0 cloning system containing the four transgenes. (**c**) Leaf of *M. truncatula* cv. Jemalong 2HA with anthocyanin patches (left) compared with an alfalfa leaf without anthocyanin production (right). (**d**) Validation of the multigenic construct by transient expression in agroinfiltrated *N. benthamiana* leaves. WT non-infiltrated leaf (left), control leaf agroinfiltrated with *DsRed* (center) and infiltrated leaf with the multigenic construct showing purple pigmentation due to anthocyanin production (right). (**e**) WT non-transformed *N. tabacum* plant with green leaves (left) and stable transformed *N. tabacum* plant with the multigenic construct showing constitutive purple color (right). (**f**) Flower of a WT non-transformed *N. tabacum* plant showing pale pink color (left) and flower of a stable transformed *N. tabacum* plant showing an intense purple color (right). Adapted from Fresquet-Corrales et al. [133], under a Creative Commons Attribution 4.0 International License (CC BY 4.0)

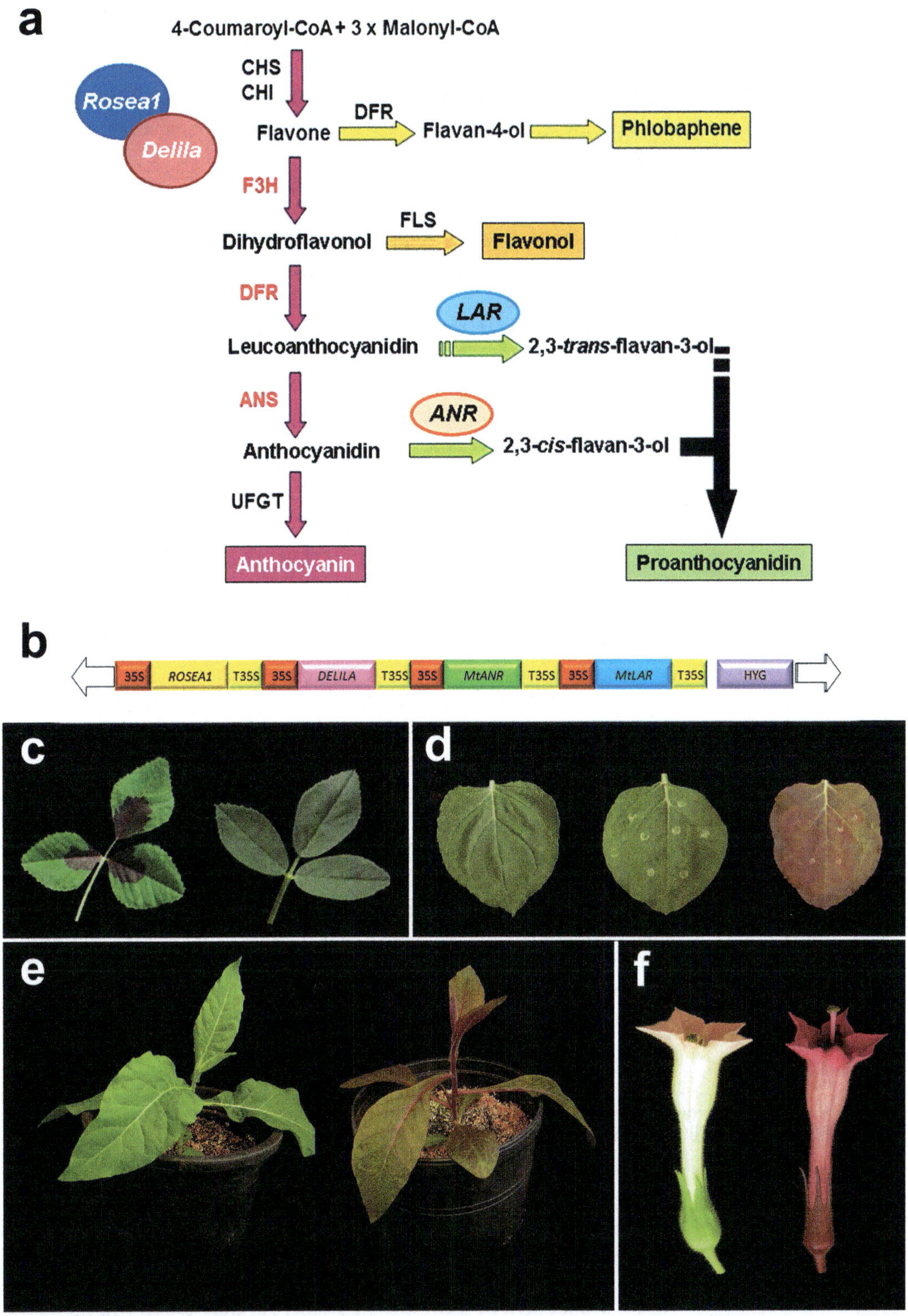

Fig. 4 Metabolic engineering to simultaneously activate the anthocyanin and proanthocyanidin pathways. (**a**) Activation of the anthocyanin pathway by means of two *A. majus* transcription factors (*AmRosea1* and *AmDelila*) and the proanthocyanidin pathway by two *M. truncatula* genes (*MtANR* and *MtLAR*). (**b**) Multigenic

linked in a multigenic construct rather than classical breeding or retransformation approaches. We generated a multigenic construct using the GoldenBraid 2.0 cloning system, a standardized DNA assembly platform developed to facilitate multigene engineering in plants by the simultaneous incorporation of several transgenes (https://gbcloning.upv.es/). The construct contained two *Antirrhinum majus* transcription factors (*AmRosea1* and *AmDelila*) [132] to upregulate the anthocyanin pathway in combination with two *M. truncatula* genes (*MtLAR* and *MtANR*) to produce the enzymes that will derivate the biosynthetic pathway to PAs production.

Transient and stable transformations of *Nicotiana benthamiana* and *Nicotiana tabacum* with the multigenic construct were, respectively, performed. Transient expression experiments in *N. benthamiana* showed the activation of the anthocyanin pathway producing a purple color in the infiltrated leaves and also the effective production of PAs (Fig. 4d). The integration capacity of the four transgenes, their respective expression levels, and their heritability in the second generation were analyzed in stably transformed *N. tabacum* plants (Fig. 4e, f). DMACA and phoroglucinolysis/HPLC-MS analyses corroborated the activation of both pathways and the effective production of PAs in T0 and T1 transgenic tobacco plants [133]. Stable transformation essays of different alfalfa genotypes with our multigenic construct are currently in progress to produce transgenic alfalfa plants with improved nutritional value to avoid pasture bloat in the ruminant livestock. The production of forage legumes with high protein content and discrete levels of condensed tannins to reduce the occurrence of bloat and to promote increased dietary protein nitrogen utilization avoiding problems of palatability is a challenge that can be achieved by metabolic engineering in a near future. In addition to well-known beneficial properties of flavonoids, recent evidence suggests that flavonoids themselves, particularly fractions rich in PAs, can significantly reduce cognitive deterioration (Alzheimer's disease) in animal model systems and may more generally promote improvements in memory acquisition, consolidation, storage, and retrieval under non-degenerative conditions [134, 135].

6 Conclusions and Future Prospects

Functional genomics in *M. truncatula* has been proven to be very useful in the study and comprehension of many aspects of legume nodulation, development, and plant secondary metabolism that could not be studied in other model plants such as *Arabidopsis*. The availability of genomic data and reverse genetic tools in this legume species allows the possibility of new discoveries in the field of plant development and especially in these processes that are almost exclusive of legumes. Functional genomics tools and phenotypic analyses of several mutants generated in *M. truncatula* have been essential to better understand the compound leaf development, the zygomorphic flower ontogeny and development, the compound inflorescence development and plant architecture, the molecular and genetic bases of nitrogen fixation, or the biosynthesis of secondary metabolites with interesting pharmacological and/or nutraceutical properties.

A better understanding of nitrogen fixation and symbiosis is fundamental for the development of a sustainable agriculture aiming at a reduction of inputs and at maintaining soil fertility. The doubling of world food production in the past four decades was contributed by a sevenfold increase of nitrogen fertilization [136]. The anthropogenic nitrogen which is mostly lost to air, water, and land affects the climate, the chemistry of the atmosphere, and the composition and function of terrestrial and aquatic ecosystems [137]. Improving the ability of plants to exploit environmental nitrogen would decrease nitrogen fertilization and its negative consequences; therefore a deep understanding of legume symbiosis with nitrogen-fixing bacteria could help the long-term goal of transferring the associative ability of legume species to nonsymbiotic crops of agronomic relevance. As a consequence functional genomics of nodulation will have an impact on reduction of intensive agriculture practices with benefits for the preservation of environment and quality of human activities [48]. The CRISPR/Cas9 system is rapidly becoming the reagent of choice for targeted mutagenesis and gene editing in crop species including legumes [127]. A modified Cas9 enzyme was shown to successfully mutate target genes in somatic cells of two legume species, soybean and *Medicago truncatula*. In addition, an online web tool for fast identification of CRISPR/*Cas9* target loci and to quickly identify a high number of potential target sites has been recently created [128, 138, 139]. These new tools would facilitate targeted mutagenesis in legumes opening the possibility to create new improved varieties through genome engineering that might be more accepted by consumers than transgenic plants carrying foreign DNA in their genomes.

Speedy developments have been witnessed in legume genomics due to advancements in next-generation sequencing and high-throughput genotyping technologies. Reference genome sequences for many model and crop legumes have been reported in the last years. The availability of the draft genome sequences and re-sequencing of elite genotypes for several important legume crops have made it possible to identify structural variations at large scale. Availability of large-scale genomic resources and low-cost high-throughput genotyping technologies are enhancing the efficiency and resolution of genetic mapping and marker-trait association studies. Deployment of molecular breeding approaches has resulted in the development of improved lines in some legume crops. In order to support genomics-driven legume improvement at a fast pace, the deployment of breeder-friendly genomics and decision support tools seems appear to be critical in breeding programs. Emerging functional genomics and informatics tools/approaches will be the key driving force for accelerating genomics-assisted breeding and ultimately ensuring nutritional and food security in developing countries [140].

Acknowledgments

We gratefully acknowledge funding from the Spanish Ministry of Economy and Competitiveness (http://www.idi.mineco.gob.es/portal/site/MICINN) grant BIO2016-75485-R.

References

1. Azani N, Babineau M, Bailey CD et al (2017) A new subfamily classification of the Leguminosae based on a taxonomically comprehensive phylogeny. The legume phylogeny working group (LPWG). Taxon 66(1):44–47
2. Young ND, Debellé F, Oldroyd GED et al (2011) The Medicago genome provides insight into the evolution of rhizobial symbioses. Nature 480(7378):520–524
3. Cannon SB (2013) The model legumes genomes. Methods Mol Biol 1069:1–14
4. Smyth DR, Bowman JL, Meyerowitz EM (1990) Early flower development in *Arabidopsis*. Plant Cell 2:755–767
5. Sommer H, Beltran JP, Huijser P et al (1990) *Deficiens*, a homeotic gene involved in the control of flower morphogenesis in *Antirrhinum majus*: the protein shows homology to transcription factors. EMBO J 9:605–613
6. Benlloch R, Navarro C, Beltrán JP et al (2003) Floral development of the model legume *Medicago truncatula*: ontogeny studies as a tool to better characterize homeotic mutations. Sex Plant Reprod 15(5):231–241
7. Tucker SC (1989) Overlapping organ initiation and common primordia in flowers of *Pisum sativum* (*Leguminosae*: *Papilionoideae*). Am J Bot 76:714–729
8. Tucker SC (2003) Floral development in legumes. Plant Physiol 131:911926
9. Ferrandiz C, Navarro C, Gomez MD et al (1999) Flower development in *Pisum sativum*: from the war of the whorls to the battle of the common primordia. Dev Genet 25:280–290
10. Kamenetsky R, Akhmetova M (1994) Floral development of *Eremurus-Altaicus* (*Liliaceae*). Israel J Plant Sci 42:227–233
11. Delaet J, Clinckemaillie D, Jansen S et al (1995) Floral ontogeny in the *Plumbaginaceae*. J Plant Res 108:289–304
12. Evans RC, Dickinson TA (1996) North-American black-fruited hawthorns. 2. Floral development of 10-stamen morphotypes in *Crataegus* section *Douglasii* (*Rosaceae*, *Maloideae*). Am J Bot 83:961–978
13. Kirchoff BK (1997) Inflorescence and flower development in the *Hedychieae* (*Zingiberaceae*) – *Hedychium*. Can J Bot 75:581–594
14. Barker DG, Bianchi S, Blondon F et al (1990) *Medicago truncatula*, A model plant for studying the molecular genetics of the *Rhizobium*-legume symbiosis. Plant Mol Biol Rep 8:40–49
15. Handberg K, Stougaard J (1992) *Lotus japonicus*, an autogamous, diploid legume species for classical and molecular genetics. Plant J 2:487–496
16. Cook DR (1999) *Medicago truncatula*--a model in the making! Curr Opin Plant Biol 2:301–304
17. Kijne J (1992) The rhizobium infection process. In: Stacey G, Burris RH, Evans HJ (eds) Biological nitrogen fixation. New York, Chapman and Hall, pp 349–398
18. Pawlowski K, Bisseling T (1996) Rhizobial and actinorhizal symbiosis: what are the shared features? Plant Cell 8:1899–1913
19. Tattersall AD, Turner L, Knox MR et al (2005) The mutant *crispa* reveals multiple roles for *PHANTASTICA* in pea compound leaf development. Plant Cell 17:1046–1060
20. Hofer J, Turner L, Moreau C et al (2009) Tendril-less regulates tendril formation in pea leaves. Plant Cell 21:420–428

21. Jing R, Knox MR, Lee JM et al (2005) Insertional polymorphism and antiquity of PDRl retrotransposon insertions in Pisum species. Genetics 171:741–752
22. Gao Z, Eyers S, Thomas C et al (2004) Identification of markers tightly linked to *sbm* recessive genes for resistance to pea seedborne mosaic virus. Theor Appl Genet 109:488–494
23. Barratt DH, Barber L, Kruger NJ et al (2001) Multiple, distinct isoforms of sucrose synthase in pea. Plant Physiol 127:655–664
24. Beveridge CA, Dun EA, Rameau C (2009) Pea has its tendrils in branching discoveries spanning a century from auxin to strigolactones. Plant Physiol 151:985–990
25. Boyer F-D, de Saint Germain A, Pillot J-P et al (2012) Strigolactone-related molecules for branching inhibition in garden pea: molecule design for shoot branching. Plant Physiol 159(4):1524–1544
26. Benlloch R, Berbel A, Ali L et al (2015) Genetic control of inflorescence architecture in legumes. Front Plant Sci 6:543
27. Berbel A, Navarro C, Ferrandiz C et al (2001) Analysis of *PEAM4*, the pea *AP1* functional homolog, supports a model for *AP1*-like genes controlling both floral meristem and floral organ identity in different plant species. Plant J 25:441–451
28. Berbel A, Navarro C, Ferrándiz C et al (2005) Functional conservation of *PISTILLATA* activity in a pea homolog lacking the PI motif. Plant Physiol 139:174–185
29. Benlloch R, d'Erfurth I, Ferrandiz C et al (2006) Isolation of *mtpim* proves Tntl a useful reverse genetics tool in *Medicago truncatula* and uncovers new aspects of AP1-like functions in legumes. Plant Physiol 142:972–983
30. Benlloch R, Berbel A, Serrano-Mislata A et al (2007) Floral initiation and inflorescence architecture: a comparative view. Ann Bot 100:659–676
31. Benlloch R, Roque E, Ferràndiz C et al (2009) Analysis of B function in legumes: PISTILLATA proteins do not require the PI motif for floral organ development in *Medicago truncatula*. Plant J 60(1):102–111
32. Roque E, Serwatowska J, Rochina MC et al (2013) Functional specialization of duplicated *AP3*-like genes in *Medicago truncatula*. Plant J 73:663–675
33. Serwatowska J, Roque E, Gómez-Mena C et al (2014) Two *euAGAMOUS* genes control C-function in *Medicago truncatula*. PLoS One 9:e103770
34. Roque E, Fares MA, Yenush L et al (2016) Evolution by gene duplication of *Medicago truncatula PISTILLATA*-like transcription factors. J Exp Bot 67(6):1805–1817
35. Vodkin L, Jones S, Gonzalez DO et al (2008) Genomics of soybean seed development. In: Stacey G (ed) Plant genetics and genomics: crops and models, Genetics and genomics of soybean, vol 2. Springer, NY, pp 163–184
36. Kwak M, Velasco D, Gepts P (2008) Mapping homologous sequences for determinacy and photoperiod sensitivity in common bean (*Phaseolus vulgaris*). J Hered 99:283–291
37. Young ND, Udvardi M (2009) Translating *Medicago truncatula* genomics to crop legumes. Curr Opin Plant Biol 12(2):193–201
38. Gepts P, Beavis WD, Brummer EC et al (2005) Legumes as a model plant family: genomics for food and feed report of the cross-legume advances through genomics conference. Plant Physiol 137:1228–1235
39. Oldroyd GE, Downie JA (2008) Coordinating nodule morphogenesis with rhizobial infection in legumes. Annu Rev Plant Biol 59:519–546
40. Xie DY, Sharma SB, Wright E et al (2006) Metabolic engineering of proanthocyanidins through co-expression of anthocyanidin reductase and the *PAP1* MYB transcription factor. Plant J 45:895–907
41. Farag MA, Huhman DV, Dixon RA et al (2008) Metabolomics reveals novel pathways and differential mechanistic and elicitor-specific responses in Phenylpropanoid and Isoflavonoid biosynthesis in *Medicago truncatula* cell cultures. Plant Physiol 146(2):387–402
42. Ha CM, Escamilla-Trevino L, Yarce JC et al (2016) An essential role of caffeoyl shikimate esterase in monolignol biosynthesis in *Medicago truncatula*. Plant J 86:363–375
43. Mouradov A, Spangenberg G (2014) Flavonoids: a metabolic network mediating plants adaptation to their real estate. Front Plant Sci 5:620
44. Yang DL, Li Q, Deng YW et al (2008) Altered disease development in the *eui* mutants and *Eui* overexpressors indicates that gibberellins negatively regulate rice basal disease resistance. Mol Plant 1:528–537
45. Gonzalez-Rizzo S, Crespi M, Frugier F (2006) The *Medicago truncatula* CRE1 cytokinin receptor regulates lateral root development and early symbiotic interaction with *Sinorhizobium meliloti*. Plant Cell 18:2680–2693
46. Lesins KA, Lesins I (1979) Genus Medicago (Leguminosae). A taxogenetic study. Dr W Yung bv Publishers, The Hague, p 228
47. Iantcheva A, Vassileva V, Ugrinova M et al (2009) Development of functional genomic platform for model legume *Medicago truncatula* in Bulgaria. Biotechnol Biotechnol Equip 23(4):1440–1443

48. Panara F, Calderini O, Porceddu A (2012) *Medicago truncatula* functional genomics – an invaluable resource for studies on agriculture sustainability. In: Meroni G, Petrera F (eds) Biochemistry, genetics and molecular biology "functional genomics". InTech. https://doi.org/10.5772/51016
49. Kang Y, Li M, Sinharoy S, Verdier J (2016) A snapshot of functional genetic studies in *Medicago truncatula*. Front Plant Sci 7:1175
50. Cook DR, VandenBosch K, de Bruijn FJ (1997) Model legumes get the nod. Plant Cell 9(3):275
51. Taylor S, Hofer J, Ian Murfet I (2001) *Stamina pistilloida*, the pea Ortholog of *Fim* and *UFO*, is required for normal development of flowers, inflorescences, and leaves. Plant Cell 13(1):31–46
52. Sagan M, Morandi D, Tarenghi E et al (1997) Selection of nodulation and mycorrhizal mutants in the model plant *Medicago truncatula* (Gaertn.) after γ-ray mutagenesis. Plant Sci 111:63–71
53. Oldroyd G (2003) Identification and characterization of nodulation-signaling pathway 2, a gene of *Medicago truncatula* involved in nod factor signaling. Plant Physiol 131(3):1027–1032
54. Schnabel E, Journet EP, de Carvalho-Niebel F et al (2005) The *Medicago truncatula SUNN* gene encodes a CLV1-like leucine-rich repeat receptor kinase that regulates nodule number and root length. Plant Mol Biol 58(6):809–822
55. Starker CG, Parra-Colmenares AL, Smith L et al (2006) Nitrogen fixation mutants of *Medicago truncatula* fail to support plant and bacterial symbiotic gene expression. Plant Physiol 140(2):671–680
56. Wang H (2006) Fast neutron bombardment (FNB) mutagenesis for forward and reverse genetic studies in plants. Global Science Books, Isleworth, pp 629–639
57. Middleton PH, Jakab J, Penmetsa RV et al (2007) An ERF transcription factor in *Medicago truncatula* that is essential for nod factor signal transduction. Plant Cell 19(4):1221–1234
58. Vernié T, Moreau S, de Billy F et al (2008) EFD is an ERF transcription factor involved in the control of nodule number and differentiation in *Medicago truncatula*. Plant Cell 20(10):2696–2713
59. Rogers C, Wen J, Chen R et al (2009) Deletion-based reverse genetics in *Medicago truncatula*. Plant Physiol 151(3):1077–1086
60. Chen J, Yu J, Ge L et al (2010) Control of dissected leaf morphology by a Cys(2)his(2) zinc finger transcription factor in the model legume *Medicago truncatula*. Proc Natl Acad Sci U S A 107(23):10754–10759
61. Peng J, Yu J, Wang H et al (2011) Regulation of compound leaf development in *Medicago truncatula* by *Fused Compound Leaf1*, a class M KNOX gene. Plant Cell 23:3929–3943
62. Murray JD (2011) Invasion by invitation: rhizobial infection in legumes. Mol Plant-Microbe Interact 24(6):631–639
63. Murray JD, Muni RRD, Torres-Jerez I et al (2011) *Vapyrin*, a gene essential for intracellular progression of arbuscular mycorrhizal symbiosis, is also essential for infection by rhizobia in the nodule symbiosis of *Medicago truncatula*. Plant J 65(2):244–252
64. Mitra RM, Gleason CA, Edwards A et al (2004) A Ca2+/calmodulin-dependent protein kinase required for symbiotic nodule development: gene identification by transcript-based cloning. Science 101(13):4701
65. Senthil-Kumar M, Mysore K (2011) New dimensions for VIGS in plant functional genomics. Trends Plant Sci 16(12):656–665
66. Grønlund M, Constantin G, Piednoir E et al (2008) Virus induced gene silencing in *Medicago truncatula* and *Lathyrus odorata*. Virus Res 135:345–349
67. Várallyay E, Lichner Z, Sáfrány J et al (2010) Development of a virus induced gene silencing vector from a legumes infecting tobamovirus. Acta Biol Hung 61(4):457–469
68. Ané JM, Kiss GB, Riely BK et al (2004) *Medicago truncatula* DMI1 required for bacterial and fungal symbioses in legumes. Science 303(5662):1364–1367
69. Kalò P, Gleason C, Edwards A et al (2005) Nodulation signaling in legumes requires NSP2, a member of the GRAS family of transcriptional regulators. Science 308(5729):1786–1789
70. Ivashuta S, Liu J, Lohar DP et al (2005) RNA interference identifies a calcium-dependent protein kinase involved in *Medicago truncatula* root development. Plant Cell 17(11):2911–2921
71. Limpens E, Mirabella R, Fedorova E et al (2005) Formation of organelle-like N2-fixing symbiosomes in legume root nodules is controlled by DMI2. Proc Natl Acad Sci U S A 102(29):10375–10380
72. Arrighi JF, Barre A, Amor BB et al (2006) The *Medicago truncatula* lysin motif-receptor-like kinase gene family includes NFP and new nodule-expressed genes. Plant Physiol 142(1):265–279
73. Huo X, Schnabel E, Hughes K, Frugoli J (2006) RNAi phenotypes and the localization of a protein::GUS fusion imply a role for *Medicago truncatula* PIN genes in nodulation. J Plant Growth Regul 25(2):156–165

74. Wasson AP, Pellerone FI (2006) Silencing the flavonoid pathway in *Medicago truncatula* inhibits root nodule formation and prevents auxin transport regulation by rhizobia. Plant Cell 18(7):1617–1629
75. Gargantini PR, Gonzalez-Rizzo S, Chinchilla D et al (2006) A CDPK isoform participates in the regulation of nodule number in *Medicago truncatula*. Plant J 48(6):843–856
76. Colditz F, Niehaus K, Krajinski F (2007) Silencing of PR-10-like proteins in *Medicago truncatula* results in an antagonistic induction of other PR proteins and in an increased tolerance upon infection with the oomycete *Aphanomyces euteiches*. Planta 226(1):57–71
77. Kevei Z, Lougnon G, Mergaert P et al (2007) 3-hydroxy-3-methylglutaryl coenzyme a reductase 1 interacts with NORK and is crucial for nodulation in *Medicago truncatula*. Plant Cell 19(12):3974–3989
78. Javot H, Penmetsa RV, Terzaghi N et al (2007) A *Medicago truncatula* phosphate transporter indispensable for the arbuscular mycorrhizal symbiosis. Proc Natl Acad Sci U S A 104(5):1720–1725
79. Messinese E, Mun JH, Yeun LH et al (2007) A novel nuclear protein interacts with the symbiotic DMI3 calcium- and calmodulin-dependent protein kinase of *Medicago truncatula*. Mol Plant-Microbe Interact 20(8):912–921
80. Rosnoblet C, Aubry C, Leprince O et al (2007) The regulatory gamma subunit SNF4b of the sucrose non-fermenting-related kinase complex is involved in longevity and stachyose accumulation during maturation of *Medicago truncatula* seeds. Plant J 51(1):47–59
81. Wan X, Hontelez J, Lillo A et al (2007) *Medicago truncatula* ENOD40-1 and ENOD40-2 are both involved in nodule initiation and bacteroid development. J Exp Bot 58(8):2033–2041
82. Zhang J, Subramanian S, Zhang Y et al (2007) Flavone synthases from *Medicago truncatula* are flavanone-2-hydroxylases and are important for nodulation. Plant Physiol 144(2):741–751
83. Floss DS, Hause B, Lange PR et al (2008) Knock-down of the MEP pathway isogene 1-deoxy-D-xylulose 5-phosphate synthase 2 inhibits formation of arbuscular mycorrhiza-induced apocarotenoids, and abolishes normal expression of mycorrhiza-specific plant marker genes. Plant J 56(1):86–100
84. Floss DS, Schliemann W, Schmidt JÃ et al (2008) RNA interferencemediated repression of MtCCD1 in mycorrhizal roots of *Medicago truncatula* causes accumulation of C27 apocarotenoids, shedding light on the functional role of CCD1. Plant Physiol 148(3):1267–1282
85. Chen SK, Kurdyukov S, Kereszt A et al (2009) The association of homeobox gene expression with stem cell formation and morphogenesis in cultured *Medicago truncatula*. Planta 230(4):827–840
86. Pii Y, Astegno A, Peroni E et al (2009) The *Medicago truncatula* N5 gene encoding a root-specific lipid transfer protein is required for the symbiotic interaction with *Sinorhizobium meliloti*. Mol Plant-Microbe Interact 22(12):1577–1587
87. de Lorenzo L, Merchan F, Laporte P et al (2009) A novel plant leucine-rich repeat receptor kinase regulates the response of *Medicago truncatula* roots to salt stress. Plant Cell 21(2):668–680
88. Kuppusamy KT, Ivashuta S, Bucciarelli B et al (2009) Knockdown of *CELL DIVISION CYCLE16* reveals an inverse relationship between lateral root and nodule numbers and a link to auxin in *Medicago truncatula*. Plant Physiol 151(3):1155–1166
89. Peleg-Grossman S, Golani Y, Kaye Y et al (2009) NPR1 protein regulates pathogenic and symbiotic interactions between rhizobium and legumes and non-legumes. PLoS One 4(12):e8399
90. Laporte P, Satiat-Jeunemaître B, Velasco I et al (2010) A novel RNA-binding peptide regulates the establishment of the *Medicago truncatula-Sinorhizobium meliloti* nitrogen-fixing symbiosis. Plant J 62(1):24–38
91. Lefebvre B, Timmers T, Mbengue M et al (2010) A remorin protein interacts with symbiotic receptors and regulates bacterial infection. Proc Natl Acad Sci U S A 107(5):2343–2348
92. Pumplin N, Mondo SJ, Topp S et al (2010) *Medicago truncatula* Vapyrin is a novel protein required for arbuscular mycorrhizal symbiosis. Plant J 61(3):482–494
93. Zhou R, Jackson L, Shadle G et al (2010) Distinct cinnamoyl CoA reductases involved in parallel routes to lignin in *Medicago truncatula*. Proc Natl Acad Sci U S A 107(41):17803–17808
94. Haney CH (2010) Plant flotillins are required for infection by nitrogen-fixing bacteria. Proc Natl Acad Sci U S A 107(1):478–483
95. Kuhn H, Kùster RN (2010) Membrane steroid-binding protein 1 induced by a diffusible fungal signal is critical for mycorrhization in *Medicago truncatula*. New Phytol 185(3):716–733
96. Zdyb A, Demchenko K, Heumann J, Mrosk C, Grzeganek P, Gòbel C et al (2011) Jasmonate biosynthesis in legume and actinorhizal nodules. New Phytol 189(2):568–579

97. Msehli SE, Lambert A, Baldacci-Cresp F et al (2011) Crucial role of (homo) glutathione in nitrogen fixation in *Medicago truncatula* nodules. New Phytol 192(2):496–506
98. Gimeno-Gilles C, Gervais ML, Planchet E et al (2011) A stress-associated protein containing A20/AN1 zing-finger domains expressed in *Medicago truncatula* seeds. Plant Physiol Biochem 49(3):303–310
99. Horchani F, Prèvot M, Boscari A et al (2011) Both plant and bacterial nitrate reductases contribute to nitric oxide production in *Medicago truncatula* nitrogen-fixing nodules. Plant Physiol 155(2):1023–1036
100. Pauly N, Ferrari C, Andrio E et al (2011) MtNOA1/RIF1 modulates *Medicago truncatula-Sinorhizobium meliloti* nodule development without affecting its nitric oxide content. J Exp Bot 62(3):939–948
101. Riely BK, He H, Venkateshwaran M et al (2011) Identification oflegume RopGEF gene families and characterization of a *Medicago truncatula* RopGEF mediating polar growth of root hairs. Plant J 65(2):230–243
102. Naoumkina M, Dixon R (2011) Characterization of the mannan synthase promoter from guar (*Cyamopsis tetragonoloba*). Plant Cell Rep 30(6):997–1006
103. Laurie ÌR, Diwadkar P, Jaudal M et al (2011) The Medicago *FLOWERING LOCUS T* homolog, *MtFTa1,* is a key regulator of Flowering time. Plant Physiol 156:2207–2224
104. Kiirika LM, Bergmann HF, Schikowsky C et al (2012) Silencing of the Rac1 GTPase MtROP9 in *Medicago truncatula* stimulates early mycorrhizal and oomycete root Colonizations but negatively affects Rhizobial infection. Plant Physiol 159(1):501–516
105. de Zélicourt A, Diet A, Marion J et al (2012) Dual involvement of a *Medicago truncatula* NAC transcription factor in root abiotic stress response and symbiotic nodule senescence. Plant J 70(2):220–230
106. O'Malley R, Ecker J (2010) Linking genotype to phenotype using the Arabidopsis unimutant collection. Plant J 61(6):928–940
107. Laffont C, Blanchet S, Lapierre C et al (2010) The compact root architecture1 gene regulates lignification, flavonoid production, and polar auxin transport in *Medicago truncatula*. Plant Physiol 153(4):1597–1607
108. Carelli M, Biazzi E, Panara F et al (2011) *Medicago truncatula* CYP716A12 is a multifunctional oxidase involved in the biosynthesis of hemolytic saponins. Plant Cell 23(8):3070–3081
109. d'Erfurth I, Cosson V, Eschstruth A et al (2003) Efficient transposition of the Tnt1 tobacco retrotransposon in the model legume *Medicago truncatula*. Plant J 34(1):95–106
110. Tadege M, Ratet P, Mysore K (2005) Insertional mutagenesis: a Swiss Army knife for functional genomics of *Medicago truncatula*. Trends Plant Sci 10(5):229–235
111. Tadege M, Wen J, He J et al (2008) Large-scale insertional mutagenesis using the Tnt1 retrotransposon in the model legume *Medicago truncatula*. Plant J 54(2):335–347
112. Cheng X, Wen J, Tadege M et al (2011) Reverse genetics in *Medicago truncatula* using Tnt1 insertion mutants. Methods Mol Biol 678:179–190
113. Cheng X, Wang M, Lee H-K et al (2014) An efficient reverse genetics platform in the model legume *Medicago truncatula*. New Phytol 201:1065–1076
114. Calderini O, Carelli M, Panara F et al (2011) Collection of mutants for functional genomics in the legume *Medicago truncatula*. Plant Genet Resour 9(2):174–176
115. Rakocevic A, Mondy S, Tirichine L et al (2009) MERE1, a low-copy-number copia-type retroelement in *Medicago truncatula active* during tissue culture. Plant Physiol 151(3):1250–1263
116. Porceddu A, Panara F, Calderini O et al (2008) An Italian functional genomic resource for *Medicago truncatula*. BMC Res Notes 1:129
117. Wang H, Chen J, Wen J et al (2008) Control of compound leaf development by *FLORICAULA/LEAFY* ortholog *SINGLE LEAFLET1* in *Medicago truncatula*. Plant Physiol 146(4):1759–1772
118. Peel GJ, Pang Y, Modolo LV (2009) The LAP1 MYB transcription factor orchestrates anthocyanidin biosynthesis and glycosylation in Medicago. Plant J 59(1):136–149
119. Zhao Q, Gallego-Giraldo L, Wang H et al (2010) An NAC transcription factor orchestrates multiple features of cell wall development in *Medicago truncatula*. Plant J 63(1):100–114
120. Vassileva V, Zehirov G, Ugrinova M (2010) Variable leaf epidermal leaf morphology in Tnt1 insertional mutants of the model legume *Medicago truncatula*. Biotechnol Biotechnol Equip 24(4):2060–2065
121. Wang D, Griffitts J, Starker C et al (2010) A nodule-specific protein secretory pathway required for nitrogen-fixing symbiosis. Science 327(5969):1126–1129
122. Zhou C, Han L, Pislariu C et al (2011) From model to crop: functional analysis of a *STAY-GREEN* gene in the model legume *Medicago truncatula* and effective use of the gene for alfalfa improvement. Plant Physiol 157(3):1483–1496
123. Tadege M, Mysore K (2011) Tnt1 retrotransposon tagging of STF in *Medicago truncatula*

reveals tight coordination of metabolic, hormonal and developmental signals during leaf morphogenesis. Mob Genet Elements 1(4):301–303
124. Uppalapati SR, Ishiga Y, Doraiswamy V et al (2012) Loss of abaxial leaf epicuticular wax in *Medicago truncatula* irg1/palm1 mutants results in reduced spore differentiation of anthracnose and nonhost rust pathogens. Plant Cell 24(1):353–370
125. Verdier J, Zhao J, Torres-Jerez I et al (2012) MtPAR MYB transcription factor acts as an on switch for proanthocyanidin biosynthesis in *Medicago truncatula*. Proc Natl Acad Sci U S A 109(5):1766–1771
126. Ratet P (2006) *Medicago truncatula* handbook. Noble Foundation
127. Curtin SJ, Voytas DF, Stupar RM (2012) Genome engineering of crops with designer nucleases. Plant Genome 5:42
128. Michno J-M, Wang X, Liu J et al (2015) CRISPR/Cas mutagenesis of soybean and *Medicago truncatula* using a new web-tool and a modified Cas9 enzyme. GM Crops Food 6(4):243–252
129. Howieson JG (2008) Nitrogen-fixing leguminous symbioses. Springer
130. Dixon RA (2010) Flavonoids and isoflavonoids: from plant biology to agriculture and neuroscience. Plant Physiol 154(2):453–457
131. Cañas LA, Fresquet S, Roque E et al (2017) Forage legumes with improved nutritional value: condensed tannins to avoid pasture bloat. In: Clemente A, Jiménez-López JC (eds) Legumes for global food security. NOVA Science Publishers, New York, pp 183–222
132. Butelli E, Titta L, Giorgio M et al (2008) Enrichment of tomato fruit with health-promoting anthocyanins by expression of select TFs. Nat Biotechnol 26:1301–1308
133. Fresquet-Corrales S, Roque E, Sarrión-Perdigones A et al (2017) Metabolic engineering to simultaneously activate anthocyanin and proanthocyanidin biosynthetic pathways in *Nicotiana spp*. PLoS One 12(9):e0184839
134. Ho L, Chen LH, Wang J et al (2009) Heterogeneity in red wine polyphenolic contents differentially influences Alzheimer's disease-type neuropathology and cognitive deterioration. J Alzheimers Dis 16(1):59–72
135. Wang J, Ferruzzi MG, Ho L et al (2012) Brain-targeted proanthocyanidin metabolites for Alzheimer's disease treatment. J Neurosci 32(15):5144–5150
136. Ollivier J, Töwe S, Bannert A et al (2011) Nitrogen turnover in soil and global change. FEMS Microbiol Ecol 78(1):3–16
137. Galloway JN, Townsend AR, Erisman JW et al (2008) Transformation of the nitrogen cycle: recent trends, questions, and potential solutions. Science 320(5878):889–892
138. Cermak T, Curtin SJ, Gil-Humanes J et al (2017) A multipurpose toolkit to enable advanced genome engineering in plants. Plant Cell 29:1196–1217
139. Cermak T, Curtin SJ (2017) Design and assembly of CRISPR/Cas9 reagents for gene knockout, targeted insertion, and replacement in wheat. Methods Mol Biol 1679: 187–212
140. Pandey MK, Roorkiwal M, Singh VK et al (2016) Emerging genomic tools for legume breeding: current status and future prospects. Front Plant Sci 7:455

Chapter 3

The *Medicago truncatula* Genome: Genomic Data Availability

David Burks, Rajeev Azad, Jiangqi Wen, and Rebecca Dickstein

Abstract

Medicago truncatula emerged in 1990 as a model for legumes, comprising the third largest land plant family. Most legumes form symbiotic nitrogen-fixing root nodules with compatible soil bacteria and thus are important contributors to the global nitrogen cycle and sustainable agriculture. Legumes and legume products are important sources for human and animal protein as well as for edible and industrial oils. In the years since *M. truncatula* was chosen as a legume model, many genetic, genomic, and molecular resources have become available, including reference quality genome sequences for two widely used genotypes. Accessibility of genomic data is important for many different types of studies with *M. truncatula* as well as for research involving crop and forage legumes. In this chapter, we discuss strategies to obtain archived *M. truncatula* genomic data originally deposited into custom databases that are no longer maintained but are now accessible in general databases. We also review key current genomic databases that are specific to *M. truncatula* as well as those that contain *M. truncatula* data in addition to data from other plants.

Key words *Medicago truncatula*, Model legume, Genome data, Genome sequence, Expression data, Databases, Symbiotic nitrogen fixation

1 Introduction

The legume family (the Fabaceae or Papilionoideae) constitutes the third largest land plant family with over 19,000 known species. Most legumes form symbiotic nitrogen-fixing root nodules with compatible soil bacteria and thus are key contributors to the global nitrogen cycle. This property makes legumes important species in many ecosystems and gives them significance as agricultural species. Legumes and legume products are important sources for human and animal protein and also for edible and industrial oils. Most economically important legumes come from one of two clades, usually referred to as the tropical and the cool season legumes. Tropical legumes include soybean (*Glycine max*) and beans (*Phaseolus* species), while the cool

Luis A. Cañas and José Pío Beltrán (eds.), *Functional genomics in* Medicago truncatula*: Methods and Protocols*, Methods in Molecular Biology, vol. 1822, https://doi.org/10.1007/978-1-4939-8633-0_3,

season legumes include the forage crop alfalfa (*Medicago sativa*), as well as pea (*Pisum sativum*) and chickpea (*Cicer arietinum*).

In the 1990s, *Medicago truncatula* and *Lotus japonicus* emerged as model legumes for fundamental research on legume biology and symbiotic nitrogen fixation [1–3]. *M. truncatula* forms indeterminate nodules typical of the cool season legumes, while *L. japonicus* forms determinate nodules characteristic of the tropical legumes. After the realization that symbiotic nitrogen fixation (SNF) shared pathways with the arbuscular mycorrhizal (AM) symbiosis, these two legumes became models for AM symbioses as well [4–7]. Genomes of both model legumes, including two different *M. truncatula* genotypes, and a number of agriculturally important crop legumes have recently been sequenced [8–17]. *M. truncatula* was chosen as a reference species because of its short life cycle, tractable diploid genetics, small genome size, ease of transformation, high levels of genetic diversity, and synteny with crop and forage legumes with much larger genomes. There are abundant genomic and genetic resources available for research in *M. truncatula*, which enrich its use as a model legume and help organize research in this important plant.

M. truncatula genomics data started to become available in the late 1990s and early 2000s. The earliest efforts were devoted to producing expressed sequence tags (ESTs) which were ordered into tentative consensus sequences (TCs) in the *M. truncatula* gene index (MtGI), finding molecular genetic markers to enable map-based cloning of symbiotic genes, production of bacterial artificial chromosome (BAC) genomic libraries accompanied by BAC end and/or survey sequences, and comparative genomics [18–28]. Many of the databases created to support these efforts have ceased to exist. These data are, however, still available in GenBank. Newer databases have collected and made available some of the older data, especially when the *M. truncatula* scientific community has expressed an interest in specific types of data to the curators who organize the current databases.

M. truncatula genomics efforts in the late 2000s and continuing to the present include the completion and release of the reference genome sequence of genotype A17 in 2011 [8], with a significant update in 2014 [9], and the release of a second reference genome sequence of genotype R108 in 2017 [17]. Next generation sequencing (NGS) efforts have propelled key studies significant for the *M. truncatula* community. These include the Medicago Hapmap project, which is resequencing diverse inbred *M. truncatula* lines to provide foundations for discovery of sequence variation. Increasingly, environmental and developmental gene expression studies use RNAseq technology.

The Internet offers a wealth of information for almost every biological domain, and finding relevant, applicable data to modern research questions can be especially daunting. Cursory online searches for plant databases are likely to find databases as general

Table 1
***M. truncatula* specific databases**

Name	Link
M. truncatula gene index (MtGI) (ESTs and TCs)	ftp://occams.dfci.harvard.edu/pub/bio/tgi/data/
Software to run MtGI	ftp://occams.dfci.harvard.edu/pub/bio/tgi/software/
MedicCyc	pathway.gramene.org/gramene/mediccyc.shtml
M. truncatula Gene Expression Atlas	mtgea.noble.org/v3/
M. truncatula A17 reference genome V3.5 assembly	www.ncbi.nlm.nih.gov/assembly/GCF_000219495.1/
M. truncatula A17 reference genome V4.0 assembly	www.ncbi.nlm.nih.gov/assembly/GCF_000219495.3
M. truncatula R108 reference genome V1.0 assembly	www.ncbi.nlm.nih.gov/assembly/GCA_002024945.1/
M. truncatula genome database	www.medicagogenome.org/
MedicMine	medicmine.jcvi.org/medicmine/begin.do
Medicago Hapmap	www.medicagohapmap.org/
Symbimics	iant.toulouse.inra.fr/symbimics/

and obtuse as the initial query, with pages dedicated to recapitulating the same information as their nearest neighbors. Here, we review *M. truncatula* genomic data availability starting with the older data and with *M. truncatula*-specific databases (Table 1) and progress to newer and more generalized databases (Table 2). We have tried to identify the databases that provide some novel, structured contribution to the crowded realm of online biological information, whether that is in the form of unique data or the form in which it is made available. We survey the current state of multiplant databases containing information related to or derived from *M. truncatula*, among other species, that we think will be especially valuable to *M. truncatula* researchers.

2 *M. truncatula* Expressed Sequence Tags (ESTs) and Tentative Consensus Sequences (TCs)

The ESTs and TCs came from random cDNAs from a variety of *M. truncatula* organs from plants grown in diverse conditions [27, 28]. These were sorted according to the libraries from which they came, for example, nodule-specific transcripts were those found exclusively in nodule cDNA libraries [27, 29]. Although the websites that made the EST and TC data available to the public (called the

Table 2
Multiplant databases/multi-database information portals

Name	Link
Legume information system	www.legumeinfo.org/home
LegumeIP	www.plantgrn.noble.org/LegumeIP/
Gramene	www.gramene.org/
EnsemblPlants	plants.ensembl.org/index.html
Phytozome	phytozome.jgi.doe.gov/pz/portal.html
KEGG Plant	www.genome.jp/kegg/genome/plant.html
Plant Reactome	plantreactome.oicr.on.ca/
National Center for Biotechnology Information Gene Expression Omnibus	www.ncbi.nlm.nih.gov/geo/
European Molecular Biology Laboratory (EMBL) Expression Atlas	www.ebi.ac.uk/gxa/home
Plant Expression Database	www.plexdb.org/
EMBL ArrayExpress	www.ebi.ac.uk/arrayexpress/
Plant Massively Parallel Signature Sequencing	mpss.danforthcenter.org/
NCBI Sequence Read Archive	www.ncbi.nlm.nih.gov/sra
PlantOrDB	bioinfolab.miamioh.edu/plantordb/interface/about-plantordb.php
PTGBase	ocri-genomics.org/PTGBase/
Plant Protein Phosphorylation Database	www.p3db.org/
PlantPIs database	plantpis.ba.itb.cnr.it/
Plant snoRNA	bioinf.scri.sari.ac.uk/cgi-bin/plant_snorna/home
PlantProm	www.softberry.com/berry.phtml?topic=plantprom&group=data&subgroup=plantprom
Plant Transcriptional Regulatory Map	plantregmap.cbi.pku.edu.cn/
Plant DNA C-values database	data.kew.org/cvalues/

TIGR Gene Index (TGI) databases that included *M. truncatula* gene index (MtGI)) no longer exist, the data are available for retrieval through FTP at ftp://occams.dfci.harvard.edu/pub/bio/tgi/data/. The software tools to mine the database are available through ftp://occams.dfci.harvard.edu/pub/bio/tgi/software/. In addition, the TCs are accessible through the MedicMine database (medicmine.jcvi.org/medicmine/begin.do; see below). The JBrowse interface at the current *M. truncatula* genome database (www.medicagogenome.org/; see below) may be used to browse the TCs in the context of the current version of the *M. truncatula* genome. For example, querying MedicMine with TC35256, described by [29] as encoding calmodulin, returns the most recent synonymous TC, TC195776, and reports that this TC corresponds to Medtr3g055570. Querying the JBrowse interface of the *M. truncatula* genome database with Medtr3g055570 and turning on the TC track display the TCs from MtGI 11.0, which in this example is TC165192, one of the synonym TCs found for TC195776 and TC35256 in the MedicMine search. The ESTs and TCs are also accessible through the Legume Information System (LIS; legumeinfo.org/organism/Medicago/truncatula; see below). There, using the JBrowse interface, one may access the *M. truncatula* EST and TC data by turning on the *Medicago truncatula* tracks under the "Transcripts"/"Dana-Farber Cancer Institute" bars after querying by gene locus. The name Dana-Farber Cancer Institute refers to the fact that the EST and TC data were housed there after it moved relocating from their first home at TIGR, which is now a part of the J. Craig Venter Institute (JCVI).

3 *M. truncatula* Bacterial Artificial Chromosome (BAC) Libraries and Their Sequences

The first *M. truncatula* genome draft was assembled from sequences of overlapping BAC clones of genomic DNA [8]. These BAC libraries were initially made to facilitate map-based cloning of symbiotic genes [24] and were thus integrated with genetic markers to form sequence-based genetic maps [25, 26, 30, 31]. BAC library production was then expanded for the *M. truncatula* genome projects. Sequences on the BACs were annotated by the International Medicago Genome Annotation Group (IMGAG) [32]. The positions of the BAC clones can be viewed in the context of the assembled *M. truncatula* genome (V4.0) using the JBrowse viewers at *M. truncatula* genome site (www.medicagogenome.org) and via the Legume Information System (legumeinfo.org/genomes/jbrowse/?data=Mt4.0). Their sequences can be retrieved at GenBank (www.ncbi.nlm.nih.gov/) or via the *M. truncatula* genome site (www.medicagogenome.org). Only a subset of the BAC clones that were in several *M. truncatula* BAC libraries were fully sequenced as part of the *M. truncatula* genome projects, with many of the remainder having partial sequence available in GenBank. These partial sequences can be used to identify overlapping BAC clones that may be useful in finding particular DNA sequences that fall into the remaining gaps in the *M. truncatula* genome [9]. To find these BACs, one can BLAST [33] the high-throughput genomic sequences (HTGS) and/or the genomic survey sequences (GSS) databases at GenBank. The query sequences for these searches should be *M. truncatula* sequences that are as close to the gap as possible. The BAC clones themselves and robot-printed filter replicas of the libraries are available at two repositories, the Clemson University Genomics Institute (www.genome.clemson.edu/) and the French Plant Genomic Resources Center (cnrgv.toulouse.inra.fr/Library/Medicago).

4 MedicCyc

The *M. truncatula* research community benefits from MedicCyc, an older plant metabolic pathway database that is entirely devoted to *M. truncatula* [34]. MedicCyc was constructed using the Pathway Tools and MetaCyc pathway databases [35, 36] and subsequent manual curation to remove non-plant pathways, rendering it a valuable resource for the plant scientific community. It includes Medicago-specific pathways, such as triterpene saponin biosynthesis, that have been specifically modified to better represent the species. Unfortunately, the database is no longer being actively updated. It is currently available through a mirror hosted at Gramene (pathway.gramene.org/gramene/mediccyc.shtml). Furthermore, MedicCyc predates the first complete sequence draft of the *M. truncatula* genome [8] and only provides annotations from UniGene [37] or Affymetrix probeset IDs for genes linked in the pathways.

5 Data from *M. truncatula* Microarrays for Gene Expression Analyses

The ESTs and TCs formed the basis for cDNA-based arrays and long oligonucleotide microarrays that became available in the 2000s for assessment of global gene expression in *M. truncatula* [38–42]. Data from the results of these studies were typically reported in the context of EST and TCs and were either not associated with databases or with databases that no longer exist. The ESTs and TCs, as well as genes annotated from genomic sequences by IMGAG [32], were the basis for the Affymetrix-based Medicago GeneChips® (www.affymetrix.com). The first chip was used to assess gene expression during rhizobia nodulation and included genes from *M. truncatula*'s rhizobial symbiont *Sinorhizobium meliloti* on the chip as well [43, 44]. The second chip, the Medicago Genome Array, included 52,700 probesets from *M. truncatula* and closely related *M. sativa* and also included probesets corresponding to *S. meliloti* [45, 46]. The data, gathered from *M. truncatula* scientific community, using the second chip has been incorporated into the *M. truncatula* Gene Expression Atlas (MtGEA) [46, 47].

6 *Medicago truncatula* Gene Expression Atlas (MtGEA)

Beginning in 2008, gene expression data obtained using the Medicago Genome Array GeneChip® from plant organs and nodulation and seed development time courses were incorporated into the *Medicago truncatula* Gene Expression Atlas (MtGEA) [46] which can be accessed via mtgea.noble.org/v3/ [47]. Since becoming available, the MtGEA has been updated with gene expression data from the Medicago scientific community, resulting in the current version, MtGEA V3 (released in Feb 2015) that contains 274 experiments encompassing 739 GeneChips (mtgea.noble.org/v3/). The gene expression probeset data in MtGEA were annotated with predicted metabolic pathway reconstruction data [34], with genes predicted to encode transporters [48] and with genes predicted to encode transcription factors. A gene network prediction program was integrated with the MtGEA to help researchers utilize the data for their research [49]. The MtGEA website contains numerous links to the data used for the probesets, with each probeset having an individual webpage with its gene sequence, gene annotation information, expression profile, information about its construction, and links to various ways to use the data.

7 *M. truncatula* A17 and R108 Genome Assemblies

The *M. truncatula* reference genome from genotype A17, published in 2011 as version Mt3.5 [8], has been updated to Mt4.0 [9]. Very recently, a reference quality genome became available for *M. truncatula* genotype R108 [17]. The genome assemblies are available at the NCBI Assembly repository at www.ncbi.nlm.nih.gov/assembly/GCF_000219495.1/, also known as MedtrA17_3.5, for Mt3.5 and at www.ncbi.nlm.nih.gov/assembly/GCF_000219495.3, also known as MedtrA17_4.0, for Mt4.0. For *M. truncatula* genotype R108, the genome assembly data is accessible via www.ncbi.nlm.nih.gov/assembly/GCA_002024945.1/, where it is called R108_V.1.0.

8 *M. truncatula* Genome Database

The *M. truncatula* A17 reference genome is hosted via a user-friendly interface by JCVI at www.medicagogenome.org/ [50]. At www.medicagogenome.org/, the genome is searchable by locus, e.g., Medtr1g009200; by keyword, e.g. transporter, kinase; by sequence using BLAST [33]; by InterPro annotation that integrates protein signatures [51]; and by gene ontology or GO annotations.

The JBrowse interface at www.medicagogenome.org/ allows one to visualize the reference *M. truncatula* genome. The visualization tools, or tracks, are listed in alphabetical order on the left side of the JBrowse viewer and can be toggled on and off. Using it, one can picture the genome in a number of ways: by assembly, by annotation, by evidence, by transposable elements (genomic features), and by data culled from the *M. truncatula* scientific community, as described in the following paragraphs. The JBrowse assembly track at www.medicagogenome.org/ allows one to visualize the data that lead to the reference genome. There are four modes to view the reference genome assembly: (1) The reference sequence browser allows visualization of DNA sequence in all six possible reading frames, if one zooms in enough. (2) Important to interpretation of the genome results is the Assembly Golden Path viewer that allows one to view the placement of the sequenced contig data underlying the reference sequence within the context of the assembled genome. (3) Turning on the BAC mapping track shows which parts of the reference sequence are backed by Sanger-quality BAC-by-BAC [20, 30] sequencing that anchored Mt3.5 [8] and which parts are assembled from Illumina whole genome shotgun sequences [9]. (4) The gap track shows the gaps in the reference sequence and the approximate gap lengths. Clicking on the names of contigs, BACs, and gaps in the JBrowse window interface brings up that section's sequence and the information about the gap size, respectively.

Annotation of the *M. truncatula* genome at www.medicagogenome.org/ includes predictions of gene loci, including those of protein coding genes, transposable elements, and tRNA models; each prediction class can be added to the JBrowse interface. Evidence for the annotations is provided and can also be toggled onto the JBrowse screen. Splice isoforms for genes in the Mt. 4.0 database were made from publicly available RNAseq data [9]; the rapidly growing RNAseq data could lead to discovery of more novel splice isoforms that may accompany the next *M. truncatula* genome release. Ab initio gene predictions using Augustus [52] and FGeneSH [53] are also available for download. Data from Sanger and 454 sequencing efforts of cDNAs as well as RNAseq data from several *M. truncatula* organs can be visualized in the context of the *M. truncatula* genome. Data from the *M. truncatula* and legume scientific communities including the *M. truncatula* Affymetrix probesets and syntenic regions from *L. japonicus* and chickpea (*Cicer arietinum*) can also be mapped onto the JBrowse interface at this *M. truncatula* genomic data portal.

9 MedicMine

MedicMine (medicmine.jcvi.org/medicmine/begin.do) is a multi-database mining portal also maintained by JCVI researchers that is primarily focused on *M. truncatula*. It is built on several public data sources [50]. MedicMine exchanges the BioMart interface used by the Gramene database for InterMine [54], which is also used by other model organism databases such as *C. elegans* WormMine [55] and the Arabidopsis ThaleMine [56]. Users can submit gene, TC, and protein identifiers of interest and build customized tables for the data of interest to them. These data can include associated gene ontology (GO) and pathway terms, derived transcripts, and brief descriptions, to more advanced data such as tissue-specific expression profiles based on RNAseq experiments. The database is not limited to gene and protein queries, as authors, pathways, and other search terms can also be searched using MedicMine to produce tables of information, and furthermore, two terms can be linked together, both directly and indirectly.

10 Medicago Hapmap

The Medicago Hapmap project (www.medicagohapmap.org/) is resequencing 384 inbred lines spanning the Medicago diversity range using Illumina next-generation technology to discover sequence variation across *M. truncatula*. Thirty of these lines, including the R108 genotype in which the *Tnt1* insertion mutant population has been developed [57], have been resequenced to a greater depth [58–60]. As of this writing, the *M. truncatula* R108 data at the Medicago Hapmap database is the R108 version 0.95 data and not yet the R108 V1.0 reference quality genome data [17]. Some of the data have appeared in genome-wide association studies (GWAS) [59, 61] and are being made available publicly through the project website while still in a prepublication state.

11 Symbimics

The Symbimics data gateway (iant.toulouse.inra.fr/symbimics/) allows researchers interested in nodulation and symbiotic nitrogen fixation to query RNAseq data for laser micro-dissected zones in *M. truncatula* nodules, where each successive zone corresponds to succeeding developmental stages of nodulation [62]. Data were collected simultaneously from both the plant and rhizobial symbiont in the nodule zones and were also produced from whole uninfected root and mature nodule tissue, and thus the nodule zone RNAseq data can be assessed for differential gene expression at the Symbimics website. In addition to data visualization, the Symbimics website allows users to download selected data from specific genes of interest or all data in the Symbimics database.

12 Multiplant and Multi-database Information Portals

The purpose of an information portal is to collect and present the immense amount of already available data in a concise, easily accessible manner and not to provide new information. In the context of plants including *M. truncatula*, these database portals often serve as the initial stop for many researchers, because they contain some of the most needed types of information, including gene and pathway annotations, sequence archives, and web interfaces to useful bioinformatics tools such as genome browsers and local alignment algorithms. In most cases, there is significant overlap between the data provided in these more inclusive information portals, though the method of their delivery varies significantly. Depending on the user, one may find an interface too simple and restrictive, while another may prefer a much more guided approach to data mining.

13 Legume Information System

The Legume Information System (LIS; legumeinfo.org/home) is a genomic data portal dedicated to legumes and focused on comparative genomics [63, 64]. LIS is an integrated system with genome and related data for 14 legume species: the two model legumes *L. japonicus* and *M. truncatula*, two wild peanut species (*Arachis duranensis* and *A. ipaensis*), two types of chickpea (*Cicer arietinum*), pigeonp0ea (*Cajanus cajan*), soybean (*Glycine max*), lupin (*Lupinus angustifolius*), common bean (*Phaseolus vulgaris*), red clover (*Trifolium pratense*), adzuki bean (*Vigna angularis*), mungbean (*V. radiata*), and cowpea (*V. unguiculata*). The GBrowse genome browser at LIS has tracks for displaying syntenic regions across legumes. These can be toggled on and off using the "Select Tracks" feature. LIS also permits users to display TCs and ESTs (from the TGI index) from *M. truncatula* as well as homologous TCs and ESTs from *G. max*, *P. vulgaris*, *L. japonicus*, and *V. unguiculata*, with links to the sequences. LIS has tools to visualize gene family and phylogenetic trees to help provide researchers with evolutionary context and possible functional information about groups of genes across legumes. LIS has links to gene families, pathway maps, genetic maps, and whole genome annotations for these legumes, in addition to a BLAST interface and links to additional legume-specific databases [64].

14 LegumeIP

LegumeIP (plantgrn.noble.org/LegumeIP/) is a database for comparative genomics and transcriptomics of legumes [65]. The current version of the platform, LegumeIP 2.0, contains genomic sequences of six legume species *M. truncatula*, *L. japonicus*, *G. max*, *P. vulgaris*, *C. arietinum*, and *C. cajan*. In addition to the legumes, the platform contains the genomic sequences of two non-legume, outgroup, reference plant species, *A. thaliana* and poplar (*Populus trichocarpa*). LegumeIP enables researchers to analyze synteny across the six legumes and outgroup species and to study gene function and genome evolution in legumes. It also enables gene family reconstruction and phylogenetic analysis across the six legumes. The database contains 391,107 protein-coding gene sequences. It also contains gene family information as well as gene expression atlases for *M. truncatula*, *L. japonicus*, *G. max*, and *C. arietinum* that users can browse, search, and retrieve using search and visualization tools available through LegumeIP 2.0.

15 Gramene and EnsemblPlants

Gramene (www.gramene.org/), a self-proclaimed "curated, open-source, integrated" data resource for crops and model plants, is a multi-database portal primarily focused on providing whole genome and pathway information [66]. As of this writing, there are 1,836,677 genes across 45 genomes available within Gramene. Genomes are provided through collaboration with Ensembl Plants (plants.ensembl.org/index.html) and are made available through the Ensembl genome browser or simply as sequence (FASTA) files [67].

Genome coordinate conversion is provided for several different plant assemblies, including the *M. truncatula* genome Mt3.5 and Mt4.0 assemblies, using the CrossMap tool [68], called "Assembly Converter" in the Ensembl webpages. Note that the *M. truncatula* genome Mt3.5 is referred to as GCA_000219495.1 and *M. truncatula* genome Mt4.0 assembly is called MedtrA17_4.0 by CrossMap, which can be confusing. The alternate names refer to the different names for these assemblies at the NCBI Assembly repository. Genome annotation files, such as the frequently used GFF3 format files [69], are also available, as they are frequently used in RNAseq alignment pipelines such as TopHat and STAR [70, 71]. EnsemblPlants, through Gramene, also provides a comparative genomics toolbox, containing tools for gene tree reconstruction and whole genome alignments for homolog or ortholog identification. Local pairwise or multiple genome alignment is also provided through a web interface employing the

BLAST algorithm [33]. Targeted data retrieval is available through Gramene Mart, providing gene annotations, variation catalogues, genetic markers, and other bulk data using the filter and attribute system of BioMart [72, 73]. More advanced users wishing to parse the extensive catalogue of data linked through Gramene can do so through Gramene's FTP servers.

16 Phytozome

Phytozome (phytozome.jgi.doe.gov/pz/portal.html), an information portal developed by the US Department of Energy's Joint Genome Institute (JGI), provides one of the largest, multiplant genome databases currently available [74]. In the current version 12.1.3, Phytozome provides 77 assembled and annotated genomes built from 74 different species, with 43 of those having been assembled and annotated through the same JGI pipeline. For several genomes, such as *Ananas comosus* (a seedless pineapple cultigen) or *Malus domestica* (table apple) that are not available in Ensembl, Phytozome is the only database that provides a web interface to BLAST the genomes directly or to analyze using a graphical genome browser. This is useful to scientists who need to compare *M. truncatula* genomic data to such data from other plant species. Data retrieval is offered through both BioMart and InterMine interfaces, making Phytozome an important resource for researchers interested in cross-species comparisons involving non-model plant species.

17 Pathway Analysis

While multi-database information portals can be useful in finding genes/proteins associated with particular biochemical pathways, there exist plant-specific databases primarily focused on applying and visualizing biochemical pathway and metabolite information that are often beyond the scope of more general databases. Much of the information in these databases may appear in general databases, as is especially the case with the ontological and pathway terms of the Kyoto Encyclopedia of Genes and Genomes (KEGG) database [75], but is only through direct interaction with the pathway analysis databases that full visual maps, and plant-specific molecular pathways can be retrieved and applied directly to user-supplied datasets. One such pathway analysis database specific for *M. truncatula* is the MedicCyc database (pathway.gramene.org/gramene/mediccyc.shtml), as previously mentioned. KEGG Plant and Plant Reactome are two multiplant pathway analysis databases.

18 KEGG Plant

KEGG Plant (www.genome.jp/kegg/genome/plant.html), part of the greater KEGG database, is the plant-specific branch of the world's largest collection of manually curated biochemical pathway maps [76]. KEGG's plant pathway maps are divided into global, regular pathway, and chemical structure and transformation subcategories. Each map consists of a series of interconnected nodes, each representative of a particular orthology term, or a reaction, or a particular chemical substrate. For many plant species, including *M. truncatula*, these terms, reactions and substrates, have been associated with particular sets of genes, allowing researchers to more easily predict biochemical pathways related to their gene(s) of interest. While interactivity within the database is limited, the sheer volume of species-specific, gene-level term association through KEGG has led to the development of several tools. For example, clusterProfiler [77] allows users to perform pathway enrichment on subsets of genes based on their associated KEGG terms. Tools such as Pathview [78] allow researchers to transform KEGG pathways into graphical heatmaps using quantification data from a variety of sources, e.g., RNAseq, in a manner reminiscent of MAPMAN [79].

19 Plant Reactome

As an alternative to KEGG, Reactome also provides a database of biological pathways. Specifically for plants, Plant Reactome (plantreactome.oicr.on.ca/) uses manually curated annotations for *Oryza sativa* (rice) to establish the links between individual reactions, substrates, and genes and extrapolates the associations to other plant species based on their homology [80]. As a newer database, the web-based interface is highly intuitive, allowing researchers to navigate various pathway maps and to overlay gene and molecule lists, expression profiles, or metabolomics data directly onto the pathways. For non-rice species, identifiers are automatically converted to their nearest rice homolog prior to being projected onto the pathways.

20 Expression Databases

High-throughput gene expression assays have become a mainstay in functional genomics research, and their data remains valuable long after the initiating studies have been completed. This is apparent in the many existing expression databases, where researchers both novice and accomplished in data mining can analyze genes of interest across thousands of experiments simultaneously [81]. The datasets in these collections are primarily divided between two different formats: RNAseq and microarray, with the former typically available in both aligned and unaligned formats.

The National Center for Biotechnology Information (NCBI) Gene Expression Omnibus (NCBI GEO)(www.ncbi.nlm.nih.gov/geo/), European Molecular Biology Laboratory (EMBL) Expression Atlas (www.ebi.ac.uk/gxa/home), and Plant Expression Database (PLEXdb) (www.plexdb.org/) are three gene expression databases that allow researchers to download fully annotated expression profiles for hundreds of species [82–84]. All three repositories offer both microarray and sequencing-based expression data, with PLEXdb focused on plants and plant pathogen datasets. While the interfaces differ, researchers can focus on expression of specific genes across experiments or can download entire experimental datasets containing thousands of gene expression profiles for different treatments or conditions or different tissues or some combination thereof. As each database is MIAME-/MINSEQE-compliant, they are frequently used repositories for many publicly available datasets, especially with scientific journals and granting agencies requiring data deposition as part of their guidelines.

NCBI GEO provides data in two primary formats: datasets and profiles. GEO datasets consist of both user-submitted and curated datasets and are integrated into NCBI's web-based analysis tools for visualization and expression analysis. This includes clustering analysis, heatmap generation, and links to related data. GEO profiles are built from curated datasets and focus on the expression of singular genes across all samples within a given set of samples. This allows a quick, visual interpretation of whether or not a gene demonstrates differential expression across different tissues, treatments, or other experimental conditions based on the dataset's origin.

Similar to NCBI GEO's profiles, the EMBL Expression Atlas provides gene expression results from different biological conditions across curated experimental datasets. Similar in size to NCBI GEO, the Expression Atlas benefits from manual curation for all datasets. Annotation to ontology terms allows users to perform enrichment for a particular subset of genes or experiments using GO, Reactome, and InterPro terms. Differential expression experimental data are only offered if they include enough replicates to allow robust statistical analysis, with the raw expression data being supplied directly from the much larger, less-curated EMBL ArrayExpress (www.ebi.ac.uk/arrayexpress/).

In addition to the expression profiles from larger repositories such as NCBI GEO and the EMBL Expression Atlas, PLEXdb and the above-mentioned *M. truncatula*-specific MtGEA offer microarray expression profiles specifically for plants [46, 47, 82]. PLEXdb contains 643 experiments across 14 plant species and several plant pathogens, which may be useful for cross-species comparisons, not available in MtGEA. In most cases, the experimental datasets are also available in larger databases (such as the EMBL ArrayExpress for the MtGEA), though the larger databases may lack a variety of plant-centric analysis and search tools, such as MtGEA's ability to recognize several of *M. truncatula*'s gene identifier formats.

For those interested specifically in small RNA and miRNA, the Plant Massively Parallel Signature Sequencing (MPSS) databases provide NGS results for several plant species [85]. In addition to RNAseq, the Plant MPSS (mpss.danforthcenter.org/) also provides datasets for small RNA, PARE-Seq, mRNA DGE, MethylC-Seq, and ChiP-Seq, which are less represented in larger databases such as NCBI and EMBL.

For researchers looking for raw sequencing data from high-throughput experiments, both NCBI and EMBL provide full access to their datasets in the form of the NCBI Sequence Read Archive (SRA) (www.ncbi.nlm.nih.gov/sra) [86] and EMBL ArrayExpress (https://www.ebi.ac.uk/arrayexpress/) [87]. Because these data repositories contain raw sequencing data, both the NCBI SRA and EMBL ArrayExpress lack the online analysis tools of their GEO and Expression Atlas counterparts. However, they allow the end user an advanced level of freedom regarding the downstream use and analysis of these experimental datasets.

21 Other Databases with Unique Data

There exist several databases focused on unique types of data, including data for *M. truncatula*, not readily available in the aforementioned information portals and beyond the scope of the annotation pipelines frequently included in most other online databases. For researchers interested in ontological associations, the PlantOrDB (bioinfolab.miamioh.edu/plantordb/interface/about-plantordb.php) provides 49,355 homologous gene families across 35 plant species and 6 algal genomes [88]. The PTGBase (ocri-genomics.org/PTGBase/) provides a collection of 129,652 genes thought to have evolved through tandem duplication mechanisms [89]. The Plant Protein Phosphorylation Database (www.p3db.org/) contains phosphosite and phosphorylation protein annotations for nine different plant species including *M. truncatula* [90]. The PlantPIs database (plantpis.ba.itb.cnr.it/) allows researchers to actively search for protease inhibitors in plants [91]. Researchers concerned with small nucleolar RNA (snoRNA) may benefit from the sequence and alignment information provided by the Plant snoRNA database (bioinf.scri.sari.ac.uk/cgi-bin/plant_snorna/home) [92]. Transcriptional regulation is heavily represented in both the PlantProm (www.softberry.com/berry.phtml?topic=plantprom&group=data&subgroup=plantprom) and Plant Transcriptional Regulatory Map (plantregmap.cbi.pku.edu.cn/)

databases, with the former providing manually curated, non-redundant promoter sequences based entirely on experimentally derived transcripts and the latter focusing on regulatory prediction using a combination of ChiP-seq, DNase-seq, genome comparisons, and experimentally derived transcription factor binding motifs [93–95]. Finally, the Plant DNA C-values database (data.kew.org/cvalues/) contains information on the amount of nuclear DNA in a plant's unreplicated gametic nucleus (C-value) for 8510 plant species [96].

Acknowledgments

We gratefully acknowledge funding from the US National Science Foundation, grants IOS-1127155, and IOS-1733470 to RD.

References

1. Barker DG, Bianchi S, London F, Dattee Y, Duc G, Essad S, Flament P, Gallusci P, Genier G, Muel X, Tourneur J, Denarie J, Huguet T (1990) *Medicago truncatula*, a model plant for studying the molecular genetics of the *Rhizobium*-legume symbiosis. Plant Mol Biol Rep 8:40–49
2. Handberg K, Stougaard J (1992) *Lotus japonicus*, an autogamous, diploid legume species for classical and molecular genetics. Plant J 2:487–496
3. Cook DR (1999) *Medicago truncatula* - a model in the making! Curr Opin Plant Biol 2:301–304
4. Oldroyd GED (2013) Speak, friend, and enter: signalling systems that promote beneficial symbiotic associations in plants. Nat Rev Microbiol 11:252–263
5. Sagan M, Morandi D, Tarneghi E, Duc G (1995) Selection of nodulation and mycorrhizal mutants in the model plant *Medicago truncatula* (Gaertn.) after γ-mutagenesis. Plant Sci 111:63–71
6. Markmann K, Parniske M (2009) Evolution of root endosymbiosis with bacteria: how novel are nodules? Trends Plant Sci 14:77–86
7. Limpens E, Bisseling T (2003) Signaling in symbiosis. Curr Opin Plant Biol 6:343–350
8. Young ND, Debelle F, Oldroyd GED, Geurts R, Cannon SB, Udvardi MK, Benedito VA, Mayer KFX, Gouzy J, Schoof H, Van de Peer Y, Proost S, Cook DR, Meyers BC, Spannagl M, Cheung F, De Mita S, Krishnakumar V, Gundlach H, Zhou S, Mudge J, Bharti AK, Murray JD, Naoumkina MA, Rosen B, Silverstein KAT, Tang H, Rombauts S, Zhao PX, Zhou P, Barbe V, Bardou P, Bechner M, Bellec A, Berger A, Berges H, Bidwell S, Bisseling T, Choisne N, Couloux A, Denny R, Deshpande S, Dai X, Doyle JJ, Dudez A-M, Farmer AD, Fouteau S, Franken C, Gibelin C, Gish J, Goldstein S, Gonzalez AJ, Green PJ, Hallab A, Hartog M, Hua A, Humphray SJ, Jeong D-H, Jing Y, Jocker A, Kenton SM, Kim D-J, Klee K, Lai H, Lang C, Lin S, Macmil SL, Magdelenat G, Matthews L, McCorrison J, Monaghan EL, Mun J-H, Najar FZ, Nicholson C, Noirot C, O'Bleness M, Paule CR, Poulain J, Prion F, Qin B, Qu C, Retzel EF, Riddle C, Sallet E, Samain S, Samson N, Sanders I, Saurat O, Scarpelli C, Schiex T, Segurens B, Severin AJ, Sherrier DJ, Shi R, Sims S, Singer SR, Sinharoy S, Sterck L, Viollet A, Wang B-B, Wang K, Wang M, Wang X, Warfsmann J, Weissenbach J, White DD, White JD, Wiley GB, Wincker P, Xing Y, Yang L, Yao Z, Ying F, Zhai J, Zhou L, Zuber A, Denarie J, Dixon RA, May GD, Schwartz DC, Rogers J, Quetier F, Town CD, Roe BA (2011) The Medicago genome provides insight into the evolution of rhizobial symbioses. Nature 480(7378):520–524 http://www.nature.com/nature/journal/v480/n7378/abs/nature10625.html#supplementary-information
9. Tang H, Krishnakumar V, Bidwell S, Rosen B, Chan A, Zhou S, Gentzbittel L, Childs KL, Yandell M, Gundlach H, Mayer KF, Schwartz DC, Town CD (2014) An improved genome release (version Mt4.0) for the model legume *Medicago truncatula*. BMC Genomics 15:312
10. Sato S, Nakamura Y, Kaneko T, Asamizu E, Kato T, Nakao M, Sasamoto S, Watanabe A, Ono A, Kawashima K, Fujishiro T, Katoh M, Kohara M, Kishida Y, Minami C, Nakayama S, Nakazaki N, Shimizu Y, Shinpo S, Takahashi C,

Wada T, Yamada M, Ohmido N, Hayashi M, Fukui K, Baba T, Nakamichi T, Mori H, Tabata S (2008) Genome structure of the legume, *Lotus japonicus*. DNA Res 15(4):227–239. https://doi.org/10.1093/dnares/dsn008

11. Schmutz J, Cannon SB, Schlueter J, Ma J, Mitros T, Nelson W, Hyten DL, Song Q, Thelen JJ, Cheng J, Xu D, Hellsten U, May GD, Yu Y, Sakurai T, Umezawa T, Bhattacharyya MK, Sandhu D, Valliyodan B, Lindquist E, Peto M, Grant D, Shu S, Goodstein D, Barry K, Futrell-Griggs M, Abernathy B, Du J, Tian Z, Zhu L, Gill N, Joshi T, Libault M, Sethuraman A, Zhang X-C, Shinozaki K, Nguyen HT, Wing RA, Cregan P, Specht J, Grimwood J, Rokhsar D, Stacey G, Shoemaker RC, Jackson SA (2010) Genome sequence of the palaeopolyploid soybean. Nature 463(7278):178–183
12. Varshney RK, Chen W, Li Y, Bharti AK, Saxena RK, Schlueter JA, Donoghue MTA, Azam S, Fan G, Whaley AM, Farmer AD, Sheridan J, Iwata A, Tuteja R, Penmetsa RV, Wu W, Upadhyaya HD, Yang S-P, Shah T, Saxena KB, Michael T, McCombie WR, Yang B, Zhang G, Yang H, Wang J, Spillane C, Cook DR, May GD, Xu X, Jackson SA (2012) Draft genome sequence of pigeonpea (*Cajanus cajan*), an orphan legume crop of resource-poor farmers. Nat Biotechnol 30:83–89
13. Bertioli DJ, Cannon SB, Froenicke L, Huang G, Farmer AD, Cannon EKS, Liu X, Gao D, Clevenger J, Dash S, Ren L, Moretzsohn MC, Shirasawa K, Huang W, Vidigal B, Abernathy B, Chu Y, Niederhuth CE, Umale P, Araújo ACG, Kozik A, Kim KD, Burow MD, Varshney RK, Wang X, Zhang X, Barkley N, Guimarães PM, Isobe S, Guo B, Liao B, Stalker HT, Schmitz RJ, Scheffler BE, Leal-Bertioli SCM, Xun X, Jackson SA, Michelmore R, Ozias-Akins P (2016) The genome sequences of *Arachis duranensis* and *Arachis ipaensis*, the diploid ancestors of cultivated peanut. Nat Genet 48:438–446
14. Varshney RK, Song C, Saxena RK, Azam S, Yu S, Sharpe AG, Cannon S, Baek J, Rosen BD, Tar'an B, Millan T, Zhang X, Ramsay LD, Iwata A, Wang Y, Nelson W, Farmer AD, Gaur PM, Soderlund C, Penmetsa RV, Xu C, Bharti AK, He W, Winter P, Zhao S, Hane JK, Carrasquilla-Garcia N, Condie JA, Upadhyaya HD, Luo M-C, Thudi M, Gowda CLL, Singh NP, Lichtenzveig J, Gali KK, Rubio J, Nadarajan N, Dolezel J, Bansal KC, Xu X, Edwards D, Zhang G, Kahl G, Gil J, Singh KB, Datta SK, Jackson SA, Wang J, Cook DR (2013) Draft genome sequence of chickpea (*Cicer arietinum*) provides a resource for trait improvement. Nat Biotechnol 31:240–246
15. Jain M, Misra G, Patel RK, Priya P, Jhanwar S, Khan AW, Shah N, Singh VK, Garg R, Jeena G, Yadav M, Kant C, Sharma P, Yadav G, Bhatia S, Tyagi AK, Chattopadhyay D (2013) A draft genome sequence of the pulse crop chickpea (*Cicer arietinum* L.). Plant J 74:715–729
16. Phytozome v9.1: Phaseolus vulgaris v1.0 phytozome.jgi.doe.gov/pz/portal.html#!info?alias=Org_Pvulgaris
17. Moll KM, Zhou P, Ramaraj T, Fajardo D, Devitt NP, Sadowsky MJ, Stupar RM, Tiffin P, Miller JR, Young ND, Silverstein KAT, Mudge J (2017) Strategies for optimizing BioNano and dovetail explored through a second reference quality assembly for the legume model, *Medicago truncatula*. BMC Genomics 18:578
18. Bell CJ, Dixon RA, Farmer AD, Flores R, Inman J, Gonzales RA, Harrison MJ, Paiva NL, Scott AD, Weller JW, May GD (2001) The *Medicago* Genome Initiative: a model legume database. Nucleic Acids Res 29:114–117
19. Lamblin A-FJ, Crow JA, Johnson JE, Silverstein KAT, Kunau TM, Kilian A, Benz D, Stromvik M, Endré G, VandenBosch KA, Cook DR, Young ND, Retzel EF (2003) MtDB: a database for personalized data mining of the model legume *Medicago truncatula* transcriptome. Nucleic Acids Res 31:196–201
20. Cannon SB, Crow JA, Heuer ML, Wang X, Cannon EKS, Dwan C, Lamblin A-F, Vasdewani J, Mudge J, Cook A, Gish J, Cheung F, Kenton S, Kunau TM, Brown D, May GD, Kim D, Cook DR, Roe BA, Town CD, Young ND, Retzel EF (2005) Databases and information integration for the *Medicago truncatula* genome and transcriptome. Plant Physiol 138(1):38–46
21. Covitz PA, Smith LS, Long SR (1998) Expressed sequence tags from a root-hair-enriched *Medicago truncatula* cDNA library. Plant Physiol 117:1325–1332
22. Gyorgyey J, Vaubert D, Jimenez-Zurdo JI, Charon C, Troussard L, Kondorosi A, Kondorosi E (2000) Analysis of Medicago truncatula nodule expressed sequence tags. Mol Plant Microbe Interact 13:62–71
23. Journet E-P, van Tuinen D, Gouzy J, Crespeau H, Carreau V, Farmer M-J, Niebel A, Schiex T, Jaillon O, Chatagnier O, Godiard L, Micheli F, Kahn D, Gianinazzi-Pearson V, Gamas P (2002) Exploring root symbiotic programs in the model legume *Medicago truncatula* using EST analysis. Nucl Acids Res 30(24):5579–5592
24. Nam YW, Penmetsa RV, Endre G, Uribe P, Kim D, Cook DR (1999) Construction of a bacterial artificial chromosome library of *Medicago truncatula* and identification of clones containing ethylene-response genes. Theor Appl Genet 98:638–646
25. Kulikova O, Gualtieri G, Geurts R, Kim DJ, Cook D, Huguet T, de Jong JH, Fransz PF,

Bisseling T (2001) Integration of the FISH pachytene and genetic maps of *Medicago truncatula*. Plant J 27:49–58

26. Choi H-K, Kim D, Uhm T, Limpens E, Lim H, Kalo P, Penmetsa VR, Seres A, Kulikova O, Bisseling T, Kiss G, Cook DR (2004) A sequence-based genetic map of *Medicago truncatula* and comparison of marker co-linearity with *Medicago sativa*. Genetics 166:1463–1502
27. VandenBosch KA, Stacey G (2003) Summaries of legume genomics projects from around the globe. Community resources for crops and models. Plant Physiol 131:840–865
28. Frugoli J, Harris J (2001) *Medicago truncatula* on the move! Plant Cell 13(3):458–463
29. Fedorova M, van de Mortel J, Matsumoto PA, Cho J, Town CD, VandenBosch KA, Gantt JS, Vance CP (2002) Genome-wide identification of nodule-specific transcripts in the model legume *Medicago truncatula*. Plant Physiol 130(2):519–537
30. Young ND, Cannon SB, Sato S, Kim D, Cook DR, Town CD, Roe BA, Tabata S (2005) Sequencing the genespaces of *Medicago truncatula* and *Lotus japonicus*. Plant Physiol 137(4):1174–1181. https://doi.org/10.1104/pp.104.057034
31. Kulikova O, Geurts R, Lamine M, Kim D-J, Cook DR, Leunissen J, de Jong H, Roe BA, Bisseling T (2004) Satellite repeats in the functional centromere and pericentromeric heterochromatin of *Medicago truncatula*. Chromosoma 113:276–283
32. Town CD (2006) Annotating the genome of *Medicago truncatula*. Curr Opin Plant Biol 9(2):122–127
33. Altschul SF, Gish W, Miller W, Meyers EW, Lipman DJ (1990) Basic local alignment search tool. J Mol Biol 215(3):403–410
34. Urbanczyk-Wochniak E, Sumner LW (2007) MedicCyc: a biochemical pathway database for *Medicago truncatula*. Bioinformatics 23:1418–1423
35. Karp PD (2002) The MetaCyc database. Nucleic Acids Res 30:59–61. https://doi.org/10.1093/nar/30.1.59
36. Karp PD, Paley S, Romero P (2002) The pathway tools software. Bioinformatics 18:S225–S232. https://doi.org/10.1093/bioinformatics/18.suppl_1.S225
37. Pontius JU, Wagner L, Schuler GD (2003) UniGene: a unified view of the transcriptome. NCBI Handbook 1:1–12
38. Hohnjec N, Vieweg MF, Pühler A, Becker A, Küster H (2005) Overlaps in the transcriptional profiles of *Medicago truncatula* roots inoculated with two different Glomus fungi provide insights into the genetic program activated during arbuscular mycorrhiza. Plant Physiol 137:1283–1301
39. Lohar DP, Sharopova N, Endre G, Penuela S, Samac D, Town C, Silverstein KAT, VandenBosch KA (2006) Transcript analysis of early nodulation events in *Medicago truncatula*. Plant Physiol 140(1):221–234
40. Kuster H, Hohnjec N, Krajinski F, El YF, Manthey K, Gouzy J, Dondrup M, Meyer F, Kalinowski J, Brechenmacher L, van Tuinen D, Gianinazzi-Pearson V, Puhler A, Gamas P, Becker A (2004) Construction and validation of cDNA-based Mt6k-RIT macro- and microarrays to explore root endosymbioses in the model legume *Medicago truncatula*. J Biotechnol 108:95–113
41. Manthey K, Krajinski F, Hohnjec N, Firnhaber C, Puhler A, Perlick AM, Kuster H (2004) Transcriptome profiling in root nodules and arbuscular mycorrhiza identifies a collection of novel genes induced during *Medicago truncatula* root endosymbioses. Mol Plant-Microbe Interact 17(10):1063–1077. https://doi.org/10.1094/mpmi.2004.17.10.1063
42. El Yahyaoui F, Kuster H, Ben Amor B, Hohnjec N, Pühler A, Becker A, Gouzy J, Vernie T, Gough C, Niebel A, Godiard L, Gamas P (2004) Expression profiling in *Medicago truncatula* identifies more than 750 genes differentially expressed during nodulation, including many potential regulators of the symbiotic program. Plant Physiol 136:3159–3176
43. Mitra RM, Shaw SL, Long SR (2004) Six nonnodulating plant mutants defective for Nod factor-induced transcriptional changes associated with the legume-rhizobia symbiosis. Proceedings of the National Academy of Sciences, USA. 0402186101
44. Barnett MJ, Toman CJ, Fisher RF, Long SR (2004) A dual-genome symbiosis chip for coordinate study of signal exchange and development in a prokaryote-host interaction. Proc Natl Acad Sci U S A 101:16636–16641
45. Tesfaye M, Silverstein KAT, Bucciarelli B, Samac DA, Vance CP (2006) The Affymetrix *Medicago* GeneChip® array is applicable for transcript analysis of alfalfa (*Medicago sativa*). Funct Plant Biol 33:783–788
46. Benedito VA, Torres-Jerez I, Murray JD, Andriankaja A, Allen S, Kakar K, Wandrey M, Verdier J, Zuber H, Ott T, Moreau S, Niebel A, Frickey T, Weiller G, He J, Dai X, Zhao PX, Tang Y, Udvardi MK (2008) A gene expression atlas of the model legume *Medicago truncatula*. Plant J 55:504–513
47. He J, Benedito VA, Wang M, Murray JD, Zhao PX, Tang Y, Udvardi MK (2009) The *Medicago*

truncatula gene expression atlas web server. BMC Bioinformatics 10:441. https://doi.org/10.1186/1471-2105-1110-1441

48. Benedito VA, Li H, Dai X, Wandrey M, He J, Kaundal R, Torres-Jerez I, Gomez SK, Harrison MJ, Tang Y, Zhao PX, Udvardi MK (2010) Genomic inventory and transcriptional analysis of *Medicago truncatula* transporters. Plant Physiol 152:1716–1730
49. Wang M, Verdier J, Benedito VA, Tang Y, Murray JD, Ge Y, Becker JD, Carvalho H, Rogers C, Udvardi M (2013) LegumeGRN: a gene regulatory network prediction server for functional and comparative studies. PLoS One 8:e67434. https://doi.org/10.61371/journal.pone.0067434
50. Krishnakumar V, Kim M, Rosen BD, Karamycheva S, Bidwell SL, Tang H, Town CD (2015) MTGD: The *Medicago truncatula* genome database. Plant Cell Physiol 56(1):e1. https://doi.org/10.1093/pcp/pcu1179
51. Hunter S, Apweiler R, Attwood TK, Bairoch A, Bateman A, Binns D, Bork P, Das U, Daugherty L, Duquenne L, Finn RD, Gough J, Haft D, Hulo N, Kahn D, Kelly E, Laugraud A, Letunic I, Lonsdale D, Lopez R, Madera M, Maslen J, McAnulla C, McDowall J, Mistry J, Mitchell A, Mulder N, Natale D, Orengo C, Quinn AF, Selengut JD, Sigrist CJA, Thimma M, Thomas PD, Valentin F, Wilson D, Wu CH, Yeats C (2009) InterPro: the integrative protein signature database. Nucleic Acids Res 37:D211–D215
52. Stanke M, Keller O, Gunduz I, Hayes A, Waack S, Morgenstern B (2006) AUGUSTUS: *ab initio* prediction of alternative transcripts. Nucleic Acids Res 34:W435–W439
53. Salamov AA, Solovyev VV (2000) Ab initio gene finding in *Drosophila* genomic DNA. Genome Res 10:516–522
54. Smith RN, Aleksic J, Butano D, Carr A, Contrino S, Hu F, Lyne M, Lyne R, Kalderimis A, Rutherford K, Stepan R, Sullivan J, Wakeling M, Watkins X, Micklem G (2012) InterMine: a flexible data warehouse system for the integration and analysis of heterogeneous biological data. Bioinformatics 28:3163–3165. https://doi.org/10.1093/bioinformatics/bts577
55. Harris TW, Baran J, Bieri T, Cabunoc A, Chan J, Chen WJ, Davis P, Done J, Grove C, Howe K, Kishore R, Lee R, Li Y, Muller HM, Nakamura C, Ozersky P, Paulini M, Raciti D, Schindelman G, Tuli MA, Auken KV, Wang D, Wang X, Williams G, Wong JD, Yook K, Schedl T, Hodgkin J, Berriman M, Kersey P, Spieth J, Stein L, Sternberg PW (2014) WormBase 2014: new views of curated biology. Nucleic Acids Res 42:42. https://doi.org/10.1093/nar/gkt1063
56. Krishnakumar V, Contrino S, Cheng CY, Belyaeva I, Ferlanti ES, Miller JR, Vaughn MW, Micklem G, Town CD, Chan AP (2017) Thalemine: a warehouse for Arabidopsis data integration and discovery. Plant Cell Physiol 58:e4. https://doi.org/10.1093/pcp/pcw200
57. Tadege M, Wen J, He J, Tu H, Kwak Y, Eschstruth A, Cayrel A, Endre G, Zhao PX, Chabaud M, Ratet P, Mysore KS (2008) Large-scale insertional mutagenesis using the *Tnt1* retrotransposon in the model legume *Medicago truncatula*. Plant J 54:335–347
58. Zhou P, Silverstein KAT, Ramaraj T, Guhlin J, Denny R, Liu J, Farmer AD, Steele KP, Stupar RM, Miller JR, Tiffin P, Mudge J, Young ND (2017) Exploring structural variation and gene family architecture with *De Novo* assemblies of 15 *Medicago* genomes. BMC Genomics 18:261
59. Curtin SJ, Tiffin P, Guhlin J, Trujillo DI, Burghardt LT, Atkins P, Baltes NJ, Denny R, Voytas DF, Stupar RM, Young ND (2017) Validating genome-wide association candidates controlling quantitative variation in nodulation. Plant Physiol 173:921–931
60. Branca A, Paape TD, Zhou P, Briskine R, Farmer AD, Mudge J, Bharti AK, Woodward JE, May GD, Gentzbittel L, Ben C, Denny R, Sadowsky MJ, Ronfort J, Bataillon T, Young ND, Tiffin P (2011) Whole-genome nucleotide diversity, recombination, and linkage disequilibrium in the model legume *Medicago truncatula*. Proc Natl Acad Sci U S A 108:E864–E870
61. Stanton-Geddes J, Paape T, Epstein B, Briskine R, Yoder J, Mudge J, Bharti AK, Farmer AD, Zhou P, Denny R, May GD, Erlandson S, Yakub M, Sugawara M, Sadowsky MJ, Young ND, Tiffin P (2013) Candidate genes and genetic architecture of symbiotic and agronomic traits revealed by whole-genome, sequence-based association genetics in *Medicago truncatula*. PLoS One 8:e65688
62. Roux B, Rodde N, Jardinaud M-F, Timmers T, Sauviac L, Cottret L, Carrère S, Sallet E, Courcelle E, Moreau S, Debellé F, Capela D, de Carvalho-Niebel F, Gouzy J, Bruand C, Gamas P (2014) An integrated analysis of plant and bacterial gene expression in symbiotic root nodules using laser-capture microdissection coupled to RNA sequencing. Plant J 77:817–837
63. Gonzales MD, Archuleta E, Farmer A, Gajendran K, Grant D, Shoemaker R, Beavis WD, Waugh ME (2005) The Legume Information System (LIS): an integrated information resource for comparative legume biology. Nucl Acids Res 33(Suppl 1):D660–D665
64. Dash S, Campbell JD, Cannon EKS, Cleary AM, Huang W, Kalberer SR, Karingula V, Rice AG, Singh J, Umale PE, Weeks NT, Wilkey AP, Farmer AD, Cannon SB (2016) Legume information

system (LegumeInfo.org): a key component of a set of federated data resources for the legume family. Nucleic Acids Res 44:D1181–D1188
65. Li J, Dai X, Liu T, Zhao PX (2012) LegumeIP: an integrative database for comparative genomics and transcriptomics of model legumes. Nucleic Acids Res 40:D1221–D1229
66. Tello-Ruiz MK, Stein J, Wei S, Preece J, Olson A, Naithani S, Amarasinghe V, Dharmawardhana P, Jiao Y, Mulvaney J, Kumari S, Chougule K, Elser J, Wang B, Thomason J, Bolser DM, Kerhornou A, Walts B, Fonseca NA, Huerta L, Keays M, Tang YA, Parkinson H, Fabregat A, McKay S, Weiser J, D'Eustachio P, Stein L, Petryszak R, Kersey PJ, Jaiswal P, Ware D (2016) Gramene 2016: comparative plant genomics and pathway resources. Nucleic Acids Res 44:D1133–D1140. https://doi.org/10.1093/nar/gkv1179
67. Kersey PJ, Allen JE, Armean I, Boddu S, Bolt BJ, Carvalho-Silva D, Christensen M, Davis P, Falin LJ, Grabmueller C, Humphrey J, Kerhornou A, Khobova J, Aranganathan NK, Langridge N, Lowy E, McDowall MD, Maheswari U, Nuhn M, Ong CK, Overduin B, Paulini M, Pedro H, Perry E, Spudich G, Tapanari E, Walts B, Williams G, Tello-Ruiz M, Stein J, Wei S, Ware D, Bolser DM, Howe KL, Kulesha E, Lawson D, Maslen G, Staines DM (2016) Ensembl genomes 2016: more genomes, more complexity. Nucleic Acids Res 44:D574–D580. https://doi.org/10.1093/nar/gkv1209
68. Zhao H, Sun Z, Wang J, Huang H, Kocher J-P, Wang L (2014) CrossMap: a versatile tool for coordinate conversion between genome assemblies. Bioinformatics 30:1006–1007
69. Stein L (2010) Generic feature format, version 3. Sequence Ontology Project, pp 1–18
70. Dobin A, Davis CA, Schlesinger F, Drenkow J, Zaleski C, Jha S, Batut P, Chaisson M, Gingeras TR (2013) STAR: Ultrafast universal RNA-seq aligner. Bioinformatics 29:15–21. https://doi.org/10.1093/bioinformatics/bts635
71. Trapnell C, Roberts A, Goff L, Pertea G, Kim D, Kelley DR, Pimentel H, Salzberg SL, Rinn JL, Pachter L (2012) Differential gene and transcript expression analysis of RNA-seq experiments with TopHat and cufflinks. Nat Protoc 7:562–578. https://doi.org/10.1038/nprot.2012.016
72. Smedley D, Haider S, Ballester B, Holland R, London D, Thorisson G, Kasprzyk A (2009) BioMart--biological queries made easy. BMC Genomics 10:22. https://doi.org/10.1186/1471-2164-10-22
73. Spooner W, Youens-Clark K, Staines D, Ware D (2012) GrameneMart: the BioMart data portal for the Gramene project. Database 2012:bar056. https://doi.org/10.1093/database/bar056
74. Goodstein DM, Shu S, Howson R, Neupane R, Hayes RD, Fazo J, Mitros T, Dirks W, Hellsten U, Putnam N, Rokhsar DS (2012) Phytozome: a comparative platform for green plant genomics. Nucleic Acids Res 40:D1178. https://doi.org/10.1093/nar/gkr944
75. Ogata H, Goto S, Sato K, Fujibuchi W, Bono H, Kanehisa M (1999) KEGG: Kyoto encyclopedia of genes and genomes. Nucleic Acids Res 27:29. https://doi.org/10.1093/nar/27.1.29
76. Kanehisa M (2016) Kegg bioinformatics resource for plant genomics and metabolomics. Methods Mol Biol 1374:55–70. https://doi.org/10.1007/978-1-4939-3167-5_3
77. Yu G, Wang L-G, Han Y, He Q-Y (2012) clusterProfiler: an R package for comparing biological themes among gene clusters. OMICS 16:284–287. https://doi.org/10.1089/omi.2011.0118
78. Luo W, Brouwer C (2013) Pathview: an R/Bioconductor package for pathway-based data integration and visualization. Bioinformatics 29:1830–1831. https://doi.org/10.1093/bioinformatics/btt285
79. Thimm O, Bläsing O, Gibon Y, Nagel A, Meyer S, Krüger P, Selbig J, Müller LA, Rhee SY, Stitt M (2004) MAPMAN: a user-driven tool to display genomics data sets onto diagrams of metabolic pathways and other biological processes. Plant J 37:914–939
80. Naithani S, Preece J, D'Eustachio P, Gupta P, Amarasinghe V, Dharmawardhana PD, Wu G, Fabregat A, Elser JL, Weiser J, Keays M, Fuentes AMP, Petryszak R, Stein LD, Ware D, Jaiswal P (2017) Plant Reactome: a resource for plant pathways and comparative analysis. Nucleic Acids Res 45:D1029–D1039. https://doi.org/10.1093/nar/gkw932
81. Guo Y, Zhao S, Bjoring M, Han L (2015) Advanced datamining using RNAseq data. Big Data Analytics in Bioinformatics and Healthcare. https://doi.org/10.4018/978-1-4666-6611-5.ch001
82. Dash S, Van Hemert J, Hong L, Wise RP, Dickerson JA (2012) PLEXdb: gene expression resources for plants and plant pathogens. Nucleic Acids Res 40:D1194. https://doi.org/10.1093/nar/gkr938
83. Edgar R (2002) Gene expression omnibus: NCBI gene expression and hybridization array data repository. Nucleic Acids Res 30:207–210. https://doi.org/10.1093/nar/30.1.207
84. Kapushesky M, Emam I, Holloway E, Kurnosov P, Zorin A, Malone J, Rustici G, Williams E, Parkinson H, Brazma A (2009) Gene expression atlas at the European bioinformatics institute. Nucleic Acids Res 38:D690–D698. https://doi.org/10.1093/nar/gkp936

85. Nakano M (2006) Plant MPSS databases: signature-based transcriptional resources for analyses of mRNA and small RNA. Nucleic Acids Res 34:D731–D735. https://doi.org/10.1093/nar/gkj077
86. Leinonen R, Sugawara H, Shumway M (2011) The sequence read archive. Nucleic Acids Res 39:D19. https://doi.org/10.1093/nar/gkq1019
87. Parkinson H, Sarkans U, Shojatalab M, Abeygunawardena N, Contrino S, Coulson R, Farne A, Lara GG, Holloway E, Kapushesky M, Lilja P, Mukherjee G, Oezcimen a RT, Rocca-Serra P, Sharma A, Sansone S, Brazma A (2005) ArrayExpress--a public repository for microarray gene expression data at the EBI. Nucleic Acids Res 33:D553–D555. https://doi.org/10.1093/nar/gki056
88. Li L, Ji G, Ye C, Shu C, Zhang J, Liang C (2015) PlantOrDB: a genome-wide ortholog database for land plants and green algae. BMC Plant Biol 15:161. https://doi.org/10.1186/s12870-015-0531-4
89. Yu J, Ke T, Tehrim S, Sun F, Liao B, Hua W (2015) PTGBase: an integrated database to study tandem duplicated genes in plants. Database 2015:bav017. https://doi.org/10.1093/database/bav017
90. Yao Q, Ge H, Wu S, Zhang N, Chen W, Xu C, Gao J, Thelen JJ, Xu D (2013) P3DB 3.0: From plant phosphorylation sites to protein networks. Nucleic Acids Res 42:D1206–D1213. https://doi.org/10.1093/nar/gkt1135
91. Consiglio A, Grillo G, Licciulli F, Ceci LR, Liuni S, Losito N, Volpicella M, Gallerani R, De Leo F (2011) PlantPIs – an interactive web resource on plant protease inhibitors. Curr Protein Pept Sci 12:448–454. https://doi.org/10.2174/138920311796391052
92. Brown JWS, Echeverria M, Qu LH, Lowe TM, Bachellerie JP, Hüttenhofer A, Kastenmayer JP, Green PJ, Shaw P, Marshall DF (2003) Plant snoRNA database. Nucleic Acids Res 31:432. https://doi.org/10.1093/nar/gkg009
93. Jin J, Tian F, Yang DC, Meng YQ, Kong L, Luo J, Gao G (2017) PlantTFDB 4.0: Toward a central hub for transcription factors and regulatory interactions in plants. Nucleic Acids Res 45:D1040–D1045. https://doi.org/10.1093/nar/gkw982
94. Shahmuradov IA, Gammerman AJ, Hancock JM, Bramley PM, Solovyev VV (2003) PlantProm: a database of plant promoter sequences. Nucleic Acids Res 31:114. https://doi.org/10.1093/nar/gkg041
95. Shahmuradov I, Abdulazimova A, Khan FZ, Solovyev V, Mustafayev N, Akbarova Y, Qamar R, Aliyev J (2012) The PlantProm DB: recent updates. Proceedings - 2012 International Conference on Biomedical Engineering and Biotechnology, iCBEB 2012. https://doi.org/10.1109/iCBEB.2012.433
96. Zonneveld BJM, Leitch IJ, Bennett MD (2005) First nuclear DNA amounts in more than 300 angiosperms. Ann Bot 96:229–244

Chapter 4

Physical Mutagenesis in *Medicago truncatula* Using Fast Neutron Bombardment (FNB) for Symbiosis and Developmental Biology Studies

Yuhui Chen and Rujin Chen

Abstract

Medicago truncatula has been selected as a model species for legume molecular genetics and functional genomics studies. With the completion of the *Medicago truncatula* cv. Jemalong A17 genome sequencing, a major challenge is to determine the function of the large number of genes in the genome. Development of diverse mutant resources is crucial for gene functional studies. In the past years, M2 seeds from over 150,000 *Medicago truncatula* mutant lines in the Jemalong A17 background have been generated coordinately at the Noble Research Institute, USA, and the John Innes Centre, UK, using fast neutron bombardment (FNB) mutagenesis. These mutant resources have been used in screening and characterization of different categories of mutants including symbiotic nitrogen fixation, nodule development, and growth and patterning of leaf, stem, and root system architecture in the legume system. Here, we describe the detail procedure that has been used for screening of mutants derived from fast neutron bombardment mutagenesis in *Medicago truncatula*.

Key words *Medicago truncatula*, FNB, Forward genetics, Screening

1 Introduction

Legumes are the third largest family of flowering plants and second only in economic values to the grass family. They contribute significantly to the food and fodder requirements of human and animals, respectively. Because legume plants can establish symbiotic interactions with the soil bacteria called rhizobia for symbiotic nitrogen fixation, they are cultivated as an integral part of sustainable agriculture for thousands of years [1]. *Medicago truncatula* has been selected by the scientific community as one of the two genetic models with *Lotus japonicus* as the other, for legume research because of its unique biology and the availability of various genetic and genomic resources. With the completion of the genome sequence of the Jemalong A17 plant, functional

Luis A. Cañas and José Pío Beltrán (eds.), *Functional genomics in* Medicago truncatula*: Methods and Protocols*, Methods in Molecular Biology, vol. 1822, https://doi.org/10.1007/978-1-4939-8633-0_4,

characterization of a large number of genes encoded in its genome is the next challenge for legume biologists. Comprehensive mutant resources are indispensable for determining gene functions. A large number of mutants of *Medicago truncatula* generated by various mutagenesis methods have been successfully isolated and characterized, helping pave the way for understanding the function of genes that play key roles in legume-specific processes.

Fast neutron, a high energy ionization mutagen, has been used in generating mutants in many plant species including *Arabidopsis* [2, 3], rice (*Oryza sativa*) [4], tomato (*Solanum lycopersicum*), soybean (*Glycine soja*; *G. max*) [5, 6], barley (*Hordeum vulgare*), and *Lotus japonicus* [7]. Compared to other mutagenesis methods, fast neutron bombardment (FNB) mutagenesis has several advantages over other methods:

1. It can generate loss-of-function mutations.
2. It can generate mutants for tandem arrayed genes. For example, in *Arabidopsis*, 19% of gene families have members in tandem repeats and in rice; the number goes as high as 29% [8]. In *Medicago truncatula*, about 33% of gene families have members in tandem repeats [9]. Due to gene redundancy, only removal of all tandem-arrayed genes, which can be achieved by FNB mutagenesis, may exhibit visible phenotypes.
3. FNB mutagenesis can also generate mutants for small genes such as those encoding small peptides. This may be difficult to achieve using other mutagenesis methods such as insertion mutagenesis using T-DNA or transposons.
4. FNB mutagenesis has been widely applied to many plant species, including those that are recalcitrant for transform, which is required for insertion mutagenesis.
5. Since there are less sequence preferences as compared to other mutagenesis methods, FNB mutagenesis induces relatively random mutations in the genome [10, 11].
6. Unlike insertion mutagenesis, FNB mutagenesis does not introduce any foreign DNA sequences into the plant genome.

Therefore, FNB-induced mutants are not GMOs (genetically modified organisms) and can be cultivated without the restrictive regulation applied to GMOs [12]. To facilitate functional studies of genes at the genome level, a comprehensive collection of mutant seeds derived from over 150,000 M_1 lines in the *M. truncatula* cv Jemalong A17 background has been generated coordinately at the Noble Research Institute and the John Innes Centre [13] using FNB mutagenesis [12, 13].

Mutation frequency can be calculated per locus within a phenotypic group and per dose so that comparisons can be made among different plant species. A comparison between barley and

Arabidopsis suggested that mutation frequencies from fast neutron bombardment mutagenesis per *cer* locus (*Eceriferum*, for bright green silique or stem) are 4.56×10^{-6}/Gy and 2.58×10^{-6}/Gy, respectively. Because the size of the barley genome (~5200 Mb) is ~40 times bigger than the size of the Arabidopsis genome, this suggested that mutation frequency induced by radiation mutagenesis is independent of the genome size [14]. It is known that mutation frequencies can vary dependent on the locus. Because the mutation frequency per locus caused by radiation mutagenesis is independent of the genome size [14], mutant populations required to reach the same mutation coverage are comparable among plants with different genome sizes. It is estimated that a population of 84,825 and 130, 397 lines will have 95% and 99% of mutation coverage, respectively, for identifying mutants of any target genes [15]. Based on this, we estimated that a population of 150,000 lines is sufficient to identify mutants of a vast majority of genes in *Medicago truncatula*. In the past 12 years, we organized annual FNB mutant screening workshops and screened on average 1000 M2 pools, representing approximately 5000 M1 lines. From these mutant screening workshops, we have isolated hundreds of putative mutants in the categories of symbiotic nitrogen fixation and leaf, root, and shoot development. For example, in 2014, we screened about 1300 M2 pools and identified 241 putative mutant lines with defects in symbiotic nitrogen fixation. Rescreening of the putative mutant lines in 2016 resulted in 118 confirmed symbiotic nitrogen fixation mutant lines. A project website (https://medicago-mutant.noble.org/mutant/) has the information for FNB mutants from the mutant screening workshops, including images and phenotypic description. Here, we describe the protocol for screening of the FNB mutant collection for mutants with visible phenotypes in different categories.

2 Materials

2.1 Plant Material

Seeds of *M. truncatula* cv. Jemalong A17 were exposed to fast neutron radiation at dosages of 30, 32.5, 35, 37.5, and 40 Gy. M1 seeds were germinated, and M1 plants were grown to set M2 seeds in greenhouse. Five M1 plants were grouped together to produce a pool of M2 seeds. M2 seeds were germinated, and M2 plants were screened for symbiotic nitrogen fixation and other categories of phenotypes as described below.

2.2 Seed Scarification and Germination

1. 15 mL falcon tubes.
2. Sulfuric acid.
3. Electronic pipette.
4. 20% bleach solution.

5. Autoclaved filter paper.
6. Plastic disposable pipette.
7. Petri dishes.
8. Laminar flow hood.
9. Dissection microscope.
10. Forceps.
11. 1.5 mL microcentrifuge tubes.
12. Glass beads.
13. 1 gallon plastic pots.
14. M830 soil mixture.

2.3 Rhizobia Inoculation

1. Rhizobia strain: *S. meliloti* strain *ABS7.*
2. TY liquid media: prepare in advance liquid TY medium. Weigh 5 g of Bacto Tryptone and 3 g yeast extract and transfer to a 1 L graduated cylinder containing about 800 mL of water. Add 10 mL of $CaCl_2$ (1 M) to the cylinder and mix for about 10 min. Adjust PH to 7.2 with 1 N NaOH. Make up to 1 L with water and autoclave at 121 °C for 20 min.
3. TY agar media: add 1.5% Type E agar to the liquid TY medium, and mix for about 10 min. The media was autoclaved at 121 °C for 20 min. The appropriated antibiotic was added after the media was cool down. Pour the media to the petri dish and make the agar plates.
4. Perlite or vermiculite.
5. Sand.
6. 162 cell trays.
7. Holed flat tray inserts.
8. Tall domes.
9. Low nitrogen solution (ISV).

This nitrogen-limited medium (ISV) is used in nodulation assays with perlite, vermiculite, or sand with manual irrigation as described in *Medicago truncatula as biological material.* (http://medicago.wisc.edu/protocols/Bio%20material.pdf).

This nitrogen-limited medium is used in nodulation assays with perlite, vermiculite, or sand with manual irrigation.

Preparation of these three components A, B, and C of the stock solution.

A

A1. KNO_3 20.2 g/L.

A2. KH_2PO_4 27.2 g/L.

A3. $CaCl_2 \times 2H_2O$ 73.0 g/L.

A4. $MgSO_4 \times 7H_2O$ 24.6 g/L.

A5. K_2SO_4 43.5 g/L.

Dissolve each component (A1–A5) separately in 1 L of deionized water.

B

$EDTA_2Na_2Fe$ 8.2 g/L of deionized water.

C

H_3BO_3 11 g.

$MnSO_4 \times H_2O$ 6.2 g.

KCl 10 g.

$ZnSO_4 \times 7H_2O$ 1 g.

(NH_4) $6Mo_7O_{24} \times 4H_2O$ 1 g.

$CuSO_4 \times 5H_2O$ 0.5 g.

H_2SO_4 0.5 mL.

Dissolwve the components in the given order in deionized water in 1 L of final volume. Heating facilitates the solubility of components.

Stock Solution

Mix in a 10 L container with deionized water up to 10 L of final volume.

250 mL of solution A1.

250 mL of solution A2.

250 mL of solution A3.

250 mL of solution A4.

250 mL of solution A5.

250 mL of solution B.

13.5 mL of Solution C.

Dilution of the stock solution for watering solution: add 200 mL of stock to 7.8 L of deionized water.

2.4 Screening and Analysis of FNB Mutants

Plants were grown in greenhouse with controlled environment, 16 h/8 h. light/dark cycle, and 150 μE/m^2/s light intensity. For the symbiotic phenotype screening, the plants were grown in 3:1 mixture of perlites and sands (or vermiculite). Three-day-old *Medicago truncatula* seedlings were inoculated with *S. meliloti* strain *ABS7*, and plants were screened their symbiotic phenotype 4 weeks post-inoculation. Leaf samples were collected from 3- to 4-week-old plants for DNA isolation and further analysis.

1. Dissection microscope.
2. Forceps.
3. 1.5 mL microcentrifuge tubes.
4. Glass beads.
5. 1 gallon plastic pots.
6. M830 soil mixture.

3 Methods

3.1 Seed Scarification and Germination

The following procedures were carried out in a laminar flow cabinet.

1. About 30 seeds from each M2 pools were poured into 15 mL conical tubes.
2. Seeds were resuspended in concentrated sulfuric acid (H_2SO_4) with intermittent agitation for 8 min. Three volumes of the acid solution per 1 volume of seeds were used to treat seeds. Make sure that seeds do not stick to the bottom or walls of the conical tubes.
3. After the sulfuric acid treatment, all excess H_2SO_4 were carefully removed with a plastic disposable pipette, and seeds were quickly rinsed with a large volume of sterilized, double distilled water (*see* **Note 1**).
4. Then, seeds were immersed in 20% bleach solution for 5 min.
5. Seeds were rinsed with prechilled, sterilized double distilled water three times (*see* **Note 2**).
6. Seeds were then transferred into a small petri dish lined with filter paper. Each petri dish was labeled with the corresponding M2 pool numbers.
7. Seeds were kept in the dark at room temperature overnight.
8. Seeds were vernalized at 4 °C for 5 days.
9. Seeds were rinsed with prechilled water once a day during vernalization.

3.2 Plant Seedlings

1. Move petri dishes with seedlings from 4 °C to room temperature, and keep them in the dark for 16–25 h.
2. Prepare in advance sufficient amount of double-autoclaved sand and perlite. Make sure that sand was washed with a generous amount of tap water before autoclaving (*see* **Note 3**).
3. Fill 162 well trays with autoclaved perlite and sand in a ratio of 3:1.
4. Transfer seedlings from petri dishes to 162 well trays. Twenty-seven seedlings from each M2 pool are transplanted so that each tray will hold 162 seedlings from six M2 pools.
5. Transfer trays that are covered with tall domes to greenhouse for 2 days. Then remove the domes and water seedlings with the low nitrogen solution every other day.

3.3 Inoculation with the Rhizobial Strain, *S. meliloti* ABS7

1. Streak a petri dish containing TY medium plus 1 M $CaCl_2$ and antibiotics (100 μg/mL Tetracycline) with glycerol stock of *S. meliloti* ABS7 strain. The plate was incubated at 28 °C for 2 days.

2. Inoculate 5 mL of TY liquid media with a single colony, and incubate in a 28 °C shaking incubator set at 280 rpm for at least 24 h.
3. Inoculate 500 mL of liquid TY medium with 500 μL of overnight starter culture, and incubate in a 28 °C shaking incubator set at 280 rpm until OD_{600} of the culture reaches to 0.750 to 1.200.
4. Precipitate the bacterial culture by centrifugation at 5000 rpm (~2348 × *g*) for 15 min at 4 °C, and carefully remove the supernatant.
5. Resuspend the pellet in the low nitrogen solution, and adjust OD_{600} of the inoculation solution to 0.02 (*see* **Note 4**).
6. Inoculate each seedling with ~1 mL of the inoculation solution
7. Repeat the inoculation process one more time in 1 week

3.4 Screening of FNB Mutant Lines

1. After 4 weeks of inoculation, plants were first recorded for visible phenotypes, including compound leaf patterning, leaf color, shoot architecture, flowering time, etc.
2. Plants were then carefully removed from the trays and washed to remove sand and perlite with tap water. Plants from each pool were kept in a disposable cup filled with tap water.
3. Nodulation phenotype of each plant was observed either visually or under a dissection microscope. Putative phenotypes such as lack of nodules (nod-), symbiotic nitrogen fixation deficiency (fix-), and hypernodulation (nod++) were recorded.
4. Images of putative mutant plants, including shoot and root, were also recorded.
5. Putative mutant plants were transplanted to 1 gallon pots filled with Metro Mix 830 (M830) professional soil mix.
6. The putative mutant plants were grown and allowed to set seeds in greenhouse.
7. Putative symbiotic nitrogen fixation and nodule development mutants will be tested in the M3 and M4 generation, and confirmed lines will be subjected to molecular and linkage analysis to identify lesions that are responsible for the mutant phenotype (*see* **Note 5**) [16].

4 Notes

1. It is very important to remove quickly all H_2SO_4 to avoid over heating that may compromise the viability of seeds. Carefully remove all excess sulfuric acid with a plastic disposal pipette,

and quickly resuspend the seeds in a large amount of water. The use of prechilled water is helpful.

2. It is highly recommended to use a laminar flow hood to treat all seeds with sulfuric acid and bleach solutions. Similarly, quickly remove the bleach solution, and extensively rinse seeds with sterile water 5–6 times.
3. Vermiculite can be used instead of perlite: sand mixture. Root growth is as efficient as in perlite: sand. However, it is more difficult to separate roots from vermiculite than from perlite: sand mixture.
4. Rhizobial cultures should be made fresh. The concentration of inoculum should not be high to avoid triggering plant defense response and resulting in poor inoculation.
5. Once the mutant phenotypes are identified, genomic DNA will be subjected to *Medicago truncatula* whole genome array-based comparative genomic hybridization (aCGH) or whole genome sequencing (WGS) to characterize lesions in the genome. Backcrossed population should be developed for co-segregation analysis. Finally, genetic complementation will be used to further confirm that the deleted genes are the cause of the phenotype.

Acknowledgments

This work was supported in part by the Noble Research Institute, LLC., the National Science Foundation (IOS-1127155) and Lanzhou University.

References

1. Graham PH, Vance CP (2003) Legumes: importance and constraints to greater use. Plant Physiol 131(3):872–877
2. Alonso JM et al (2003) Five components of the ethylene-response pathway identified in a screen for weak ethylene-insensitive mutants in Arabidopsis. Proc Natl Acad Sci U S A 100(5): 2992–2997
3. Silverstone AL, Ciampaglio CN, Sun T (1998) The Arabidopsis RGA gene encodes a transcriptional regulator repressing the gibberellin signal transduction pathway. Plant Cell 10(2): 155–169
4. Li X, Lassner M, Zhang Y (2002) Deleteagene: a fast neutron deletion mutagenesis-based gene knockout system for plants. Comp Funct Genomics 3(2):158–160
5. Bolon YT et al (2011) Phenotypic and genomic analyses of a fast neutron mutant population resource in soybean. Plant Physiol 156(1): 240–253
6. Men AE et al (2002) Fast neutron mutagenesis of soybean (*Glycine soja L.*) produces a supernodulating mutant containing a large deletion in linkage group H. Genom Lett 1(3): 147–155
7. Hoffmann D, Jiang Q, Men A, Kinkema M, Gresshoff PM (2007) Nodulation deficiency caused by fast neutron mutagenesis of the model legume Lotus japonicus. J Plant Physiol 164(4):460–469
8. Goff SA et al (2002) A draft sequence of the rice genome (Oryza sativa L. ssp. japonica). Science 296(5565):92–100

9. Young ND et al (2011) The Medicago genome provides insight into the evolution of rhizobial symbioses. Nature 480(7378):520–524
10. Li X, Zhang Y (2002) Reverse genetics by fast neutron mutagenesis in higher plants. Funct Integr Genomics 2(6):254–258
11. Wu JL et al (2005) Chemical- and irradiation-induced mutants of indica rice IR64 for forward and reverse genetics. Plant Mol Biol 59(1):85–97
12. Wang H, Li G, Chen R (2006) Fast neutron bombardment (FNB) induced deletion mutagenesis for forward and reverse genetic studies in plants. In: da Silva JT (ed) Floriculture, ornamental and plant biotechnology: advances and topical issues, 1st edn. Global Science Books, Isleworth, pp 629–639
13. Rogers C, Wen J, Chen R, Oldroyd G (2009) Deletion-based reverse genetics in Medicago truncatula. Plant Physiol 151(3):1077–1086
14. Koornneef M, Dellaert LW, van der Veen JH (1982) EMS- and relation-induced mutation frequencies at individual loci in Arabidopsis thaliana (L.) Heynh. Mutat Res 93(1): 109–123
15. Li X et al (2001) A fast neutron deletion mutagenesis-based reverse genetics system for plants. Plant J 27(3):235–242
16. Chen Y, Wang X, Shunfei L, Wang H, Li S, Chen R (2017) An array-based comparative genomic hybridization platform for efficient detection of copy number variations in fast neutron-induced Medicago truncatula mutants. J Vis Exp 129

Chapter 5

Targeting Induced Local Lesions IN Genomes (TILLING) in *Medicago truncatula*

Myriam Sánchez, Christine Le Signor, Gregoire Aubert, Brigitte Darchy, Karine Gallardo, and Richard D. Thompson

Abstract

TILLING is a reverse genetics strategy that combines the high density of point mutations provided by traditional chemical mutagenesis with rapid screening of DNA pools from a mutagenized population for induced mutations (McCallum et al., Nat Biotechnol 18:455–457, 2000). This high-throughput technique allows the identification of point mutations in any gene of interest.

Key words TILLING, Reverse genetics, *Medicago truncatula*, EMS chemical mutagenesis

1 Introduction

With the availability of next-generation sequencing, genome sequences of most crop species are now available. To identify gene function, a number of reverse genetic approaches have been described for *M. truncatula*, including TILLING. The advantages of TILLING include the fact that it is non-GMO, that a saturating population is easily attained, and that point mutations are relatively well-tolerated, allowing reasonable population sizes to be used and an allelic series of mutations at a given locus to be isolated. The method we describe is straightforward to set up and screenings are routine. For a higher throughput, more recent screening techniques employing NGS should be considered [2]. Once a mutation has been identified at a given locus, mutant lines obtained by TILLING have to be backcrossed to a non-mutated genotype to reduce the background of mutations at other loci.

Luis A. Cañas and José Pío Beltrán (eds.), *Functional genomics in* Medicago truncatula*: Methods and Protocols*, Methods in Molecular Biology, vol. 1822, https://doi.org/10.1007/978-1-4939-8633-0_5,

2 Materials

2.1 EMS Mutagenesis of Medicago Seeds

1. Seeds of *Medicago truncatula* (about 20,000 seeds were necessary to generate a 5000 M1 plant population) (*see* [3] and **Note 1**).
2. EMS (SIGMA M0880, highly toxic) is a mutagenic, teratogenic, and suspected carcinogenic compound to be handled under a fume hood for chemical volatiles, with appropriate personal protection (refer to EMS safety data sheet).
3. Sulfuric acid 95–98%.
4. A stainless steel tea ball.
5. Deionized water.
6. Micropipette.
7. 100 mL Schott glass bottle with a leak-proof cap with silicone sealing ring.
8. Petri dishes and blotter.
9. Rotator.
10. Fume hood.

2.2 DNA Preparation and Pooling of Individuals

1. 1.1 mL polypropylene tubes in strips of eight, with caps, in a 96-well rack.
2. Stainless steel beads (diameter 3 mm).
3. Liquid nitrogen.
4. TissueRuptor or a TissueLyser.
5. Qiagen DNeasy 96 Plant Kit.
6. Absolute ethanol and 70% ethanol.
7. Micropipettes.
8. Sterile tips.
9. 65 °C water bath.
10. Vortex mixer.
11. Optional: pipetting robot

2.3 PCR Amplification of Target Gene

1. PCR machine (thermocycler).
2. 1.5 mL microcentrifuge tubes.
3. PCR plate.
4. PCR Sealing Sheets.
5. PCR reagents: 10× PCR buffer, 2.5 mM dNTPs set, TAQ DNA polymerase, and Milli-Q (MQ) water.
6. A computer connected to the Internet
7. Target gene's outer primers.

8. Target gene's inner labeled primers.
9. 2% agarose gel.
10. Agarose gel electrophoresis system.
11. 1× TAE buffer.
12. DNA ladder.
13. DNA loading buffer: (1.5 g of Ficoll, 25 g of urea, 10 mL of 0.5 M EDTA, bromophenol blue and xylene cyanol, MQ water up to 100 mL).
14. Optional: infrared imaging system (Odyssey by LI-COR).

2.4 Endonuclease Digestion

1. Cel I nuclease.
2. 10× reaction buffer: 0.1 M Hepes pH 7.5, 0.1 M $MgSO_4$, 0.02% Triton X-100, 0.1 M KCl, and MQ water.
3. Storage and dilution buffer: 50 mM Tris HCl pH 8, 100 mM KCL, 10 μM $ZnCl_2$, 0.01% Triton X-100, 50% glycerol, 20 μg/mL acetylated BSA, and MQ water.
4. 150 mM EDTA.

2.5 Denaturing Gel Detection and Image Analysis

1. Gloves.
2. Safety glasses.
3. Nonabrasive tissues.
4. 25 cm plates: front plate (notched) and back plate (rectangular).
5. One set of 0.25 mm spacers.
6. 0.25 mm casting comb.
7. One set of rail assemblies.
8. Casting plates.
9. Casting stand (optional).
10. Concentrated laboratory detergent solution.
11. Deionized water.
12. Ethanol (70–100%).
13. 6% manual sequencing gel solution bis-acrylamide ratio 19:1 (ready to use formulation contains 7 M urea (denaturant) and 1× TBE buffer).
14. TEMED.
15. 10% APS.
16. LI-COR 4300 DNA Analysis Systems.
17. Upper buffer tank.
18. Upper buffer tank lid.
19. Electrode leads
20. Lower buffer tank.

21. Lower buffer tank lid.
22. 1 L of 1× TBE buffer.
23. Formamide loading buffer: (deionized formamide 5 mL, 0.1 mL of 0.5 M EDTA pH 8, 1 mg bromophenol blue, and MQ water up to 20 mL).
24. 96-tooth membrane paper comb.
25. Membrane comb-loading tray.
26. Multichannel pipettes, 0.5–10 μL.
27. Image analysis software.

3 Methods

3.1 EMS Mutagenesis of Medicago Seeds

1. Put a batch of maximum 6000 *Medicago truncatula* seeds in a tea ball (approximately 25 g). Immerse the tea ball + seeds in pure sulfuric acid, for a maximum of 3 min to scarify seeds. Then immerse rapidly the tea ball in a large volume of distilled water (otherwise risk of burning seeds), and rinse at least three times. After scarification, the seeds are packed in cotton gauze closed with staples.
2. Dilute EMS to the chosen dose in deionized water. Pipette the sample using filter tips, close hermetically the bottle, and shake gently until the product has completely dissolved (the EMS is viscous and dissolves quite poorly in water) (*see* **Notes 2–4**).
3. Put the seeds packed in cotton gauze in the EMS solution. The bottles are closed preferably with red PBT type cap, leak proof with a silicone sealing ring. Closed bottles are gently shaken (50 rpm) on a rotary shaker under the fume hood overnight (24 h soaking).
4. Remove the EMS solution, and dispose it in a dedicated bin (to be stored under the fume hood until evacuation; *see*-**Note 4–5**).
5. The bottle is refilled with distilled water, closed and shaken by inversion three times and re-emptied. Three fast washes are followed by 11 washes, each of 30 min under rotary shaking. Always dispose of wash liquid in the dedicated bin.
6. Seeds are then collected in petri dishes on blotters soaked with distilled water and placed in a vernalization chamber (8 °C in the dark) for about 48 h.
7. Seeds are sown in 0.5 L pots filled with inert disinfected medium (i.e., pouzzolane) in a greenhouse and covered with a bed of coarse sand. Plants were automatically watered with a nutrient solution of 3.5 N/3.1 P/8.6 K. The temperature was above 19 °C during the night and below 30 °C during the day;

complementary artificial lighting was provided to attain a 16-h day. Given the loss at emergence, at least two seeds are sown per pot. It is preferable that after sowing, the greenhouse temperature rises gradually to a minimum of 19 °C.

8. Each plant is bagged to prevent cross-fertilization and, at maturity, harvested individually and seeds extracted from pods. Store M2 seed lots at 4 °C, keeping humidity at ≤30–40%.
9. After imbibition and vernalization, sow four seeds per M2 seed lot in 0.5 L pots filled with inert medium. Before flowering, three plants are discarded (*see* **Note 6**), and leaf material of the remaining plant is collected for DNA extraction; it is self-fertilized and harvested. M3 seed lots were harvested and stored in cold chamber under controlled humidity.

3.2 DNA Preparation and Pooling of Individuals

1. Collect 50–100 mg leaf material from each M2 plant in a 1.1 mL polypropylene tube (single tube or in strip of 8) with two stainless steel beads (diameter 3 mm), and put it in a rack. Samples are organized according to a predefined order in racks of 96 tubes.
2. Immerse the racks into liquid nitrogen to freeze the samples. Disrupt and homogenize samples using Qiagen TissueLyser (1 min 30 to 30 Hz). Then store the samples at −80 °C until needed.
3. Carry out the DNA purifications using the Qiagen DNeasy 96 Plant Kit (*see* **Note 7**).
4. Dilute aliquots from each well plate 20 fold, keeping the initial order.
5. To increase the throughput, pool eight diluted DNAs into each well of a new set of 96-well plates. Take care of the new sample order. Pool plates are ready to use for the following steps (*see* **Note 8**).

3.3 PCR Amplification of Target Gene and Heteroduplex Formation

1. For screening a given gene, choose a specific DNA sequence with a length of around 1 kb. Design two primer pairs, one for the outer fragment and one for the inner fragment to be amplified. The second, internal set of primers is labeled with infrared fluorochromes: IRD 700 and IRD 800 fluorescent labels, respectively (Fig. 1, *see also* **Note 9**).
2. Perform the first PCR amplification with unlabeled outer primers in a total volume of 11 μL. For one 96-well PCR plate, allocate 5 μL of pooled DNA in each well keeping the same order, and add 6 μL of PCR master mix. Prepare the master mix in a sterile 1.5 mL microcentrifuge tube. Each reaction contains 1.1 μL of 10×Taq DNA polymerase buffer with 15 mM MgCl2, 1 μL of 2.5 mM dNTP set, 0.3 μL of each of

Fig. 1 Target is amplified by PCR with gene-specific primers (1), followed by a nested PCR with internal gene-specific labeled primers (2)

the 10 μM primer pair, 0.5 U of Taq DNA polymerase, and sterile water up to 6 μL.

3. Seal the 96-well PCR plate, put on ice, and set up the thermocycling program for the PCR: 94 °C for 3 min, 39 cycles of (94 °C for 45 s, Tm of the primers used for 45 s, 72 °C for 1 min/Kb), 72 °C for 5 min, and final hold at 22 °C.
4. Put the 96-well PCR plate into the thermocycler and run the program.
5. When the run is finished, verify the amplification of some samples by electrophoresis of the PCR products on 2% agarose gel in the presence of a DNA size marker. Store the PCR plate at −20 °C.
6. Check the yield and the specificity of the PCR by observing a unique DNA band on the electrophoresis gel, and compare it to the size marker.
7. If this first PCR is ok, dilute it by adding 100 μL of sterile water to each well (*see* **Note 10**).
8. Perform the second PCR amplification with inner labeled primers, using the same protocol. In a new 96-well PCR plate, allocate 5 μL of the diluted PCR product, keeping the same order, and add 6 μL of PCR master mix. Prepare the master mix in a sterile 1.5 mL microcentrifuge tube. Each reaction contains 1.1 μL of 10×Taq DNA polymerase buffer with 15 mM MgCl2, 1 μL of 2.5 mM dNTP set, 0.3 μL of each of the 10 μM primer pair, 0.5 U of Taq DNA polymerase, and sterile water up to 11 μL final (*see* **Note 11**).
9. The 96-well plate is then sealed hermetically with a film supplied for this purpose, put the plate into the thermocycler, and run the program as follows: 94 °C for 3 min, 39 cycles of (94 °C for 45 s, Tm of the primers used for 45 s, 72 °C for 1 min/Kb), and 72 °C for 5 min. To favor the formation of heteroduplexes, further denaturation and reannealing steps are included at the end of this PCR program as follows: 99 °C for 10 min, 69 cycles of 70 °C for 20 s with 0.3 °C decrease in temperature per cycle, final hold 22 °C.
10. After the run, check some samples by electrophoresis on a 2% agarose gel.

11. Verify the products and the labeling of the PCR by scanning the electrophorese gel using an infrared imaging system (Odyssey by LI-COR).

3.4 Endonuclease Digestion: Cel I

1. Digest the sample with the mismatch cleavage enzyme Cel I. Add 9 μL of Cel I digestion mix to the total volume of the PCR products previously obtained. Prepare the digestion mix in a sterile 1.5 mL microcentrifuge tube. Each reaction contains, 2 μL of 10×Cel I reaction buffer, 0.6 μL of Cel I (diluted 1/10, purified according to [4]), and sterile water up to 20 μL final.
2. Seal the plate and incubate it for 40 min at 45 °C. Perform the incubation using a thermocycler.
3. Stop the reaction by adding 3 μL of 150 mM EDTA to each well (*see* **Note 12**).

3.5 Denaturing Gel Detection and Image Analysis

1. Prepare a loading plate. Allocate 1 μL of each sample and add 5 μL of DNA loading buffer.
2. Heat the samples at 95 °C for 3 min using a thermocycler, and place immediately on ice.
3. Load 1 μL of sample by capillarity absorption onto a 96-tooth membrane paper comb using a membrane comb-loading tray and a multichannel pipette.
4. Put the comb containing samples in contact with the front of the polyacrylamide gel.
5. Run the electrophoresis using LI-COR 4300 DNA Analysis System at: voltage, 1500 V; current, 40 mA power, 40 W; temperature, 45 °C; and scan speed, 3.
6. Collect image from the LI-COR 4300 in TIFF format, and analyze visually using image analysis software. An example is shown in Fig. 2.

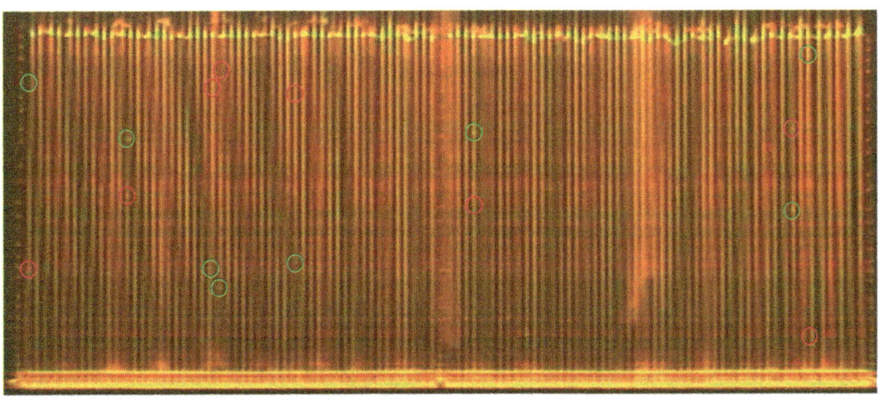

Fig. 2 Example of gene TILLING screen on eightfold pooled Medicago DNA. The gel image represents the superposed image from IRD700 and IRD800 channels

7. Identify novels bands due to the Cel I action by comparing all the samples. The bands must be complementary between the gel images at 700 and 800 nm, i.e., add up in size to that of the undigested PCR fragment. Note the numbers of the positive lanes on the gel, and return to the plan of the plate loaded to identify the mutant DNA pools.

3.6 Identification of the Mutated Individual

1. Identify the mutated individual applying the same technology (nested PCR amplification, Cel I cleavage, denaturing gel detection) to each DNA which composed the positive pool.
2. Confirm the mutation on the target gene by DNA sequencing.

3.7 Select Best Mutation Using Online Tools

1. Add the mutations to a database and cross-reference with the plant line identifiers.
2. Map and analyze the effects of mutations identified to the encoded protein using web tools like PARESNP (Project Aligned Related Sequences and Evaluate SNPs) and SIFT (Sort Intolerant From Tolerant) (*see* **Note 13**). An example of PARESNP output depicting the sites and codon types of predicted mutations for a given gene is shown in Fig. 3.
3. Continue the analysis by the characterization of mutant lines identified from M3 seeds. The chosen lines will be genotyped and phenotyped over several generations of backcrosses (*see* **Note 14**).

3.8 Prepare Plant Material for Phenotype Validation

1. After seed vernalization step, grow the M3 plants from M3 seeds retrieved from the TILLING collection together with A17 reference line.
2. Harvest leaf material as soon as possible.
3. Proceed to DNA extraction as described above.
4. Use a genotyping method (TILLING method as described above or Sanger sequencing after PCR amplification or dCaps marker; see below).
5. Eliminate plants that do not carry mutant allele in greenhouse.
6. At flowering, proceed to crosses between mutant and reference line. Always bring pollen from mutant line to emasculated flower of A17. Tag the flower and wrap it in Parafilm®.

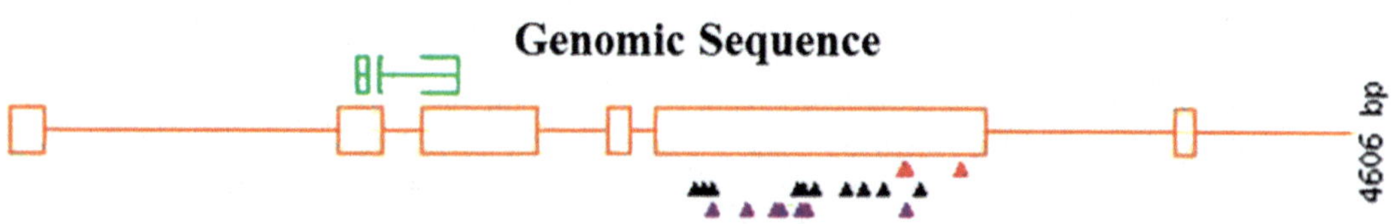

Fig. 3 PARSESNP output. Graphical representation of mutations identified. The orange boxes represent coding sequence. The purple and the black triangles represent silent and missense mutations, respectively. The red triangles represent nonsense mutations

7. At maturity, harvest tagged pods.
8. After seed vernalization, grow first backcrossed plants (BC1) and reference line A17 in greenhouse. Harvest leaf material, and proceed to genotyping as before to select only plants carrying mutation.
9. Proceed to second backcross by bringing BC1 pollen directly on emasculated flower of reference line A17. Tag the flower.
10. At maturity, harvest tagged pods.
11. After seed vernalization, grow second backcrossed plants (BC2) in greenhouse. Harvest leaf material, and proceed to genotyping as before to select only plants carrying mutation.
12. At maturity, harvest all pods on the plants resulting from selfing of BC2.
13. After seed vernalization, grow a sample of second backcross selfed plants (BC2S1). Harvest leaf material, and genotype these segregating plants, including ¼ homozygous mutant plants, ¼ homozygous WT plants, and ½ heterozygous plants. This segregating population is the appropriate material for recording WT and mutant phenotypes in the aim of gene function validation.

3.9 Development of CAPS (Cleaved Amplified Polymorphic Sequences) Markers for SNP Genotyping

1. Design CAPS or dCAPS primers using the bioinformatics tool: dCAPs Finder 2.0 (http://helix.wustl.edu/dcaps/dcaps.html) [5].
2. Enter, respectively, the sequence of the 55–60 DNA bases bordering the SNP for the two genotypes (e.g., 27 bases-SNP) into the "wild-type sequence" and "mutant sequence" boxes, respectively. Never exceed 60 bases in total.
3. Enter a number of mismatches starting from zero, and increase according to the output generated. The output from zero shows the presence of a CAPS marker. If a CAPS marker is not generated by the presence of the SNP, search with one mismatch for generating a derived CAPS (dCAPS) marker. Increase the number of mismatches in each run until potential dCAPS markers are proposed. Each run produces several optional primers in its output.
4. As an option, add a tail at the 5′ end of the primer (CACGACGTTGTAAAACGAC). This will allow to display a bigger size difference between the restriction fragments corresponding to the different alleles.
5. Define a second primer, respectively, reverse or forward, so as to amplify a 250–500 bp fragment in combination with the first primer overlapping the SNP. The fragment to be amplified should not contain any further restriction sites for the enzyme chosen. This primer must have a similar Tm to that of the first primer (calculated without the tail).

6. Test the markers by performing PCR amplification on the DNAs of the two genotypes. Proceed as described above using 5 μL of each diluted DNA tenfold in 0.2 mL PCR tubes. Add 1.1 μL of 10×Taq DNA polymerase buffer with 15 mM $MgCl_2$, 1 μL of 2.5 mM dNTP set, 0.3 μL of each of the 10 μM primer pair, 0.5 U of Taq DNA polymerase, and sterile water up to 11 μL final. Close the tubes and run the PCR on a thermocycler as follows: 94 °C for 3 min, 39 cycles of (94 °C for 45 s, Tm of the primers used for 45 s, 72 °C for 1 min/Kb), 72 °C for 5 min, and final hold at 22 °C.
7. Verify the amplification by electrophoresis of 5 μL of PCR product on a 2% agarose gel.
8. If satisfactorily amplified, incubate 5 μL of PCR product in a new 0.2 mL PCR tube, with 1 μL of 10× buffer (supplied with the restriction enzyme), 0.5–1 μL of restriction enzyme, and 3.5–3 μL of deionized water, for 12 h at the optimum temperature required for the restriction enzyme used (37 °C for the majority of enzymes, unless otherwise indicated).
9. Check the digestion products by performing an electrophoresis on a 3% agarose gel. Compare electrophoretic profiles for both genotypes.

4 Notes

1. Even though *M. truncatula* is predominantly self-fertilizing, some residual cross-pollination may occur. Consequently, plants were raised in insect-proof glasshouses and covered with perforated plastic bags to keep them isolated from their neighbors. To ensure the quality of the TILLING population, parental plants used to produce the M0 seed lot were analyzed for genetic uniformity using a set of 16 simple sequence repeat (SSR) markers distributed over the eight chromosomes (SSR markers and genetic map provided by T. Huguet, ENSAT Toulouse, France; [6]). We checked that no heterogeneity appeared in banding patterns.
2. The EMS vial is transported in a triple pack: bottle + hermetic bag + closed metal box filled with moisture-absorbing granules. Only the plastic bag and the bottle are placed under the fume hood during handling. The metal box is never in contact with the product.
3. All vials and utensils that come into contact with the EMS are to be disposed of in specific bins (for reprocessing) on which the name and approximate concentration of the product is indicated.
4. To determine the EMS dose to apply, preliminary tests are needed in small volumes to test the sensitivity of the seed lot

to the EMS [3, 7]. An EMS concentration range from 0 (water control) to 0.3% for *Medicago truncatula* is sufficient to determine the optimum dosage that is the highest dose that allows an emergence rate similar to the control and fertility close to 50%. In general, according to the seed lots, doses from 0.15 to 0.20% are retained (*see* [3]).

5. *EMS protocol for pea seeds*: In pea, the criterion of emergence rate is not valid; whatever the doses tested (up to 0.7%), the rate of emergence is greater than 90% of the control. In this species, we select the dose essentially based on plant fertility [7]. Prepare the soaking solution in a 1 L bottle (Schott type), with hermetic cap (PBT, leak-proof equipped with a silicone seal). The bottle is filled with approximately 900 pea seeds (example for Caméor genotype with 1000-seed weight around 180 g) and 450 mL of EMS solution. After 15 h of soaking (50 rpm) on a rotary shaker and under the fume hood, there is almost no liquid left. Then proceed to washes as described for *Medicago truncatula*. No vernalization is needed, and seeds can be sown directly in pots under greenhouse conditions.
6. Before flowering, a phenotype scoring was done on each M1 family. An ontology adapted to *M. truncatula* was defined, based on those already developed for pea ([7, 8]). Of all M2 families grown and phenotyped, 40% showed a visible phenotype. The most commonly observed phenotypes were related to leaf and cotyledon color, plant size, leaf size, and leaflet shape, as reported for pea and lotus. All data are available on www.thelegumeportal.net.
7. Proceed according to the *Qiagen DNeasy Plant Handbook*, June 2015. "Purification of Total DNA from Frozen or Lyophilized Plant Tissue"—DNeasy 96 Protocol, p35–41. Tissues are collected plant to plant, and DNAs extracted individually. The DNA collection of the EMS *Medicago truncatula* mutant population represents 44 racks each of 96 samples.
8. The use of a pipetting robot facilitates the preparation of DNA pools in the plates. Each pool plate contains 768 individuals of M2 families. The plates containing pure DNA samples are stored at −80 °C. The plates containing diluted DNA and DNA pools are stored at −20 °C.
9. For primer design, take into account all available information about gene structure and function. The use of the web tools "Codons Optimized to Detect Deleterious Lesions" (CODDLE) allows one to take into account the gene model (intron and exon positions) and conserved domains of the encoded protein.
10. The volume of water added depends on the PCR yield; it may be smaller or larger, but routinely we add about 100 μL

of sterile water. This dilution may be made in a new 96-well PCR plate.

11. Labeled primers are photosensitive, so minimize exposure to light of the primers' stock and the PCR plate for the rest of the experiment.
12. Samples can be desalted using a Sephadex G-50 separation column (optional).
13. Just over one-half of the induced mutations in the coding sequences are predicted to result in a defective protein, mainly missense (50%) or nonsense (5%) [9]. Additional loss-of-function mutations can result from mutation of splice junctions.
14. Two backcrosses to the reference line A17 will remove approximately 75% of the background EMS mutations. Homozygous mutant lines derived from these backcrosses will be sufficiently pure for use in genetic and phenotypic analyses. The allelic series of mutations provided by TILLING are powerful tools for functional validation of candidate genes.

References

1. McCallum CM, Comai L, Greene EA, Henikoff S (2000) Targeted screening for induced mutations. Nat Biotechnol 18:455–457. https://doi.org/10.1038/74542
2. Henry IM, Nagalakshmi U, Lieberman MC, Ngo KJ, Krasileva KV, Vasquez-Gross H, Akhunova A, Akhunov E, Dubcovsky J, Tai TH, Comai L (2014) Efficient genome-wide detection and cataloging of EMS-induced mutations using exome capture and next-generation sequencing. Plant Cell 26:1382–1397. https://doi.org/10.1105/tpc.113.121590
3. Le Signor C, Savois V, Aubert G, Verdier J, Nicolas M, Pagny G, Moussy F, Sanchez M, Baker D, Clarke J, Thompson R (2009) Optimizing TILLING populations for reverse genetics in Medicago truncatula. Plant Biotechnol J 7:430–441. https://doi.org/10.1111/j.1467-7652.2009.00410.x
4. Oleykowski CA, Bronson Mullins CR, Godwin AK, Yeung AT (1998) Mutation detection using a novel plant endonuclease. Nucleic Acids Res 26:4597–4602
5. Neff MM, Turk E, Kalishman M (2002) Web-based primer design for single nucleotide polymorphism analysis. Trends Genet 18:613–615
6. Pierre J-B, Huguet T, Barre P, Huyghe C, Julier B (2008) Detection of QTLs for flowering date in three mapping populations of the model legume species Medicago truncatula. Theor Appl Genet 117:609–620. https://doi.org/10.1007/s00122-008-0805-4
7. Dalmais M, Schmidt J, Le Signor C, Moussy F, Burstin J, Savois V, Aubert G, Brunaud V, de Oliveira Y, Guichard C, Thompson R, Bendahmane A (2008) UTILLdb, a Pisum sativum in silico forward and reverse genetics tool. Genome Biol 9:R43. https://doi.org/10.1186/gb-2008-9-2-r43
8. Perry JA, Wang TL, Welham TJ, Gardner S, Pike JM, Yoshida S, Parniske M (2003) A TILLING reverse genetics tool and a web-accessible collection of mutants of the legume Lotus japonicus. Plant Physiol 131:866–871. https://doi.org/10.1104/pp.102.017384
9. Markiewicz S, Subtil A, Dautry-Varsat A, Fischer A, de Saint Basile G (1994) Detection of three nonsense mutations and one missense mutation in the interleukin-2 receptor gamma chain gene in SCIDX1 that differently affect the mRNA processing. Genomics 21:291–293. https://doi.org/10.1006/geno.1994.1265

Chapter 6

T-DNA Insertional Mutagenesis and Activation Tagging in *Medicago truncatula*

Francesco Panara, Ornella Calderini, and Andrea Porceddu

Abstract

The development of plant genetic transformation techniques has greatly enhanced our capacity to investigate and understand gene function. Since T-DNA constructs insert randomly in genomes, in principle, it is possible to construct a population of individuals harboring one or more T-DNA inserted in any region of the genome. Such populations can be screened following two approaches: (1) given a mutant phenotype, one could find the gene subtending the phenotypic alteration (forward approach), or (2) given a gene of interest, one could identify the phenotypic effect of its expression perturbation (reverse approach).

Activation tagging is an application of T-DNA mutagenesis aimed at obtaining gain-of-function mutations. This can be achieved by introducing enhancer sequences randomly in the target genome via a T-DNA shuttle and then analyzing the genomic regions flanking the insertion sites in individuals showing phenotypic alterations. In this chapter, we describe the detailed procedure to obtain and screen an activation-tagged population in *Medicago truncatula*.

Key words Activation tagging, I-PCR, *Medicago truncatula*, *Agrobacterium*-mediated plant transformation, pSKI074, Adapter-ligation-mediated PCR, Genome walking

1 Introduction

One of the most effective ways to investigate gene function in model organisms is based on the generation of large tagged mutant populations exploiting T-DNAs or transposons as mutagens. An insertion of genetic elements within or in proximity of a gene often determines its functional disruption, and the corresponding gene function is inferred by the phenotypic characterization of the (loss-of-function) mutant. This strategy has been largely adopted in many plant model species including *Medicago truncatula* [1–3]. However, not all genes can be studied via loss-of-function mutations as, for example, in the case of genes providing essential function (lethal mutations) or functionally redundant genes.

As a complementary approach, activation tagging was proposed to obtain gain-of-function mutations. This method is based

Luis A. Cañas and José Pío Beltrán (eds.), *Functional genomics in* Medicago truncatula*: Methods and Protocols*, Methods in Molecular Biology, vol. 1822, https://doi.org/10.1007/978-1-4939-8633-0_6,

on the use of a genetic element harboring one or several enhancer sequences that when inserted in proximity of a gene may activate it ectopically and/or highly increase its expression. Advantages of this technique are the generation of mutations with dominant phenotypes already observable in T_0 lines and the possibility to obtain phenotypes from functionally redundant genes or other genes non-tractable with loss-of-function approaches and the presence of a DNA tag. Loss-of-function mutations are still possible within the activation tagging approach as in classical insertional mutagenesis when the T-DNA insertion introduce a disablement: an example is described in [4]. Hybrid loss/activation tagging approach can be established by developing activation tagging collection from mutant genotypes and screening for suppression/complementation of the mutant phenotype [5, 6].

Elements used in activation tagging include a T-DNA with strong constitutive enhancers reading toward one of its borders as the one used in [5]. Other strategies using the maize transposon system engineered with 35S enhancers, including En-I [7], En/Spm [8], and Ac/Ds [9], have also been established. Recently the FOX hunting system (Full-length cDNA Over-eXpressing gene hunting system) has also emerged [10]. In addition, Tnt1 retrotransposon, used in insertional mutagenesis, harbors enhancer elements within long terminus repeats that can determine upregulation of an adjacent gene [11] (Panara et al., unpublished).

We used an activation tagging approach based on the pSKI074 vector developed by Detlef Weigel [5] to generate a mutant collection in *Medicago truncatula*. The collection was described in [12]. Here we describe methods and protocols used for creation and routine screening of this population.

2 Materials

2.1 Plant Material and Bacterial Strains

1. Seeds of the *Medicago truncatula* ecotype R108–1 C3 [13].
2. SURE electroporation competent *Escherichia coli* (Agilent™, USA).
3. TOP 10 chemically competent *Escherichia coli* (Thermo-Fisher Scientific™, USA) (*see* **Note 1**).
4. EHA105 *Agrobacterium tumefaciens* strain [13].

2.2 Growth Media and Solutions

1. LB medium: 10 g/L tryptone, 5 g/L yeast extract, 10 g/L NaCl, pH 7.5. For solid medium: 15 g/L Bacto™ Agar.
2. YEB medium: 5 g/L beef extract, 1 g/L yeast extract, 5 g/L peptone, 5 g/L sucrose, 2 mL of a 1 M solution of magnesium sulfate, pH 7.2. For solid medium: 15 g/L Bacto™ Agar.
3. SH3a medium: 100 mL/L N6 major solution, 1 mL/L SH minor solution, 1 mL/L SH vitamins solution, 20 mL/L

EDFS stock solution, 100 mg/L myoinositol, 30 g/L sucrose, 4 mg/L 2,4-dichlorophenoxyacetic acid, 0.5 mg/L 6-benzylaminopurine, pH 5.8 (*see* **Note 2**). For solid medium: 3 g/L phytagel.

4. SH9 medium: 100 mL/L N6 major solution, 1 mL/L SH minor solution, 1 mL/L SH vitamins solution, 20 mL/L EDFS stock solution, 100 mg/L myoinositol, 20 g/L sucrose, pH 5.8 (*see* **Note 2**). For solid medium: 9 g/L Kalys agar (Kalys SA, France).
5. ½ SH9 medium: 50 mL/L N6 major solution, 0.5 mL/L SH minor solution, 0.5 mL/L SH vitamins solution, 10 mL/L EDFS stock solution, 50 mg/L myoinositol, 10 g/L sucrose, pH 5.8 (*see* **Note 2**). For solid medium: 7 g/L Kalys agar.
6. N6 major solution: 1.85 g/L $MgSO_4 \cdot 7H_2O$, 28.3 g/L KNO_3, 4.63 g/L $(NH_4)_2SO_4$, 1.66 g/L $CaCl_2 \cdot 2H_2O$, 4 g/L KH_2PO_4.
7. SH minor solution: 10 g/L $MnSO_4 \cdot H_2O$, 5 g/L H_3BO_3, 1 g/L $ZnSO_4 \cdot 7H_2O$, 1 g/L KI, 0.1 g/L $NaMoO_4 \cdot 2H_2O$, 0.2 g/L $CuSO_4 \cdot 2H_2O$, 0.1 g/L $CoCl_2 \cdot 6H_2O$.
8. SH vitamin solution: 5 g/L nicotinic acid, 5 g/L thiamine hydrochloride, 5 g/L pyridoxine hydrochloride.
9. EDFS stock solution: 7 g/L Ethylenediaminetetraacetic acid ferric sodium salt (EDFS).
10. Ampicillin stock solution: dilute100 mg/mL salt in bdH_2O, filter sterilize, and store at −20 °C in 0.5–1 mL aliquots.
11. Kanamycin stock solution: dilute 100 mg/mL in bdH_2O, filter sterilize and store at −20 °C in 0.5–1 mL aliquots.
12. Rifampicin stock solution: dilute 100 mg/mL in DMSO (Dimethyl sulfoxide) and store at −20 °C in 0.5–1 mL aliquots.
13. Augmentin: we used it in the powder format; individually packed bags contain 1 g of suspendible powder (875 mg amoxicillin, 125 mg clavulanic acid).

Bacterial media [1, 2] refer to standard protocols; plant media/solutions [3–9] are reported in [13].

2.3 The pSKI074 Vector

pSKI074 is a binary plasmid expressing the *nptII* gene under mannopine synthase promoter that confers kanamycin resistance in plants and contains the pBluescript KS(+) sequence and tetramerized CaMV 35S enhancer sequences close to T-DNA right border (Fig. 1a, b). The plasmid sequence pSKI074 is flanked by restriction enzyme sites that can be used for rescue of T-DNA and flanking plant sequences from transformed plants. The restriction enzymes KpnI, EcoRI, and HindIII can be used for rescue of sequences adjacent to the right T-DNA border, and BamHI, SpeI, and NotI can be used for left border rescue. pSKI074 vector is available at

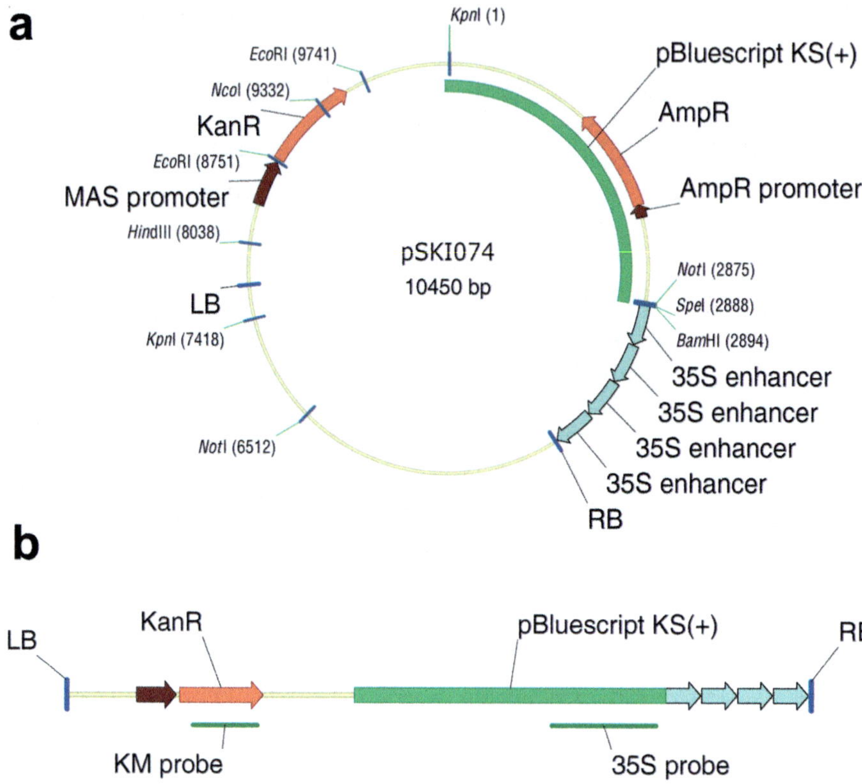

Fig. 1 pSKI074 vector map (**a**) and of the linearized T-DNA with indication of Southern blot probes positions (**b**)

Addgene (Plasmid #11571). The four CaMV 35S enhancer repeats in pSKI074 vector are unstable in *E. coli* and *A. tumefaciens* if stored at 4 °C for extended time [5]; thus particular care must be adopted in the various steps of its manipulation and use.

2.4 Other Reagents and Solutions

1. Glycerol.
2. Absolute and 70% ethanol.
3. Leaf sterilization solution: 5% sodium hypochlorite (*see* **Note 3**), 0.05% TWEEN® 20 in sterile bdH_2O.
4. Phenol.
5. Chloroform.
6. Sodium acetate 3 M, pH 5.2.
7. TAE buffer: 40 mM Tris, 20 mM acetic acid, and 1 mM EDTA, pH 8.0.
8. Plasmid purification kit: we used a commercial plasmid mini-prep kit, PureLink™ Quick Plasmid Miniprep Kit, Thermo Fisher Scientific™, USA, (*see* **Note 4**).
9. High-yielding DNA extraction kit: Illustra Nucleon PhytoPure Genomic DNA Extraction Kit (RPN8511 from GE Healthcare™, UK).

10. Restriction enzymes and T4 DNA ligase were purchased from New England Biolabs™, USA.
11. Blotting membrane (Hybond N+ nylon membrane) and reagents were purchased from Amersham™, GE Healthcare™, UK.
12. Nick Translation Kit, GE Healthcare™, UK.
13. TaKaRa Ex Taq®, Takara Bio USA™, USA.

2.5 Equipment

1. Seed scarifier: a plastic Petri dish, fine sandpaper (180 is good), and glue are necessary to assemble a seed scarifier as shown in Fig. 2.
2. Refrigerators (4, −20, −80 °C).
3. Incubators (28–37 °C) with and without shaking.
4. Ice generator.
5. Lab grade water generator.
6. Refrigerated centrifuges.
7. UV-visible spectrophotometer.
8. Standard benchtop pHmeter.
9. Biorad™ MicroPulser Electroporator and compatible cuvettes.
10. Standard laboratory autoclave.
11. Plant growth chamber with light, temperature, and humidity control.
12. In vitro culture chamber with light and temperature control.
13. Plant in vitro culture and bacteria handling devices and disposables.
14. Horizontal laminar flow hood equipped with Bunsen burner.
15. Vertical laminar flow hood equipped with Bunsen burner.
16. Glass vacuum desiccator jar connected to a vacuum pump.

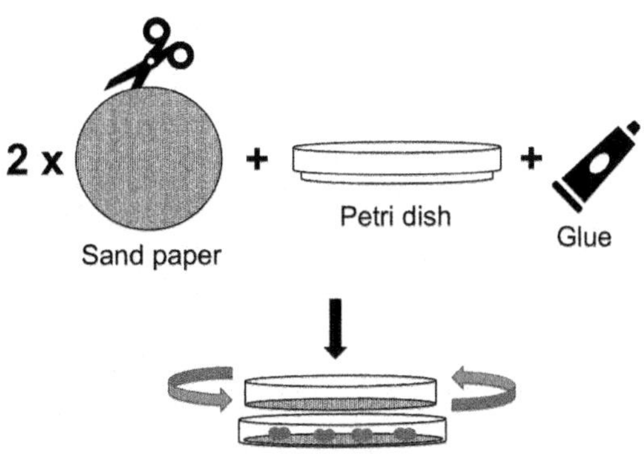

Fig. 2 Schematic representation of a simple hand-made seed scarifier. Sandpaper pieces are stuck within the lid and below the bottom of a Petri dish and used as shown with rotary movements

17. Standard laboratory glassware, plasticware, and liquid handling devices.
18. Programmable thermal cycler and electrophoresis equipment.
19. Plant potting and growth equipment.

3 Methods

3.1 First Manipulation of pSKI074

When obtained from Addgene, the plasmid (pSKI074) is sent as a bacterial stab at room temperature. In this condition pSKI074 is unstable; thus, it is highly recommendable to proceed immediately with its conservation as a glycerol stock as follows.

1. Pick bacteria from stab and spread on an LB agar plate containing 50 mg/L ampicillin to obtain single colonies.
2. Pick a single colony to inoculate 50 mL liquid LB containing 50 mg/L ampicillin in an Erlenmeyer flask, and incubate overnight at 37 °C with shaking (200 rpm).
3. Use an aliquot (3–5 mL) of the overnight culture for plasmid extraction. Use the remaining culture to create glycerol stocks: add 500 μL of 50% glycerol to every 500 μL of culture in 2 mL cryovials and gently mix. Freeze the tubes at −80 °C.
4. Check the plasmid by PCR for the integrity of the 35S enhancer repeats using primer T7, 5′ TAA TAC GAC TCA CTA TAG GG 3′ and IK007, 5′ ACC CGC CAA TAT ATC CTG 3′ (*see* Table 1). An amplicon of 1.46 kb is obtained from intact vectors.

3.2 Production of Agrobacterium tumefaciens EHA105 Electrocompetent Cells

The plasmid can be introduced in EHA105 *Agrobacterium* strain cells by electroporation. EHA105 cells can be made electrocompetent with the procedure detailed below.

1. Inoculate 100 mL of LB added with 100 mg/L rifampicin with 100 μL from a 5 mL pre-inoculum culture. Put at 28 °C with 150 rpm shaking.
2. Let it grow until OD_{600} reaches 0.4–0.8, and then transfer to two 50 mL sterile tubes and put in ice for 30′.
3. Spin tubes at 4000 rcf for 10′ in a refrigerated centrifuge. Discard supernatant and resuspend each pellet in 25 mL of sterile 10% glycerol.
4. Repeat the previous step resuspending first in 12.5, then in 6 and finally in 1.5 mL of sterile 10% glycerol.
5. Transfer in two 2 mL tubes, spin at 4000 rcf for 10′ in a refrigerated centrifuge and resuspend in 500 μL of sterile 10% glycerol.
6. Freeze 50 μL aliquots in liquid nitrogen and store at −80 °C.

Table 1
List of primers designed on the pSKI074 T-DNA sequence

Primer name	Sequence (5′–3′)	Position(s)	Orientation
T7	taatacgactcactataggg	5581–5600	Direct
M13 forward	tgtaaaacgacggccagt	5557–5574	Direct
IK007	acccgccaatatatcctg	7024–7041	Complementary
Right1	cgttcaagatgcctctaccg	5864–5883 6203–6222 6542–6561 6881–6900	Direct
Right2	agacgttccaaccacgtctt	5936–5955 6275–6294 6614–6633 6953–6972	Direct
Right3	gacaggatatattggcggg	7022–7040	Direct
Left1	tatcttccacacgtgaaaatgc	557–578	Complementary
Left2	tttttatttcatccgacatgg	484–504	Complementary
KMFW	gcccggttctttttgtcaag	1202–1221	Direct
KMRV	gcgataccgtaaagcacgag	1792–1811	Complementary
35SFW	ataataccgcgccacatagc	4587–4606	Direct
35SRV	aacgtcgtgactgggaaaac	5538–5557	Complementary

Positions and orientations are referred to T-DNA sequence (7045 bp from 5′, left border to 3′, right border)

3.3 Introduction of pSKI074 in *Agrobacterium tumefaciens* EHA105 by Electroporation

To transfer the pSKI074 plasmid in electrocompetent EHA105, do the following:

1. Melt a 50 μL aliquot of electrocompetent EHA105 cells in ice.
2. Add 1 μL of pSKI074 plasmid miniprep, and transfer to 0.1 cm ice-cold electroporation cuvettes.
3. Perform electroporation with a Biorad™ MicroPulser electroporator using the preset *Agrobacterium* program (2.2 kV).
4. Add immediately 1 mL LB to the cuvette. Recover cells with a pipette and transfer to a sterile 2 mL tube.
5. Incubate 4 h at 28 °C with shaking (180 rpm).
6. Streak 100 μL on an LB plate with 50 mg/L rifampicin and 50 mg/L ampicillin and incubate at 28 °C. In 2 days isolated colonies should be visible.
7. Check for pSKI074 presence and integrity by colony PCR using the primers described above.

3.4 Preparation of Agrobacterium tumefaciens EHA105 Glycerol Stocks and Starter Cultures

A colony can be used to prepare glycerol stocks as well as starter cultures for plant transformation.

1. Inoculate a single colony into 5 mL of liquid YEB medium containing 50 mg/L ampicillin and 50 mg/L rifampicin, and grow overnight at 28 °C on a rotary shaker at 200 rpm.
2. Keep an aliquot (0.5 mL) of the overnight culture, and inoculate into 50 mL of liquid YEB medium containing the same antibiotics, and grow at 28 °C with vigorous shaking until the OD_{600} reaches approximately 0.6–1.0.
3. Pellet 2 mL of culture and perform a PCR check for the integrity of the 35S enhancer repeats using a standard colony PCR protocol picking the *Agrobacterium* pellet.
4. Add 500 μL of 50% glycerol to every 500 μL of the remaining culture in 2 mL cryovials and gently mix. Freeze the tubes at −80 °C. These stocks will be used as starters for plant transformation cultures.

3.5 Cultivation and Genetic Transformation of M. truncatula R108-1 Genotype

1. Scarify seeds of *M. truncatula* R108-1 in order to create small abrasions on the seed cuticle and favor seed germination (2–4 rotary movements are sufficient with the scarifier described in Fig. 2).
2. Put scarified seeds on Petri dishes containing solidified water (10–15% agar in bdH_2O) in the dark at 4 °C.
3. When radicles reach ~20 mm length (~1 week), transfer seedlings in pots containing a vermiculite: soil (1:2) mixture under growth room conditions (16 h light/8 h dark, 23 ± 2 °C, and 60 ± 5% relative humidity). In five to eight weeks plants can be used for in vitro transformation experiments.
4. The day before plant transformation, take a frozen *Agrobacterium* aliquot stock (*see* Subheading 3.4), and inoculate it into 50 mL of sterile liquid YEB medium containing 50 mg/L ampicillin and 50 mg/L rifampicin. Grow at 28 °C with vigorous shaking until the OD_{600} reaches approximately 0.6–1.0. Spin down 2 mL of culture, and check the integrity of pSKI074 by PCR on *Agrobacterium* pellet as described above. Collect the remaining cells by centrifugation at 5000 rcf for 5 min at 4 °C and re-suspend at $OD_{600} = 0.4$ in sterile water; this suspension will be used to infect ~100 *Medicago* explants.
5. Leaf sterilization. Harvest ~50 healthy, round-shaped leaflets from *Medicago* plants, and put them in a 50 mL sterile tube with sterile bdH_2O. Wash in 70% ethanol inverting the capped tube for 30 s under a laminar flow hood. Discard 70% ethanol and wash three times with sterile bdH_2O. Discard water and add leaf sterilization solution (*see* Subheading 2.4). Invert

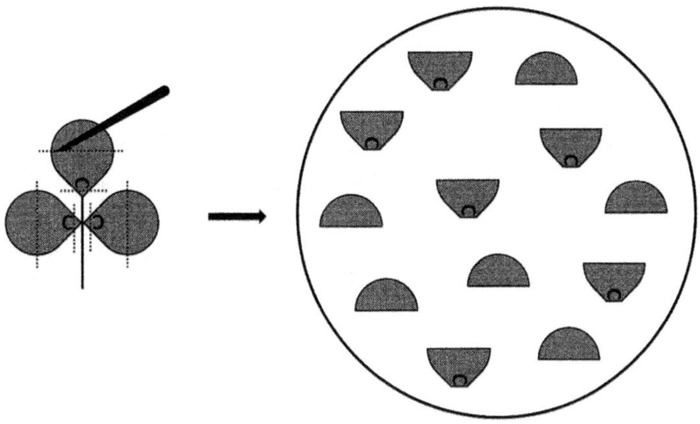

Fig. 3 Representation of leaf explant preparation and disposition on Petri dish containing agarized growth medium

every 2′ and discard sterilization solution after 10′. Wash three times with sterile bdH_2O and keep sterilized leaflets in water under laminar flow hood.

6. Cut the leaflets in halves with a sterile scalpel as shown in Fig. 3, and place them in the solution with *Agrobacterium*. Mix gently to separate the plant pieces. Apply vacuum to the leaf explants in the solution with the *Agrobacterium* for 20′ at 650 psi (*see* **Note 5**). Release slowly the vacuum and place the closed recipient at room temperature for 1 h to allow the tissue to recover from the infiltration procedure. The vacuum infiltration step can be bypassed determining a slight reduction in transformation efficiency.
7. Co-cultivation. Transfer the leaf explants to solid SH3a medium without antibiotics. The abaxial side of the leaf explant should be in contact with the medium. Seal the plates with parafilm and incubate for a maximum of 2 days in the dark in the plant growth culture room (24 °C). Check frequently the plates as the *Agrobacterium* should not overgrow the leaf explants.
8. Callogenesis. Prepare a suspension of 1 g/L Augmentin in water. Rinse explants in the Augmentin suspension and transfer to clean Augmentin suspension three times. Dry explants touching a piece of sterile filter paper before transferring them to plates with SH3a medium with 500 mg/L Augmentin (to select against the agrobacteria) and with 40 mg/L kanamycin to select for the transformed cells. Place the sealed plates in the dark in the growth chamber (24 °C) for 5–6 weeks and check regularly for contaminations. Transfer to fresh medium every 2 week. At the end of this phase 1–2 cm brownish calli should be formed.

9. Embryogenesis. Transfer *calli* to SH9 medium with 40 mg/L kanamycin, and place them in the light in the growth chamber (24 °C, 16 h photoperiod). If bacterial growth contaminates the plate at this stage, add 250 mg/L Augmentin to SH9 medium. Transfer to fresh medium every 3 weeks. The Augmentin can be added for 2 or 3 weeks and then can be removed if no further bacterial growth is visible. During this stage calli will develop proembryos (Fig. 4a) that will turn into true embryos (Fig. 4b–d) and later form leaflets (Fig. 4e). In this phase, separate regularly well-formed embryos from calli to favor further development.
10. Rooting. When leaflets are present, transfer plantlets to magenta boxes containing ½ SH9 to induce rooting (Fig. 4f). Transfer to fresh medium every 4 weeks if necessary.
11. Transfer to green house. Plants with well-developed leaves and roots (Fig. 4g) can be transferred to ex vitro non-sterile conditions. This step is performed gradually as plants need to adapt to an environment not saturated with humidity. Transfer plants in small pots containing a vermiculite soil (1:2) mixture after washing away agarized medium residues. Cap the pots with a transparent plastic bag (i.e., an upside-down 1.5–2 L freezer bag fixed with an elastic to the pot). Keep pots in a tray and water directly from the tray. Progressively open the top of the bag making cuts with scissors. Within a couple of weeks, the bag can be completely removed (*see* **Note 6**).
12. Check regenerated plants by PCR using primers KMFW, 5′ GCC CGG TTC TTT TTG TCA AG 3′ and KMRV, 5′ GCG ATA CCG TAA AGC ACG AG 3′ on *nptII* sequence (*see* **Note 7**). Positive plants represent the T_0 generation and can show phenotypes in case of dominant mutations (*see* **Note 8**). Pods from fertile mutants can be stored and/or used to obtain generation T_1. R108 seeds maintain a good fertility rate for 5–10 years.

3.6 Southern Blot

Southern blot analysis is used to verify the number of T-DNA insertions in transformed plants. The principle behind this approach is to digest genomic DNA with a frequent cutter enzyme whose cutting site is present only once in the T-DNA sequence. After digestion, electrophoretic separation, and blotting, a DNA probe complementary to a part of the T-DNA sequence is used to hybridize with fragments containing T-DNA portions. Theoretically it results in one signal from each insert. In our experience we successfully used KpnI as restriction enzyme and PCR probes amplified with KMFW, 5′ GCC CGG TTC TTT TTG TCA AG 3′ and KMRV, 5′ GCG ATA CCG TAA AGC ACG AG 3′ on the *nptII* gene in 5′ respect to KpnI site and 35SFW, 5′ ATA ATA CCG CGC CAC ATA GC 3′ and 35SRV, 5′ AAC GTC GTG ACT GGG

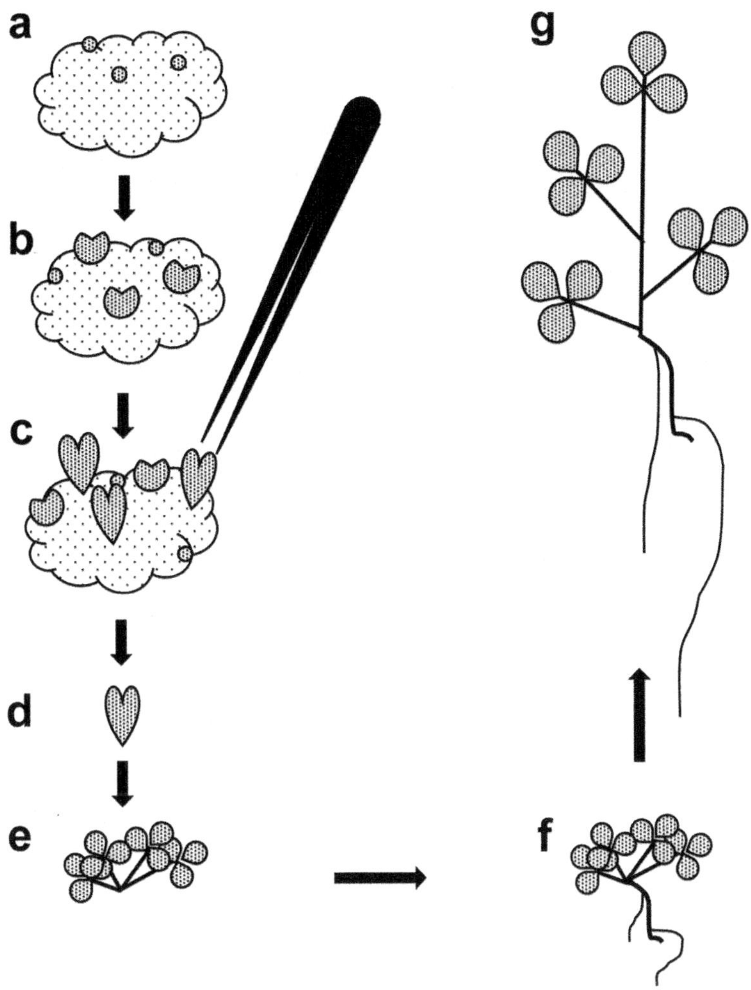

Fig. 4 Outline of the various steps of ***Medicago*** in vitro culture. Callus with pro-embryos (**a**) that later develop into heart-shaped (**b**) and torpedo-shaped embryos (**c**). Mature embryos are separated from callus (**d**) and start to grow and differentiate into plantlets (**e**). Later rooting is induced (**f**) and, when well-developed leaves and roots are present, plants can be adapted to ex vitro conditions (**g**)

AAA AC 3′ in 3′ respect to KpnI site. The KM probe is 610 bp long; the 35S probe is 971 bp. Position of probes on T-DNA is shown in Fig. 1b.

1. DNA extraction. The use of a high-yelding method is suggested to isolate genomic DNA from the mutated plant.
2. Digest about 10 μg of genomic DNA with KpnI-HF® (New England Biolabs™, USA) in the appropriate buffer. Incubate at 37 °C overnight (the use of standard KpnI enzyme may result

in star activity; thus, if high fidelity version is not available, reduce restriction time).

3. Purify restricted DNA fragments by adding 2.5 volumes of absolute ethanol and 1/10 volume of sodium acetate 3 M pH 5.2. Mix and precipitate at −20 °C for 1 h amd then spin 20′ at full speed in a refrigerated benchtop mini centrifuge. Discard supernatant and wash twice with 500 μL of 70% ethanol. After the second wash, discard supernatant, and let the pellet air-dry at room temperature. Resuspend digested DNA in 20–40 μL bdH_2O. Effective restriction and recovery can be checked by running 2–4 μL of purified DNA on a stained 0.6–1% TAE agarose gel.
4. Separate DNA fragments by electrophoresis loading the remaining purified DNA on an unstained 0.8% TAE agarose gel running at 35 Volt/cm for ~20 h.
5. Blot gel onto Hybond N+ nylon membrane following manufacturer protocol for gel preparation to transfer (depurination, denaturation, and neutralization), capillary blotting, fixation, and hybridization.
6. In the hybridization step, use labeled probes (use either the one in 5′ or in 3′ of the KpnI site) previously obtained by PCR amplification followed by nick translation in the presence of [α-^{32}P]-dCTP or DIG-dUTP (if facilities for manipulation of radioactive substances are not available).
7. Signal detection. After washing expose the filter to an X-ray film in a light shielded cassette. Expose for 3–5 days at −80 °C. Develop the X-ray film in a dark room using the appropriate reagents and supplier instructions. Alternatively, hybridization signals (both radioactive and chemiluminescent) from blotted filters can be detected by devices for digital acquisition.
8. Each band that produced a signal represents an insertion. The filter can be stripped and rehybridized with the second probe; the same number of bands should be obtained with both probes.

3.7 Isolation of the Mutated Gene

Isolation of the tagged gene can be performed with various techniques whose effectiveness and can vary from plant to plant. Since T-DNA should influence genes that are in the proximity of the insertion site, the approach of choice is to isolate the sequences flanking the insertion sites and use this information to identify the genomic region with both in silico and molecular techniques. We describe T-DNA rescue, inverse PCR, adapter-ligation-mediated PCR, Genome walker, and TAIL methods that all have been used successfully on *Medicago truncatula* mutants. Sequence information can be also used to make probes/primers for screening of *Medicago truncatula* genomic libraries such as those available at INRA-CNRGV for genotype A17 (http://cnrgv.toulouse.inra.fr).

If the recovered sequence is present in multiple copies in the genome, the sequence homology search may be inconclusive. In these cases, sequence information about the neighboring insertion site can be extended by using the genome walker approach as described below with some modifications.

To help flanking sequence analysis, a genome assembly is available for *M. truncatula* cv. A17 [14], and recently an assembly was released also for genotype R108–1 [15].

3.7.1 T-DNA Rescue

Plasmid (pBluescript KS+) sequence in pSKI074 allows the recovery of T-DNA and portion of the flanking plant genome region due to restriction, re-ligation, and transformation in competent *E. coli* cells. Restriction enzymes that leave plasmid sequence intact and cut only on one side between pBluescript KS+ and either the left (Fig. 5a) or the right border (Fig. 5b) can be used. Tagged

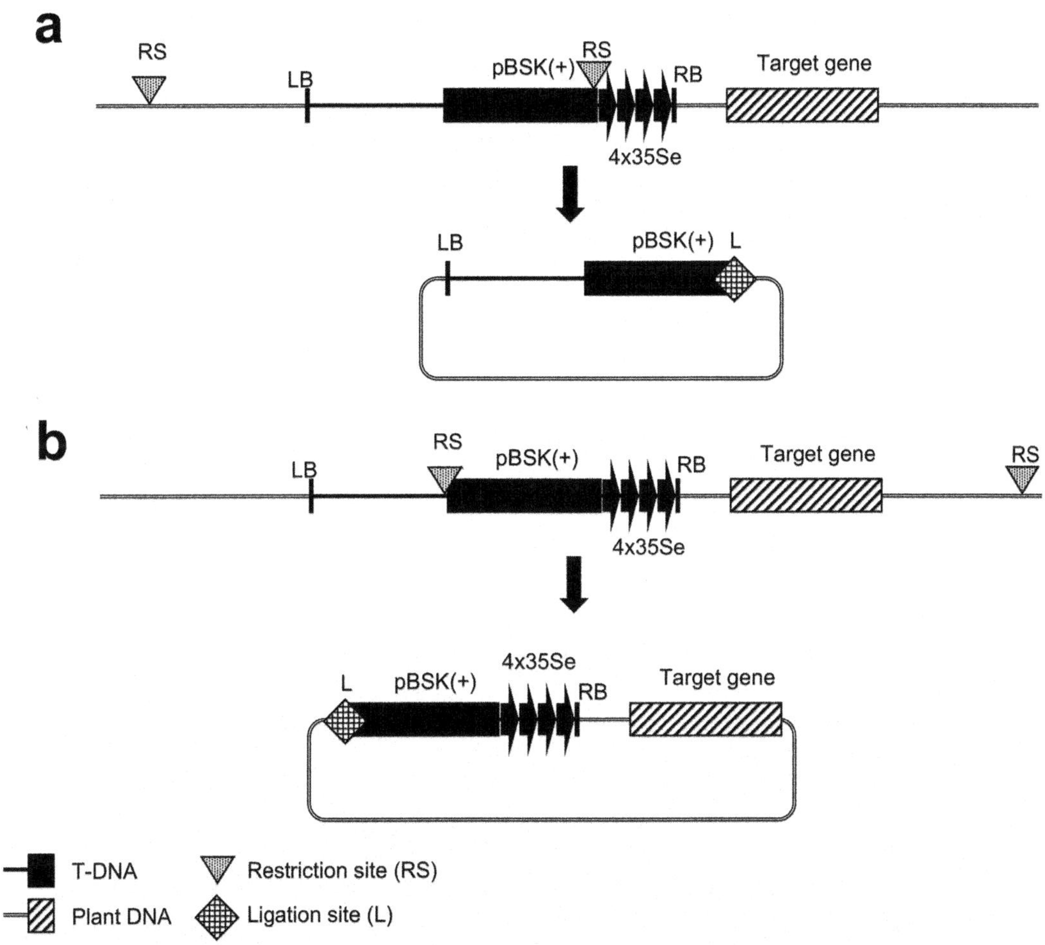

Fig. 5 Representation of T-DNA rescue procedure. Suitable restriction enzymes do not cut within the pBluescript sequence. Self-ligated fragments are functional plasmids that incorporate flanking plant sequences on the left (**a**) or on the right border of the T-DNA (**b**)

genes are expected to be in proximity of the right border of the T-DNA that harbors the 35S enhancers.

1. DNA restriction. Digest 1–3 μg of genomic DNA extracted from the mutant plant with an appropriate restriction enzyme (KpnI, EcoRI, and HindIII for rescue of sequences adjacent to the right T-DNA border and BamHI, SpeI, and NotI for left border rescue) and relative buffer in 50 μL of reaction volume.
2. Extraction of restricted DNA with phenol-chloroform. Add an equal volume (50 μL) of phenol and mix gently; spin 1′ at full speed in a benchtop microcentrifuge at room temperature. After centrifugation two phases are present, the upper aqueous and the lower organic phase. Recover the upper phase with a pipette, and repeat the same steps using 50 μL of chloroform instead of phenol.
3. Purify the recovered aqueous phase containing restricted DNA fragments by adding 2.5 volumes of absolute ethanol and 1/10 volume of sodium acetate 3 M pH 5.2. Mix and put at −20 °C for 1 h, and then spin 20′ at full speed on a refrigerated benchtop mini centrifuge. Discard supernatant and wash twice with 500 μL of 70% ethanol. After the second wash, discard supernatant, and let the pellet air-dry at room temperature. Resuspend digested DNA in 20–40 μL bdH_2O.
4. Ligate restricted DNA overnight using 1 μL (400 U) T4 DNA ligase and the appropriate buffer at 16 °C in a volume of 250 μL. Ligation in a high volume promotes attachment of fragment ends and consequently formation of circularized, functional plasmids.
5. Precipitate ligated DNA as in **step 3** and resuspend in 5 μL. Transform 1 μL by electroporation into recombination-deficient *E. coli* SURE cells to maximize stability of the multimerized CaMV 35S enhancers.
6. Thaw a 40 μL aliquot of electrocompetent *E. coli* SURE cells in ice.
7. Add 1 μL of resuspended DNA and transfer to 0.1 cm prechilled electroporation cuvettes.
8. Perform electroporation with a Biorad™ MicroPulser electroporator using the preset Ec1 program (1.7 kV).
9. Add immediately 1 mL LB to the cuvette. Recover cells with a pipette and transfer to a sterile 2 mL tube.
10. Incubate 1 h at 37 °C with shaking (220 rpm).
11. Spin 3′ at 3000 g, remove 800 μL supernatant, and resuspend pellet in the remaining (~240 μL).
12. Streak 120 μL on a LB plate with 100 mg/L ampicillin and incubate overnight at 37 °C.

13. Check colonies by PCR using 35SFW-35SRV primer pair.
14. Inoculate 5 mL LB containing 100 mg/L ampicillin picking positive colonies and grow overnight at 37 °C. Extract plasmid and sequence using primers going outward T-DNA borders.

3.7.2 Inverse PCR

Inverse PCR allows the simultaneous isolation of both left and right genomic DNA portions adjacent to the T-DNA inserted in the genome. Digestion of the genomic DNA using restriction enzymes that do not cut in the T-DNA sequence (Fig. 6a) generates fragments that include both T-DNA and flanking host DNA. Self-ligated restriction fragments can be amplified using primers reading outward from the T-DNA generating amplicons that contain the sequence of the insertion site (Fig. 6b).

1. Digest 200 ng of genomic DNA with BsiWI and BsrGI restriction enzymes in 20 μL with the 3.1 buffer for 2 h at 37 °C. Inactivate the enzymes incubating 20′ at 80 °C.
2. Incubate the digested DNA overnight at 16 °C with 0.5 μL (200 U) T4 DNA ligase and the appropriate buffer in a total volume of 200 μL (high volume promotes fragment circularization vs. fragment to fragment ligation as in T-DNA rescue).
3. Perform a first PCR reaction using primers designed for the right (Right1, 5′ CGT TCA AGA TGC CTC TAC CG 3′) and the left (Left1, 5′ TAT CTT CCA CAC GTG AAA ATG C 3′) borders of the T-DNA (*see* **Note 9**); mix 10 μL of ligation to 0.1 μL TaKaRa Ex Taq®, 2 μL 10× Ex Taq buffer, 1.6 μL dNTPs (2.5 μM each), 0.5 μL of each primer (Right1 and Left1 from 10 μM stock), 5.3 μL bdH_2O. Use the following amplification program: 2′ at 94 °C, 30 cycles of 20″ at 94 °C, 20″ at 68 °C, 3′ at 72 °C, and a final 5′ at 72 °C step.
4. Dilute the products of first PCR reaction 1:100 and amplify in a nested PCR to increase the concentration of the desired fragments using Right2, 5′ AGA CGT TCC AAC CAC GTC TT 3′, and Left2, 5′ TTT TTA TTT CAT CCG ACA TGG 3′ primers. Mix 2 μL of 1:100 diluted first PCR to 0.1 μL TaKaRa Ex Taq®, 2 μL 10× Ex Taq buffer, 1.6 μL dNTPs (2.5 μM each), 0.5 μL of each primer (Right2 and Left2 from 10 μM stock), 13.3 μL bdH_2O. Use the following amplification program: 2′ at 94 °C, 30 cycles of 20″ at 94 °C, 20″ at 60 °C, 3′ at 72 °C, and a final 5′ at 72 °C step.
5. Run both first and second PCR product on a 1.5% agarose TAE gel. Amplicons can be gel purified and/or cloned using any commercial TA cloning kit and transformed into chemically competent TOP10 cells.

3.7.3 Adapter-Ligation-Mediated PCR

Inverse PCR has the major advantage of giving knowledge of both the flanking regions of the T-DNA insertion sites, but it requires the presence of two restriction site sequences, one for each side of

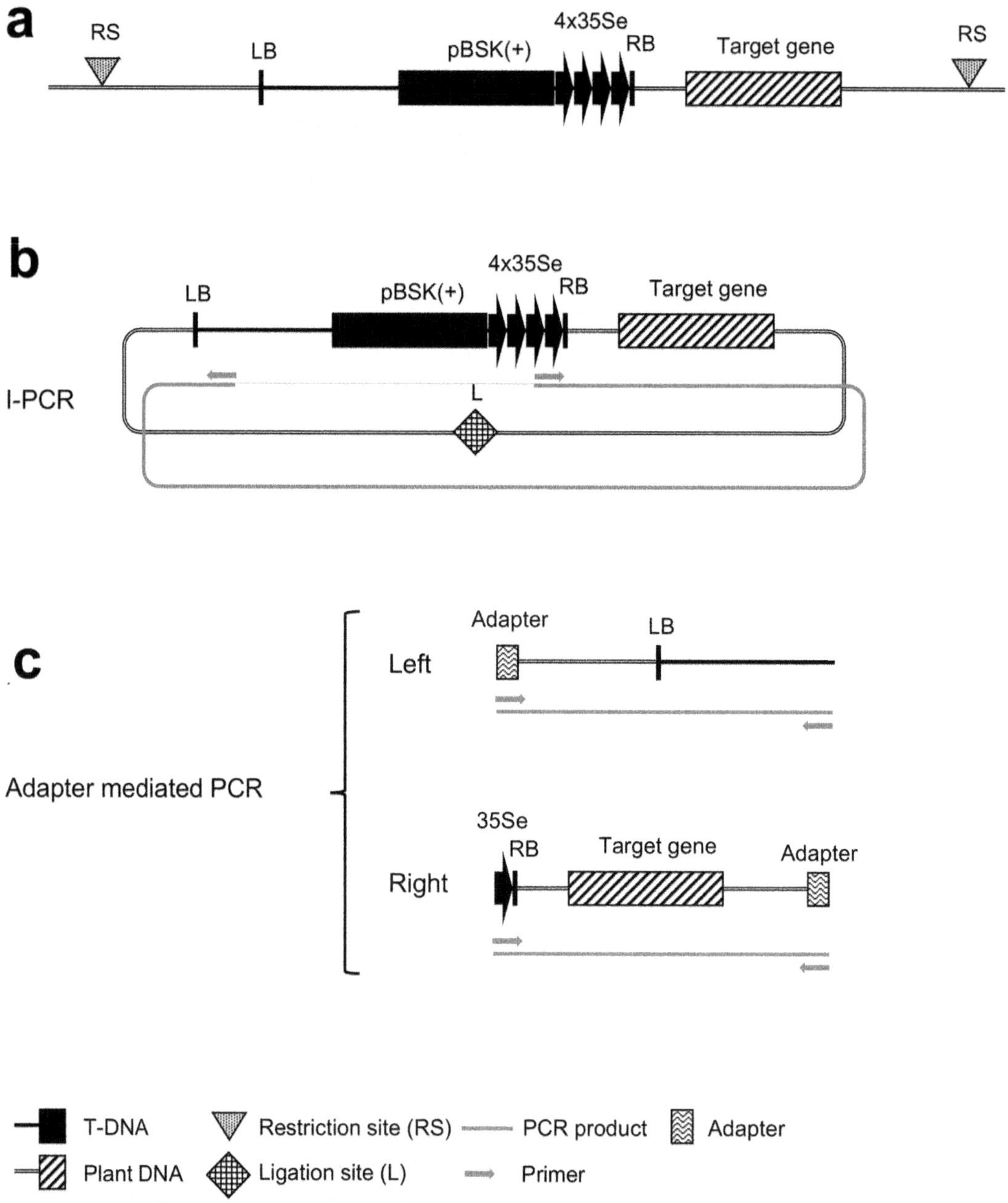

Fig. 6 Representation of PCR-based insertion site recovery techniques. On the top is represented T-DNA insertion within plant genome and position of suitable restriction sites (**a**). I-PCR template is generated by restriction and circularization of fragments bearing the complete T-DNA and both left and right plant flanking regions (**b**). Ligation of synthetic adapters as in adapter-ligation-mediated PCR or in the genome walker methods generates PCR templates that can be amplified using primers on T-DNA and adapter sequences (**c**)

the insertion. An additional method for finding unknown genomic DNA sequences adjacent to a known sequence is genome walking. Many genome walking techniques and kits are commercially available. Here we describe the protocol that we successfully used to amplify flanking regions in *M. truncatula* insertional mutants and that is derived from S-SAP (sequence-specific amplification polymorphism) [16, 17]. In this method DNA is digested and ligated to appropriate adapters. Fragments containing T-DNA ends can be amplified by PCR using primers designed on T-DNA and on the adapter (Fig. 6c).

The EcoRI adapter is obtained by annealing the following oligos: EcoAdapter1, 5′ CTC GGA AGC CTC AGT CGT AGA CTG CGT ACC 3′ and EcoAdapter2, 5′ AAT TGG TAC GCA GTC TAC GAC TGA GGC T 3′ (*see* **Note 10**). EcoAdapter2 can be phosphorylated in 5′ in order to favor ligation reaction and amminated in 3′ to avoid its extension in PCR.

1. Restriction and ligation of genomic DNA can be performed in a single reaction as ligation of the adapter does not recreate the EcoRI recognition site. In a total volume of 50 μL, mix 500 ng of genomic DNA to 2.5 μL of 1 M NaCl; 5 μL of BSA 0.5 mg/mL; 0.5 U EcoRI, 0.17 μL of T4 DNA ligase; 5 μL of 10×T4 DNA ligase buffer; and 1 μL EcoRI Adapter 5 μM. Keep to final volume with bdH_2O. Incubate for 2 h at 37 °C.
2. Dilute restriction-ligation reaction 1:10, and use it as a template in the following PCR reaction: 5 μL of diluted template, 5 μL of 10× reaction buffer, 1.5 μL of 50 mM $MgCl_2$, 0.1 μL Taq polymerase, 2 μL dNTPs (2.5 mM each), 1 μL of 10 mM Right1 (or Left1) primer, 1 μL of primer Eco0, 5′ GAC TGC GTA CCA ATT C 3′, 34.4 μL bdH_2O. Use the following touchdown PCR program as it favors amplification of fragments with Right1 (or Left1) primers that have higher melting temperatures compared to Eco0 in the first cycles. First 13 cycles: 30″ at 94 °C, 30″ at 65 °C, 3′ at 72 °C decreasing annealing temperature by 0.7 °C at each cycle. Twenty-two more cycles with 30″ at 94 °C, 30″ at 56 °C, 3′ at 72 °C, and a final extension at 72 °C for 7′.
3. Dilute the first amplification product 1:20 and use as template in a nested PCR reaction as follows: 5 μL of diluted template, 5 μL of 10× reaction buffer, 1.5 μL of 50 mM $MgCl_2$, 0.1 μL Taq polymerase, 2 μL dNTPs (2.5 mM each), 1 μL of 10 mM Right2 (or Left2) primer, 1 μL of primer Eco + N, 5′ GAC TGC GTA CCA ATT CN 3′ (primer Eco + A, Eco + C, Eco + G, Eco + T can be also used separately in four reactions to facilitate amplified fragment isolation), 34.4 μL bdH_2O. Use the following amplification program: 35 cycles with 30″ at 94 °C, 30″ at 58 °C, 3′ at 72 °C, and a final extension at 72 °C for 7′.
4. Run both first and second PCR product on a 1.5% agarose TAE gel. Amplicons can be gel purified and/or cloned using any commercial TA cloning kit and transformed into chemically competent TOP10 cells.

3.7.4 Genome Walker Method

This method is described in [18] and uses the same concept of adapter-ligation-mediated PCR (Fig. 6c) with several modifications. The main differences are in the use of a specially designed adapter and primers that can be used with any blunt end producing restriction enzyme that does not cut between the T-DNA specific primers and the T-DNA ends.

The adapter is obtained by annealing the following oligos: GWAdapter1, 5′ GTA ATA CGA CTC ACT ATA GGG CAC GCG TGG TCG ACG GCC CGG GCT GGT 3′ and GWAdapter2, 5′ ACC AGC CC 3′ (*see* **Note 10**). GWAdapter2 is phosphorylated in 5′ in order to favor ligation reaction and amminated in 3′ to avoid its extension in PCR.

1. Perform digestion of genomic DNA with an appropriate restriction enzyme that creates blunt-ended fragments. Here we report the use of RsaI that recognizes a 4-base site: 5′ GT ^ AC 3′. The use of a frequent cutter increases the chance to have a restriction site close to the T-DNA insertion. Mix 500 ng of genomic DNA to 1 μL RsaI and the appropriate buffer in a total volume of 30 μL. Incubate at 37 °C for 2 h. Incubation can be extended overnight with this enzyme.
2. Since RsaI cannot be heat inactivated, a phenol-chloroform purification is recommended (*see* Subheading 3.7.1). Alternatively, commercially available spin-column-based kits can be used. Resuspend purified digested DNA in 20–40 μL bdH_2O. Effective restriction and recovery can be checked running 2–4 μL of purified DNA on a 1% agarose TAE gel.
3. Perform adapter ligation as follows: in a total volume of 50 μL, mix 20 μL of purified restriction to 0.2 μL (100 U) of T4 DNA ligase, 5 μL of 10X T4 DNA Ligase Buffer, 5 μL Adapter 5 μM, keep to final volume with bdH_2O. Incubate overnight at 16 °C.
4. Dilute restriction-ligation reaction 1:10, and use it as a template in the following PCR reaction: 5 μL of diluted template, 5 μL of 10× reaction buffer, 1.5 μL of 50 mM $MgCl_2$, 0.1 μL Taq polymerase, 2 μL dNTPs (2.5 mM each), 1 μL of 10 μM Right1 (or Left1) primer, 1 μL of 10 μM AP1 primer, 5′ GTA ATA CGA CTC ACT ATA GGG C 3′, 34.4 μL bdH_2O. Use the following touchdown PCR program as it favors amplification of fragments with Right1 (or Left1) primers that have higher melting temperatures with respect to AP1 in the first cycles. First 16 cycles: 30″ at 94 °C, 30″ at 68 °C, 3′ at 72 °C decreasing annealing temperature by 0.5 °C at each cycle. Do 20 more cycles with 30″ at 94 °C, 30″ at 60 °C, 3′ at 72 °C, and a final extension at 72 °C for 7′.

5. Dilute the first amplification product 1:20 and use as template in a nested PCR reaction as follows: 5 μL of diluted template, 5 μL of 10× reaction buffer, 1.5 μL of 50 mM $MgCl_2$, 0.1 μL Taq polymerase, 2 μL dNTPs (2.5 mM each), 1 μL of 10 μM Right2 (or Left2) primer, 1 μL of 10 μM AP2 primer, 5′ ACT ATA GGG CAC GCG TGG T 3′, 34.4 μL bdH_2O. Use the following amplification program: 35 cycles with 30″ at 94 °C, 30″ at 60 °C, 3′ at 72 °C, and a final extension at 72 °C for 7′.
6. Run both first and second PCR product on a 1.5% agarose TAE gel. Amplicons can be gel purified and/or cloned using any commercial TA cloning kit and transformed into chemically competent TOP10 cells.

3.7.5 Tail-PCR

Thermal asymmetric interlaced PCR is a technique that does not require restriction and ligation steps; thus, it is not dependent on restriction site presence. The method is described in Chapter 7 and does not require any special modification except for the use of Left1, Left2 or Right1, Right2 as T-DNA-specific primers.

3.8 Segregation Analysis

Activation tagging may give rise to dominant mutation which is expected to give a phenotypic segregation ration of 75% mutant phenotype and 25% wild-type phenotype. Sometimes heterozygous individuals may show incomplete dominance, and thus phenotypic segregation ratios of 25% (homozygote-dominant phenotype), 50% (heterozygote-intermediate phenotype), and 25% (wild-type phenotype) are expected.

Once cloned the T-DNA flanking sequences, the segregation of T-DNA, and the phenotype should be verified. In Fig. 7, a scheme of primer design and amplification results is given for such analyses.

3.9 Expression Analysis

Expression levels of the activated target gene are verified by quantitative real-time PCR. It was shown that the presence of the tetramerized 35S enhancer sequence may not induce constitutive expression of the target gene in all tissues of the transformed plant [5]; thus, we suggest to perform the analysis of transcriptional activity on several organs, i.e., leaf and root. Choice of tissues/organs can be varied based on the phenotype of the mutated plant, expression profile of target gene in the wild type, and literature data about homolog genes.

3.10 Overexpression of Tagged Gene(S) (See Note 11)

Because the enhancer activity could influence the expression of adjacent genes, both the segregation analysis and gene expression studies may be not sufficient to identify the casual mutation. In these cases, it is advisable to assess the dependence of the mutant phenotype on the tagged gene by additional approaches. Rescued T-DNA

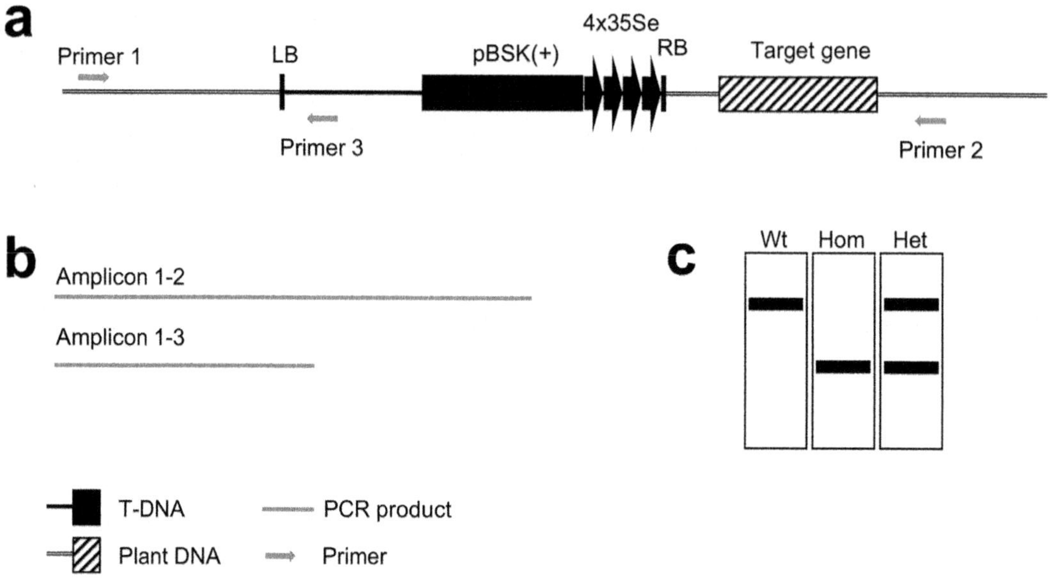

Fig. 7 Representation of segregation analysis. On the top is represented T-DNA insertion within plant genome and position primers: primer 1 and 2 are opportunely designed 1–2 kb distant on plant DNA; primer3 is designed on T-DNA close to the left border (Left 2 primer can be used for this analysis) (**a**). Amplicon 1–2 will be generated only in the absence of T-DNA whose extension would require nonstandard PCR conditions; amplicon 1–3 will be generated only in the presence of T-DNA (**b**). Electrophoresis of PCR products will result in three possible profiles: Wt (wild type, only amplicon 1–2 is generated), Hom (homozygous, only amplicon 1–2 is generated), Het (heterozygous, both 1–2 and 1–3 amplicons are generated) (**c**)

that includes 35S enhancer and the complete target gene can be subcloned into a binary vector and used to transform wild-type plants. Alternatively, the activated gene can be cloned in pMN19 (Addgene id #11572) or pMN20 (Addgene id #11573), binary vectors that harbor tetramerized 35S enhancer sequences adjacent to a multicloning site that can be used to transform wild-type plant. Finally, wild-type plants can be transformed with a construct where the mutated gene is cloned under the full 35S promoter. If the phenotype under study is obtained by an alteration of the expression of the transferred gene, its overexpression should reproduce the phenotypic alteration of the tagged line.

4 Notes

1. As an alternative to commercial chemically competent cells, we suggest XL1 Blue *E. coli* cells (Agilent technologies, USA) made chemically competent using a published protocol [19] that provides excellent efficiency when 18 ° C is used as growth temperature of the batch culture.
2. For plant cultivation media and solutions adjust pH with 1 M KOH.

3. Sodium hypochlorite can be substituted with commercial bleach. Just pay attention to chloride title and possibly to absence of additives (i.e., soap, perfumes, etc.).
4. Commercial kits and molecular biology reagents that we used successfully are reported. In many cases they can be substituted with equivalent products from other manufacturers or by standard lab protocols.
5. Glass vacuum desiccator jar should be cleaned with ethanol under laminar flow hood to reduce possible contaminations.
6. After the transfer to wet soil and during the time that plants are still bagged, consider to check soil humidity every few days and water again if necessary avoiding excess moisture. Established plants will be evident after a few days by their erect position. Plants that produce new leaves are reasonably ready for complete bag removal.
7. Any couple of primers designed within the T-DNA sequence can be used for positive plant check. Just avoid oligos too close to borders as loss of nucleotides in these positions may happen upon T-DNA integration.
8. Sometimes plants regenerated from in vitro culture display altered morphology that is not determined by T-DNA insertion. These morphological alterations are not heritable and may interest shape and growth pattern of shoot, root, and leaves. Thus, T_0 phenotypes should be confirmed in T_1.
9. Primers on the right border anneal in four different positions due to the 35S enhancer repetitions. This did not adversely affect the effectiveness of the proposed protocols because shorter amplicons are favored in both PCR and cloning steps. Nevertheless, to reduce fragment complexity, a further PCR amplification step can be introduced before cloning using primer Right3 designed just after the last enhancer sequence (*see* Table 1).
10. Do usual 10 μM oligo dilutions. Mix an equal amount of both oligos in a total volume of 50 μL. Denature using a thermal cycler, a water bath, or a heated block for 5′–10′ at 95 °C. Leave at room temperature for 3–4 h for annealing of complementary oligos then store at −20 °C. Double-stranded adapter concentration is 5 μM.
11. It is important to bear in mind that since with the experiments described in Subheading 3.10, it is not always possible to reproduce exactly the same level of alteration of gene expression level as in the tagged line; the phenotypic severity of the overexpressing lines could be slightly different from that of the original activation tagging line.

References

1. Panara F, Calderini O, Porceddu A (2012) Medicago truncatula functional genomics – an invaluable resource for studies on agriculture sustainability. In: InTech (ed) Funct Genomics. pp 131–153
2. Yarce JCS, Lee H-K, Tadege M, Ratet P, Mysore KS (2013) Forward genetics screening of *Medicago truncatula* Tnt1 insertion lines. Methods Mol Biol 1069:93–100
3. Cheng X, Wang M, Lee HK, Tadege M, Ratet P, Udvardi M, Mysore KS, Wen J (2014) An efficient reverse genetics platform in the model legume *Medicago truncatula*. New Phytol 201:1065–1076. https://doi.org/10.1111/nph.12575
4. Carelli M, Biazzi E, Panara F, Tava A, Scaramelli L, Porceddu A, Graham N, Odoardi M, Piano E, Arcioni S, May S, Scotti C, Calderini O (2011) *Medicago truncatula* CYP716A12 is a multifunctional oxidase involved in the biosynthesis of hemolytic saponins. Plant Cell 23:3070–3081. https://doi.org/10.1105/tpc.111.087312
5. Weigel D, Ahn JH, Blazquez MA, Borevitz JO, Christensen SK, Fankhauser C, Ferrandiz C, Kardailsky I, Malancharuvil EJ, Neff MM, Nguyen JT, Sato S, Wang ZY, Xia Y, Dixon RA, Harrison MJ, Lamb CJ, Yanofsky MF, Chory J (2000) Activation tagging in Arabidopsis. Plant Physiol 122:1003–1013. https://doi.org/10.1104/pp.122.4.1003
6. Kim A-R, Min J-H, Lee K-H, Kim CS (2017) PCA22 acts as a suppressor of atrzf1 to mediate proline accumulation in response to abiotic stress in Arabidopsis. J Exp Bot 68:1797–1809. https://doi.org/10.1093/jxb/erx069
7. Marsch-Martinez N, Greco R, Van Arkel G, Herrera-Estrella L, Pereira A (2002) Activation tagging using the En-I maize transposon system in Arabidopsis. Plant Physiol 129:1544–1556. https://doi.org/10.1104/pp.003327
8. Schneider A, Kirch T, Gigolashvili T, Mock HP, Sonnewald U, Simon R, Flügge UI, Werr W (2005) A transposon-based activation-tagging population in *Arabidopsis thaliana* (TAMARA) and its application in the identification of dominant developmental and metabolic mutations. FEBS Lett 579:4622–4628. https://doi.org/10.1016/j.febslet.2005.07.030
9. Wilson K, Long D, Swinburne J, Coupland G (1996) A dissociation insertion causes a semidominant mutation that increases expression of TINY, an Arabidopsis gene related to *APETALA2*. Plant Cell 8:659–671. https://doi.org/10.1105/tpc.8.4.659
10. Ichikawa T, Nakazawa M, Kawashima M, Iizumi H, Kuroda H, Kondou Y, Tsuhara Y, Suzuki K, Ishikawa A, Seki M, Fujita M, Motohashi R, Nagata N, Takagi T, Shinozaki K, Matsui M (2006) The FOX hunting system: an alternative gain-of-function gene hunting technique. Plant J 48:974–985. https://doi.org/10.1111/j.1365-313X.2006.02924.x
11. Pauls PK, Kunert K, Huttner E, Grandbastien MA (1994) Expression of the tobacco Tnt1 retrotransposon promoter in heterologous species. Plant Mol Biol 26:393–402. https://doi.org/10.1007/BF00039548
12. Porceddu A, Panara F, Calderini O, Molinari L, Taviani P, Lanfaloni L, Scotti C, Carelli M, Scaramelli L, Bruschi G, Cosson V, Ratet P, de Larembergue H, Duc G, Piano E, Arcioni S (2008) An Italian functional genomic resource for *Medicago truncatula*. BMC Res Notes1:129.https://doi.org/10.1186/1756-0500-1-129
13. Chabaud M, Ratet P, Araújo SDS, Sofia A, Lopes R, Duque A, Harrison M, Barker DG (2007) *Agrobacterium tumefaciens*-mediated transformation and *in vitro* plant regeneration of *M. truncatula*. *Medicago truncatula* handbook, http://www.noble.org/MedicagoHandbook/. pp 1–34
14. Young ND, Debellé F, Oldroyd GED, Geurts R, Cannon SB, Udvardi MK, Benedito VA, Mayer KFX, Gouzy J, Schoof H, Van de Peer Y, Proost S, Cook DR, Meyers BC, Spannagl M, Cheung F, De Mita S, Krishnakumar V, Gundlach H, Zhou S, Mudge J, Bharti AK, Murray JD, Naoumkina MA, Rosen B, Silverstein KAT, Tang H, Rombauts S, Zhao PX, Zhou P, Barbe V, Bardou P, Bechner M, Bellec A, Berger A, Bergès H, Bidwell S, Bisseling T, Choisne N, Couloux A, Denny R, Deshpande S, Dai X, Doyle JJ, Dudez A-M, Farmer AD, Fouteau S, Franken C, Gibelin C, Gish J, Goldstein S, González AJ, Green PJ, Hallab A, Hartog M, Hua A, Humphray SJ, Jeong D-H, Jing Y, Jöcker A, Kenton SM, Kim D-J, Klee K, Lai H, Lang C, Lin S, Macmil SL, Magdelenat G, Matthews L, McCorrison J, Monaghan EL, Mun J-H, Najar FZ, Nicholson C, Noirot C, O'Bleness M, Paule CR, Poulain J, Prion F, Qin B, Qu C, Retzel EF, Riddle C, Sallet E, Samain S, Samson N, Sanders I, Saurat O, Scarpelli C, Schiex T, Segurens B, Severin AJ, Sherrier DJ, Shi R, Sims S, Singer SR, Sinharoy S, Sterck L, Viollet A, Wang B-B, Wang K, Wang M, Wang X, Warfsmann J, Weissenbach J, White DD, White JD, Wiley GB, Wincker P, Xing Y, Yang L, Yao Z, Ying F,

Zhai J, Zhou L, Zuber A, Dénarié J, Dixon RA, May GD, Schwartz DC, Rogers J, Quétier F, Town CD, Roe BA (2011) The Medicago genome provides insight into the evolution of rhizobial symbioses. Nature 480:5–9. https://doi.org/10.1038/nature10625
15. Moll KM, Zhou P, Ramaraj T, Fajardo D, Devitt NP, Sadowsky MJ, Stupar RM, Tiffin P, Miller JR, Young ND, Silverstein KAT, Mudge J (2017) Strategies for optimizing BioNano and dovetail explored through a second reference quality assembly for the legume model, Medicago truncatula. BMC Genomics 18:578. https://doi.org/10.1186/s12864-017-3971-4
16. Porceddu A, Albertini E, Barcaccia G, Marconi G, Bertoli F, Veronesi F (2002) Development of S-SAP markers based on an LTR-like sequence from *Medicago sativa* L. Mol Gen Genomics 267:107–114. https://doi.org/10.1007/s00438-002-0643-z
17. Ratet P, Porceddu A, Tadege M, Mysore KS (2006) Insertional mutagenesis in Medicago truncatula using Tnt1 retrotransposon. *Medicago truncatula* handbook
18. Siebert PD, Chenchik A, Kellogg DE, Lukyanov KA, Lukyanov SA (1995) An improved PCR method for walking in uncloned genomic DNA. Nucleic Acids Res 23:1087–1088. https://doi.org/10.1093/nar/23.6.1087
19. Inoue H, Nojima H, Okayama H (1990) High efficiency transformation of *Escherichia coli* with plasmids. Gene 96:23–28. https://doi.org/10.1016/0378-1119(90)90336-P

Chapter 7

Tnt1 Insertional Mutagenesis in *Medicago truncatula*

Hee-Kyung Lee, Kirankumar S. Mysore, and Jiangqi Wen

Abstract

Legumes play irreplaceable roles in sustainable agriculture due to their unique capability of fixing gaseous nitrogen in the atmosphere and turning into plant-usable ammonium through interaction with rhizobia. With the completion of genome sequencing of several model and non-model legumes, it is highly desirable to generate mutant populations for characterizing gene functions in genome-wide scales. In the past decade, we have generated a near-saturated insertional mutant population in the model legume *Medicago truncatula* using the tobacco-derived *Tnt1* retrotransposon at Noble Research Institute. The mutant population was generated through callus induction, subculture, and regeneration from a starting transgenic line harboring three homozygous copies of *Tnt1* insertion. The population consists of 21,700 regenerated lines that encompass more than 500,000 *Tnt1* insertions. Based on the genome size, average gene length, and random insertion nature of *Tnt1*, this mutant population covers about 90% of genes in the *M. truncatula* genome. Due to the convenience of known *Tnt1* sequence, the mutant population is highly feasible for both forward and reverse genetics. Over the past 12 years, we have distributed more than 9000 mutant lines to 203 research groups in 24 countries.

Key words Insertion mutagenesis, *Medicago truncatula*, Regeneration, Tissue culture, *Tnt1*

1 Introduction

The legume family, which has approximately 700 genera with more than 18,000 species [1], is critical for sustainable agriculture by supplying non-fertilizer nitrogen. Food legumes (pulses), including common bean, soybean, pea, chickpea, broad bean, pigeon pea, cowpea, and lentil, are widely grown worldwide. Seeds and shoots of many legume species are rich sources of dietary proteins, oils, fibers, and/or other beneficial supplemental metabolites for humans and livestock [2]. Legumes account for approximately one third of the world's primary crop production, human dietary proteins, and processed vegetable oils.

The evolution of mutualistic symbioses with nitrogen-fixing bacteria—rhizobia—allows legumes to directly capture atmospheric dinitrogen (N2) and turn into ammonium to support plant growth in an environment-friendly way [3]. This mutually

Luis A. Cañas and José Pío Beltrán (eds.), *Functional genomics in* Medicago truncatula*: Methods and Protocols*, Methods in Molecular Biology, vol. 1822, https://doi.org/10.1007/978-1-4939-8633-0_7,

beneficial association (nodulation) fixes 40–60 million tons of nitrogen globally each year by cultivated legumes [4]. Investigations into the genetic regulation and manipulation of legume nodulation are of significant importance in agriculture.

While forage and pulse legumes are primary targets of most applied research, many of these species have complex genomes (e.g., alfalfa and pea) and/or are recalcitrant to transformation (e.g., common bean and pea); therefore, they are not good systems for functional genomics research and genetic manipulation. In this situation, *Medicago truncatula* was selected as a model or reference species for legume research [5, 6]. To facilitate functional genomics studies in *M. truncatula*, we generated a near-saturated *Tnt1* insertional mutant population in the *Medicago* subspecies R108 at Noble Research Institute through tissue culture and regeneration [7, 8]. The population consists of more than 21,700 *Tnt1* mutant lines, which harbor an estimate of 500,000 randomly distributed *Tnt1* insertions in the *Medicago* genome.

Tnt1, one of the well-characterized plant long terminal repeat (LTR) retrotransposons, is a 5.3-Kb-long autonomous copia-like transposable element [9]. *Tnt1* sequence encodes a capsid-related protein, a protease, an integrase, a reverse transcriptase, and ribonuclease H and contains a 610 bp LTR flanking each end [10]. *Tnt1* is very active in *M. truncatula* during tissue culture but is stable during seed-to-seed propagation [11]; thus, we choose *Tnt1* to initiate a near-saturation mutagenesis in *M. truncatula*. In this chapter, we will describe the procedure of *Tnt1* mutant generation using the tissue culture and regeneration approach in *M. truncatula*.

2 Materials

2.1 Plant Materials

Starting material: Transgenic line *Tnt88-7-7* was generated by d'Erfurth et al. [11]. This specific line is in T2 generation and contains three homozygous *Tnt1* insertions in the genome. We obtained seeds of *Tnt88-7-7* from Dr. Pascal Ratet.

2.2 Media and Solutions

2.2.1 Stock Solutions

1. SH minor solution: Add 350 mL water to a 1 L beaker. To the beaker, add 10 g of manganese sulfate hydrate ($MnSO_4{\cdot}H_2O$), 5 g of boric acid (H_3BO_3), 1 g of zinc sulfate heptahydrate ($ZnSO_4{\cdot}7H_2O$), 1 g of potassium iodide (KI), 0.1 g of sodium molybdate dihydrate ($Na_2MoO_4{\cdot}2H_2O$), 0.2 g of copper (II) sulfate pentahydrate ($CuSO_4{\cdot}5H_2O$), and 0.1 g of cobalt (II) chloride hexahydrate ($CoCl_2{\cdot}6H_2O$). When all solutes are fully dissolved, make up to 400 mL. Autoclave the solution at 121 °C for 15 min. Store at 4 °C.
2. SH vitamin solution: Add 350 mL water to a 1 L beaker. To the beaker, add 5 g of nicotinic acid, 5 g of thiamine hydrochloride (vitamin B1), and 5 g of pyridoxine hydrochloride

(vitamin B6). Make up to 400 mL. Autoclave the solution at 121 °C for 15 min. Store at 4 °C.

3. N6 major solution: Add 2.5 L of water to a 5 L beaker. To the beaker, add 18.5 g of manganese sulfate hydrate ($MnSO_4{\cdot}H_2O$), 283 g of potassium nitrate (KNO_3), 46.3 g of ammonium sulfate ($(NH_4)_2SO_4$), and 40 g of potassium dihydrogen phosphate (KH_2PO_4). Stir until all solutes are fully dissolved. In a 1 L beaker, add 500 mL water and then 16.6 g of calcium chloride dihydrate ($CaCl_2{\cdot}2H_2O$). Heat to dissolve the solute. Mix the CaCl solution with the first part of the solution. Make up to 4 L. Autoclave the solution at 121 °C for 15 min. Store at 4 °C.
4. BAP (1 mg/mL): Add 50 mg of 6-benzylaminopurine (BAP) to 50 mL water. The solution is stored at 4 °C after filter sterilization.
5. 2, 4-D (2 mg/mL): Add 20 mg of 2, 4-dichlorophenoxyacetic acid (2, 4-D) to 10 mL ethanol. The solution is stored at 4 °C after filter sterilization.
6. Cefotaxime (250 mg/mL): Add 5 g of cefotaxime to a 50 mL Falcon tube. Add 15 mL water, and mix to dissolve the solute. Make up to 20 mL. The solution is stored at 4 °C after filter sterilization.
7. EDFS (7 g/L): Add 7 g of ethylenediaminetetraacetic acid iron (III) sodium (EDFS) to 1 L water. Stir until the solute is fully dissolved. Autoclave and store at 4 °C.
8. 5 M KOH: Dissolve 30.55 g of KOH in 80 mL water, and then make up to 100 mL. Store at room temperature.

2.2.2 Medium Preparation

1. SHMab medium: To a 2 L beaker, add 500 mL water. Then add 160 mL of N6 major, 80 mL of EDFS, 1.6 mL of SH minor, 1.6 mL of SH vitamin, 2 mL of BAP, 180.8 μL of 2, 4-D, 0.4 g of myoinositol, and 120 g of sucrose. Mix until all solutes are dissolved. Make up to 1 L. Adjust the pH to 5.8 with 5 M KOH, and then add 3 g of Phytogel, mix, and autoclave. When the autoclaved solution is cooled to approximate 60 °C, add 1 mL of cefotaxime, mix, and pour about 20 mL of SHMab medium into each 100 × 25 mm deep Petri dish.
2. SHMb medium: To a 2 L beaker, add 500 mL water. Then add 160 mL of N6 major, 80 mL of EDFS, 1.6 mL of SH minor, 1.6 mL of SH vitamin, 2 mL of BAP, 0.4 g of myoinositol, and 120 g of sucrose. Mix until all solutes are dissolved. Make up to 1 L. Adjust the pH to 5.8 with 5 M KOH, and then add 3 g of Phytogel, mix, and autoclave. When the autoclaved solution is cooled to approximate 60 °C, add 1 mL of cefotaxime, mix, and pour about 20 mL of SHMab medium into each 100 × 25 mm deep Petri dish.
3. SHM medium: To a 2 L beaker, add 500 mL water. Then add 160 mL of N6 major, 80 mL of EDFS, 1.6 mL of SH minor,

1.6 mL of SH vitamin, 0.4 g of myoinositol, and 120 g of sucrose. Mix until all solutes are dissolved. Make up to 1 L. Adjust the pH to 5.8 with 5 M KOH, and then add 3 g of Phytogel, mix, and autoclave. When the autoclaved solution is cooled to approximate 60 °C, add 1 mL of cefotaxime, mix, and pour about 20 mL of SHMab medium into each 100 × 25 mm deep Petri dish.

4. SHMp medium: To a 2 L beaker, add 500 mL water. Then add 160 mL of N6 major, 80 mL of EDFS, 1.6 mL of SH minor, 1.6 mL of SH vitamin, 0.4 g of myoinositol, and 80 g of sucrose. Mix until all solutes are dissolved. Make up to 1 L. Adjust the pH to 5.8 with 5 M KOH, and then add 3 g of Phytogel, mix, and autoclave. When the autoclaved solution is cooled to approximate 60 °C, pour about 20 mL of SHMab medium into each 100 × 25 mm deep Petri dish.

3 Methods

3.1 *Tnk88-7-7* Plant Material Preparation

3.2.1 Seed Scarification and Germination

1. Thirty to forty *Tnk88-7-7* seeds are treated using sandpaper for 2 min to remove seed coat.
2. Sandpaper-treated seeds are sterilized in 30% Clorox with one drop of Triton X-100 for 8 min.
3. Seeds are washed three times with autoclaved water.
4. Sterilized seeds are placed on 100 × 25 mm deep Petri dishes with autoclaved filter paper. The seeds are fully covered with water overnight at 4 °C. The excessive water is removed with a pipette the next morning.
5. After 1 week of cold treatment, seeds begin to germinate.

3.1.1 Explants Preparation

1. Germinated *Tnk88-7-7* seedlings are transplanted into a 36-cell tray filled with soil. After 4 weeks of growth in the cells, the seedlings are transferred to 1-gallon pots in such a way that each pot contains one plant. After growing for 6–8 weeks in pots, individual plants are ready to be cut as leaf explants.
2. From each individual plant, three leaflets with petioles are cut and collected into a 1000 mL beaker. The edges of each leaflet are removed to make a square or rectangular shape for better callus induction. Petioles are cut into short segments. 50 mL screw-cap tubes are filled with 30 mL of distilled water for collecting the cut explant samples (at least 40–50 samples per tube). The leaf samples fill tubes halfway. Bio-Rad sieve caps are used to change solutions.
3. Explants are first sterilized with 70% ethanol for 2 min with frequent inverting and shaking of tubes. After removing the ethanol solution, the explants are treated with 30% Clorox

(with two drops of Tween-20) for 12 min on a shaker. After treatment with the second solution, explants are washed with autoclaved water for at least four times until no bubbles in the tubes. Repeated cutting is carried out two times a week for 1 month from the same individual plants.

3.2 Callus Induction

1. After washing, the sterilized explants are placed on SHMab medium plates. Each plate has 10–15 explants (Fig. 1a). The stacked plates are covered with aluminum foil to prevent light. Plates are then placed in a growth chamber at 24 °C. After 1 week on the SHMab medium, explants begin to lose chlorophyll and show signs of callus induction (Fig. 1b).
2. The explants are transferred onto new plates with fresh medium every 2 weeks for three times. At the time of first transfer, callus can be observed on some explants (Fig. 1c, arrows), and the explants are greenless (Fig. 1c). At the time of second transfer, more calli are observed around most explants (Fig. 1d, arrows). Explants that have no callus formation at this stage are discarded.

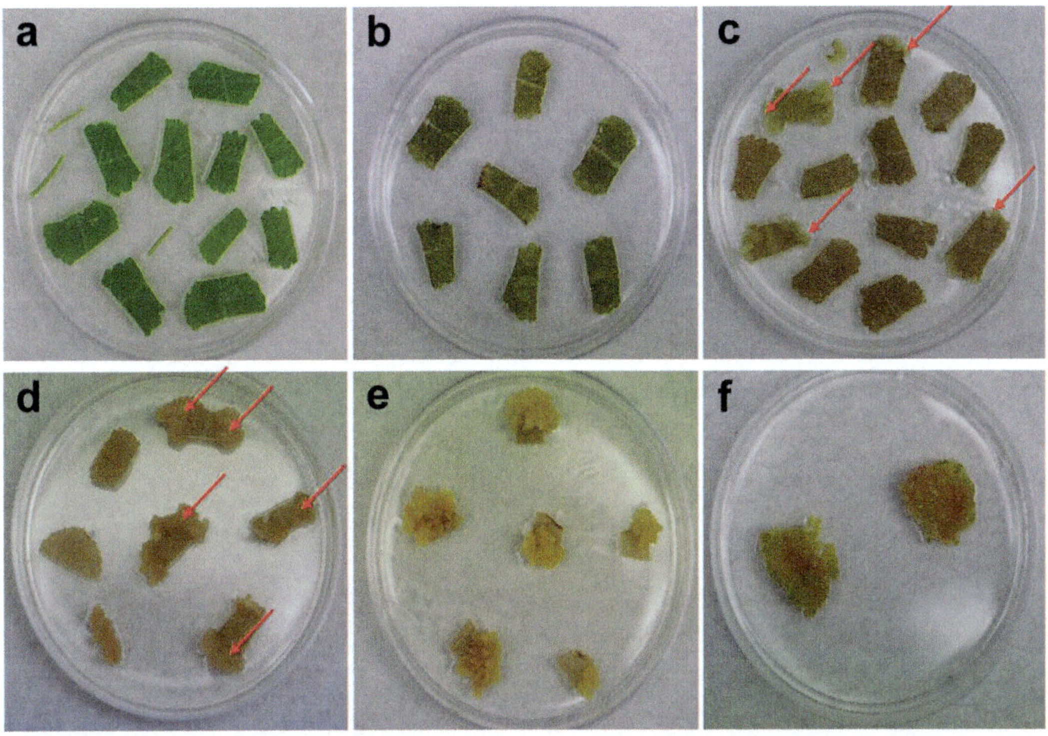

Fig. 1 Representative photos of callus induction steps. (**a**) Explants on SHMab medium for 0 days. (**b**) Explants on SHMab medium for 1 week. (**c**) Explants on SHMab medium for 2 weeks. (**d**) Explants on SHMab medium for 4 weeks. (**e**) Explants on SHMab medium for 6 weeks. (**f**) Callus on SHMb medium for 2 weeks

3. After the third transfer, callus grows well and turns into yellowish color (Fig. 1e). If several calli are produced from one leaf explant, each callus is separated on the new medium plate (*see* **Note 1**)

3.3 Embryogenesis and Embryo Development

1. Bright yellowish callus from the last SHMab medium is transferred onto SHMb medium for embryogenesis. Starting from this stage, medium plates with callus are no longer stacked and placed directly under light in a growth chamber at 24 °C.
2. Callus is maintained on SHMb medium for 2 weeks. At the end of 2 weeks, callus starts to turn green, suggesting embryogenesis is occurring (Fig. 1f).
3. Callus is then transferred onto SHM medium for embryo development. Embryonic callus is transferred onto fresh SHM medium every 2 weeks for two times. At the end of this stage, callus turns to dark green (Fig. 2a).

3.4 Plant Regeneration and Mutant Generation

1. Dark green callus from the last SHM medium is pressed to a flat shape and transferred onto SHMp medium for shoot and root development. After 1 week on SHMp medium, callus turns to brownish (Fig. 2b).

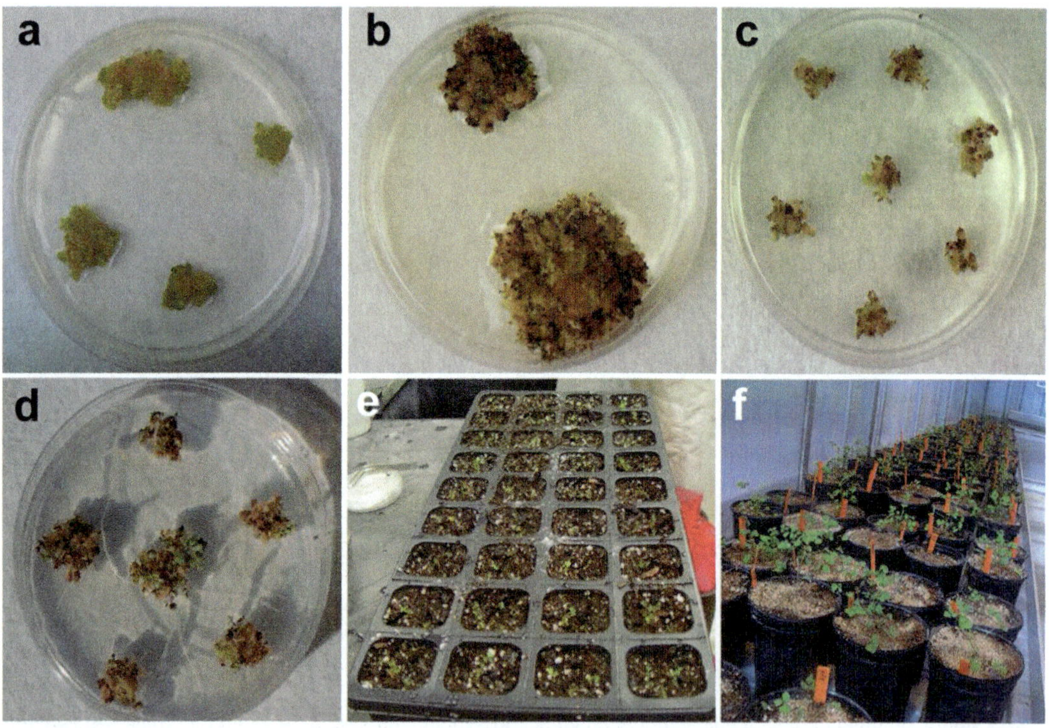

Fig. 2 Representative photos of plantlet regeneration steps. (**a**) Callus on SHM medium for 4 weeks. (**b**) Callus on SHMp medium for 1 week. (**c**) One callus from **b** is split into seven small pieces. (**d**) Callus on SHMp medium for 2–3 weeks, showing plantlets are generated. (**e**) Regenerated plantlets in 36-cell tray. (**f**) Plants in 1-gallon pots in the greenhouse

2. Each callus is split into several small pieces and one callus per plate (Fig. 2c). After 2–3 weeks on SHMp medium, most callus produces plantlets (Fig. 2d). One plantlet from each callus is kept (*see* **Note 2**) and numbered from NF0001.
3. Regenerated plantlets are transplanted into a 36-cell tray filled with soil (Fig. 2e). After transplanting, the tray is fully soaked and covered with a transparent plastic dome in a growth room. Plantlets in the tray are allowed to adapt for 4 weeks before they are moved to the greenhouse.
4. After 4-week adaptation, regenerated plants are transferred into 1-gallon pots with soil filled to 75% of the pot volume (Fig. 2f). The pots are placed on benches in a fully air-conditioned greenhouse room with a regime of 16-h light and 8-h dark.
5. When plants grow for 2 weeks in pots in the greenhouse room, a trifoliate leaf is collected from each plant for genomic DNA isolation (*see* **Note 3**).
6. The plants in pots are allowed to grow to maturation and seeds from each plant are harvested.

The entire mutant regeneration procedure is summarized in a flowchart (Fig. 3).

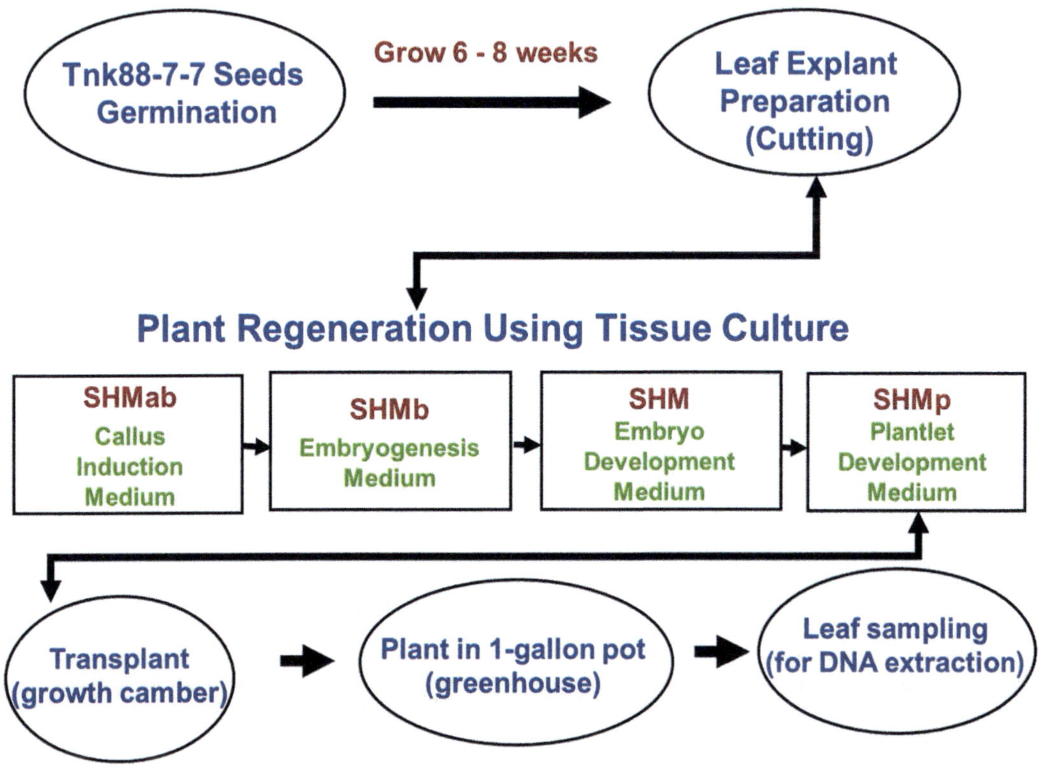

Fig. 3 A flowchart to show the main steps of the entire tissue culture and regeneration procedure described in the chapter

4 Notes

1. *Tnt1* transposition occurs at the very early stage of callus induction. Each callus represents a specific transposition event. To avoid mixed-up *Tnt1* insertions, callus separation at early stages is critical.
2. Sometime several plantlets are generated from one callus. To avoid duplicate/redundant mutant lines in the mutant population, only one plantlet from each callus is kept in the *Tnt1* insertional mutant population.
3. We normally collect three copies of leaf tissues from each regenerated line. One copy is used for DNA isolation. The other two copies are kept in a −80 °C freezer as backup samples for DNA isolation.

Acknowledgments

This work was supported by Noble Research Institute and in part by NSF plant genome grants (DBI 0703285 and IOS 1127155).

References

1. Doyle JJ, Luckow MA (2003) The rest of the iceberg: legume diversity and evolution in a phylogenetic context. Plant Physiol 131:900–910
2. Wang TL, Domoney C, Hedley CL, Casey R, Grusak MA (2003) Can we improve the nutritional quality of legume seeds? Plant Physiol 131:886–891
3. Hernández G, Valdés-López O, Ramírez M, Goffard N, Weiller G, Aparicio-Fabre R, Fuentes SI, Erban A, Kopka J, Udvardi MK, Vance CP (2009) Global changes in the transcript and metabolic profiles during symbiotic nitrogen fixation in phosphorus-stressed common bean plants. Plant Physiol 151:1221–1238
4. Smil V (1999) Nitrogen in crop production: an account of global flows. Glob Biogeochem Cycles 13:647–662
5. Bell CJ, Dixon RA, Farmer AD, Flores R, Inman J, Gonzales RA, Harrison MJ, Paiva NL, Scott AD, Weller JW, May GD (2001) The Medicago genome initiative: a model legume database. Nucleic Acids Res 29:114–117
6. May GD, Dixon RA (2004) Medicago truncatula. Curr Biol 14:R180–R181
7. Tadege M, Wen J, He J, Tu H, Kwak Y, Eschstruth A, Cayrel A, Endre G, Zhao PX, Chabaud M, Ratet P, Mysore KS (2008) Large-scale insertional mutagenesis using the *Tnt1* retrotransposon in the model legume *Medicago truncatula*. Plant J 54: 335–347
8. Cheng X, Wang M, Lee HK, Tadege M, Ratet P, Udvardi M, Mysore KS, Wen J (2014) An efficient reverse genetics platform in the model legume *Medicago truncatula*. New Phytol 201:1065–1076
9. Grandbastien MA, Spielmann A, Caboche M (1989) *Tnt1*, a mobile retroviral-like transposable element of tobacco isolated by plant cell genetics. Nature 337:376–380
10. Paszkowski J (2015) Controlled activation of retrotransposition for plant breeding. Curr Opin Biotechnol 32:200–206
11. d'Erfurth I, Cosson V, Eschstruth A, Lucas H, Kondorosi A, Ratet P (2003) Efficient transposition of the *Tnt1* tobacco retrotransposon in the model legume *Medicago truncatula*. Plant J 34:95–106

Chapter 8

Transient Posttranscriptional Gene Silencing in *Medicago truncatula*: Virus-Induced Gene Silencing (VIGS)

Joanna Serwatowska, Ole Søgaard Lund, and Ida Elisabeth Johansen

Abstract

Successful application of virus-induced gene silencing for functional genomics requires a virus vector that can initiate a systemic infection of the host plant. Agroinoculation of the pea early browning virus vectors pCAPE1 and pCAPE2 can establish infection in several genotypes of *Medicago truncatula* and can reduce target gene RNA levels to an extent that allows investigation of gene function.

Key words Pea early browning virus, PEBV, Functional genomics, Reverse genetics, Virus vector

1 Introduction

Virus-induced gene silencing (VIGS) was established in *Medicago truncatula* by [1] using pea early browning virus (PEBV) binary vectors pCAPE1 and pCAPE2 and applied to study the function of duplicated *MtAGAMOUS* genes [2]. pCAPE1 carries the viral replication and movement functions, and pCAPE2 carries the coat protein and a fragment of the gene, which is targeted for silencing. PEBV also induces silencing in *Nicotiana benthamiana* [3], *Lathyrus odorata*, *L. sativus* [1], and *Pisum sativum* [4]. In *P. sativum*, PEBV VIGS has been applied in the analysis of meristematic genes [5–7] and symbiosis with rhizobia and mycorrhiza [8, 9]. Details of the experimental setup are described in [10].

This protocol describes ligation-free cloning into pCAPE2, transformation into *Agrobacterium*, and agroinoculation of *Medicago*. Subsequent analysis will depend on the purpose of the experiment and is not described here.

Luis A. Cañas and José Pío Beltrán (eds.), *Functional genomics in* Medicago truncatula*: Methods and Protocols*, Methods in Molecular Biology, vol. 1822, https://doi.org/10.1007/978-1-4939-8633-0_8,

2 Materials

2.1 Plants and Microorganisms

M. truncatula cv. Jemalong A17, R108, SA1306, SA1326, SA1335, or other genotype that is systemically infected by PEBV [1]. Plants are inoculated 1 month after germination.

Electrocompetent *Agrobacterium tumefaciens* GV3101, GV3850, or similar

Chemically competent *Escherichia coli* XL1-blue, DH5α, TOP10, or similar

2.2 pCAPE VIGS Vectors and Sequencing Primer

1. Virus vectors pCAPE1 (HQ687213.1) and pCAPE2-USER can be obtained for research purposes, by contacting Ida Elisabeth Johansen at University of Copenhagen, Denmark (email: ejoh@plen.ku.dk).
2. The pCAPE2-USER sequence is obtained by replacing the *Nco*I/*Eco*RI *PsPDS* fragment of pCAPE2-PsPDS (HQ687214.1) with the sequence <u>CCATGG</u>CTGAGGCA ATTTAAATACCCTCAGC<u>GAATTC</u> (*Nco*I and *Eco*RI sites are underlined). For details of the USER cassette, *see* Fig. 1a.
3. Sequencing primer CAPE2-FW 5′-TCTAAACGGACGTTG GGGTTTTGCTA.

2.3 Growth Media

1. Luria-Bertani broth (LB) medium, liquid: 10 g tryptone, 5 g yeast extract, and 10 g NaCl per liter liquid medium. Adjust pH to 7.0 using 1 M NaOH. Sterilize by autoclaving. For plates, add 15 g agar per liter before autoclaving.
2. YEP medium, liquid: 10 g Bacto Peptone, 10 g yeast extract, and 5 g NaCl per liter liquid medium. Adjust pH to 7.0 with 1 M NaOH. Sterilize by autoclaving. For plates, add 15 g agar per liter before autoclaving.
3. Infiltration medium: 10 mM NaCl, 1.75 mM $CaCl_2$.
4. Acetosyringone 100 mM in DMSO.

2.4 Kits, Reagents, and Instruments

1. Plant RNA extraction kit: Spectrum Plant Total RNA Kit (Sigma Aldrich), RNeasy Plant Mini Kit (Qiagen), or similar.
2. cDNA synthesis kit: iScript cDNA synthesis kit (Bio-Rad) or similar.
3. DNA gel purification kit: Monarch DNA Gel Extraction Kit (NEB) or similar.
4. PCR for USER cloning: PfuTurbo Cx Hotstart DNA polymerase (Agilent Technologies), which overcomes uracil stalling and is compatible with uracil-containing primers.
5. Restriction enzyme *Swa*I (NEB), nicking endonuclease *Nt.BbvCI* (NEB), USER enzyme (NEB).
6. NEB buffer 3.1, CutSmart Buffer, PCR reaction buffer, or equivalent.

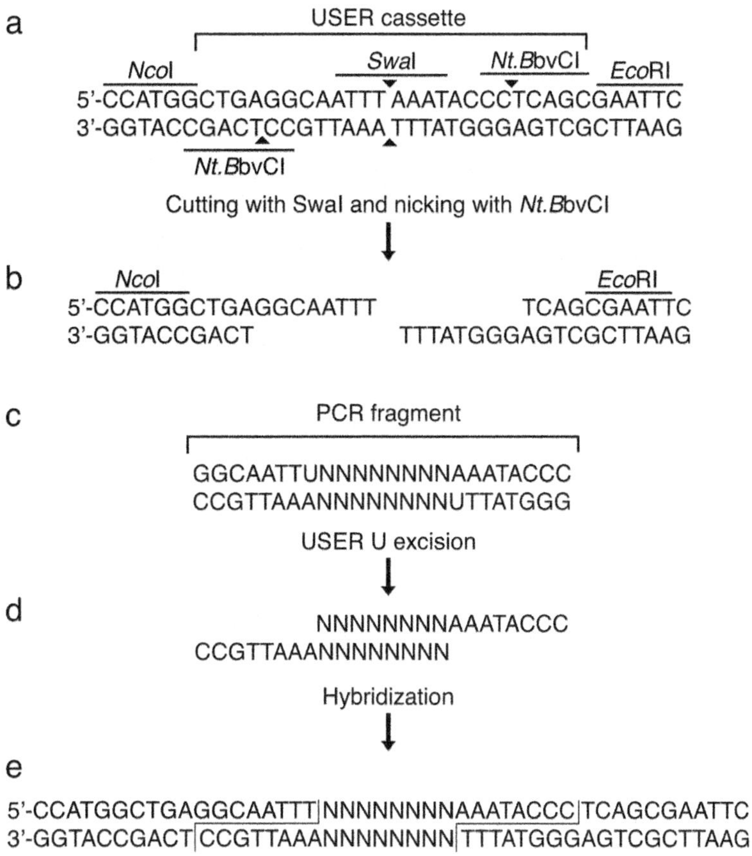

Fig. 1 Insertion of PCR products into pCAPE2-USER. (**a**) USER cassette flanked by *Nco*I and *Eco*RI restriction sites. (**b**) Staggered ends are created by cutting with *Swa*I followed by nicking with *Nt.Bb*vCI. (**c**) PCR-generated insert with uracil-containing primers. (**d**) Staggered ends are created by uracil excision with USER enzyme. (**e**) Hybridization product of staggered ends of vector and insert are transformed into competent *E. coli* without ligation

7. Template of the gene of interest (cDNA or cloned).
8. USER primers diluted to 5 pmol/μL.
9. 10 mM deoxynucleotide solution mix (NEB).
10. Antibiotics: Rifampicin (*Agrobacterium*), kanamycin (pCAPE), gentamicin (GV3101), ampicillin, or carbenicillin (GV3850) (Sigma Aldrich).
11. Electroporation equipment: Gene Pulser II (Bio-Rad), electroporation cuvettes 1 mm, 90 μL (VWR).

3 Methods

3.1 Preparation of pCAPE2-USER

1. Transform pCAPE2-USER into XL1-blue, DH5α, TOP10, or similar competent *E. coli* strain. Plate on LB-agar supplemented with 50 μg/mL kanamycin. Grow 16 h at 37 °C (*see* **Note 1**).

2. Select individual colonies, and amplify in 15–20 mL LB with 50 μg/mL kanamycin in 100 mL culture flasks at 37 °C, 200 rpm for 12–16 h.
3. Purify the plasmid using a spin column-based commercial kit. Measure concentration, and check the sequence of the USER cassette with primer CAPE2-FW.
4. Digest 10–20 μg plasmid with 20 units *Swa*I in a total volume of 200 μL 1 × NEB3.1 buffer. Incubate at 25 °C overnight. Add additional ten units *Swa*I, and incubate at 25 °C for 1 h. Check that the plasmid is linearized completely by agarose gel electrophoresis of 1–2 μL of the digestion. pCAPE2-USER is 9541 bp long.
5. Precipitate the linearized DNA by adding 20 μL 3 M sodium acetate pH 5.5 and 600 μL 96% ethanol, vortex, and incubate at −20 °C for 1 h. Pellet by centrifugation at 15,000 × *g* for 20 min. Aspirate the ethanol, and wash the pellet with 70% ethanol followed by centrifugation at >15,000 × *g* for 5 min. Remove all ethanol, and let the pellet dry.
6. Dissolve the DNA pellet in 176 μL H_2O. Add 20 μL 10× CutSmart Buffer and 40 units *Nt.Bbv*CI. Incubate for 2 h at 37 °C.
7. Purify the nicked vector (Fig. 1b) using a spin column-based purification kit, and elute in 50 μL. Determine DNA concentration and quality by measuring absorbance at 280 and 260 nm, and dilute the nicked vector to 50 ng/μL.

3.2 Selection and Primer Design for VIGS Fragments

1. Design primers to amplify two to three fragments (200–500 bp) of the gene of interest (GOI) with USER primers having 5′ extensions 5′-GGCAATTU (FW primer) and 5′-GGGTATTU (RV primer) (Fig. 1c) (*see* **Note 2**).
2. Silencing of *phytoene desaturase* (*PDS*) is well suited as a positive control because it results in photobleaching of the leaves [4, 11]. A 476 bp fragment derived from *M. truncatula PDS* mRNA [1] can be amplified from cDNA with USER primers 5′GGCAATTUCTAAACTTCATCAATCCTGATG and GGGTATTUCAAGTTACTGACATGTCGGCAT (*see* **Note 3**). The resulting plasmid is named pCAPE2-MtPDS. Inoculation to *M. truncatula* induces photobleaching as shown in (Fig. 2a).
3. A negative control is constructed by amplification of a fragment of *GUS*, *GFP*, or other inserts, which is not expected to affect gene expression in the host plant. Primers are designed with 5′ extensions 5′-GGCAATTU (FW primer) and 5′-GGGTATTU (RV primer).
4. To assay infectivity of PEBV in *M. truncatula*, the complete coding region of *GUS* (Fig. 2b, c), *GFP*, or other reporters is inserted in pCAPE2 using the *Nco*I and *Eco*RI restriction sites.

Fig. 2 Pea early browning virus (PEBV)-mediated silencing in *M. truncatula*. (**a**) *PDS* bleaching phenotypes in uninoculated upper leaves of *M. truncatula* SA1326 35 days postinoculation (dpi) with pCAPE1/pCAPE2-MtPDS 35. (**b**, **c**) In situ histochemical staining demonstrating GUS activity in *M. truncatula* inoculated with pCAPE1/pCAPE2-GUS. (**b**) SA1326 inoculated leaf 5 dpi and (**c**) uninoculated upper leaf 21 dpi. (**d**) WT flower of *M. truncatula*. (**e**) Flower of *M. truncatula* plants inoculated with pCAPE1/pCAPE2-MtAGa/MtABb carrying fragments of MADS-box genes *AGAMOUSa* and *AGAMOUSb*

3.3 Templates for Amplification of VIGS Fragments

1. If the gene of interest is cloned in a plasmid vector, approximately 10–100 ng plasmid DNA is used as template (*see* **Note 4**).
2. If the cDNA of the gene of interest is not cloned, then use a plant RNA extraction kit to extract RNA from 50 to 100 mg tissue expressing the GOI. Prepare cDNA using a cDNA synthesis kit. Follow the manufacturers' instructions for the commercial kits. Use 1 μL cDNA as template for a control PCR to check cDNA quality. *MtPDS* primers from Subheading 3.2, **step 2** can be used for this purpose.

3.4 Preparing USER PCR Products

1. Perform PCR with USER primers and PfuTurbo Cx Hotstart DNA polymerase following the manufacturers' instructions. Check the PCR reaction by running 1 μL on agarose gel.
2. If the USER PCR is done on cDNA or plasmid DNA that does not contain kanamycin resistance and there is only one PCR product of the expected size, the PCR product can be used directly in USER cloning.

3. If the USER PCR is done on plasmid DNA that contains kanamycin resistance or there are additional PCR amplification products, the entire PCR reaction is run on an agarose gel; the correct fragment is excised and purified with a DNA gel purification kit. Run 1 μL of the eluted purified product on an agarose gel, and determine the concentration.

3.5 Insertion of USER VIGS Fragments into pCAPE2- USER

1. Prepare a 10 μL USER reaction by mixing 50–100 ng USER PCR product in 6.25 μL H_2O, with 2 μL 50 ng/μL linearized PEBV USER vector, 1 μL CutSmart Buffer (NEB) or PCR reaction buffer, and 0.75 μL 1unit/μL USER enzyme (Fig. 1d, e).
2. Incubate for 15 min at 37 °C and then for 15 min at 25 °C.
3. Transform chemically competent *E. coli* by adding 2 μL USER reaction to 25 μL competent cells that are thawed on ice. Mix gently (not by vortexing), and incubate for 15 min on ice. Heat shock 30 seconds at 42 °C, followed by 2 min on ice. Add 100 μL LB medium, and incubate at 37 °C for 1 h at 100 rpm. Plate 40 and 80 μL transformation on two LB-agar plates with 50 μg/mL kanamycin. Incubate at 37 °C overnight.
4. Amplify plasmids from single colonies (four should be sufficient) by inoculating a culture tube with 2 mL LB with 50 μg/mL kanamycin. Incubate overnight at 37 °C at 200 rpm. Spin 1.5 mL cell culture down in a microcentrifuge tube, and purify plasmid with a spin column-based kit.
5. Check plasmid by restriction digestion on 1 μL. An *Nco*I/*Eco*RI double digest will excise the USER PCR fragment. Send plasmid with the correct digestion pattern for sequencing with CAPE2-FW.

3.6 Transformation of Agrobacterium with pCAPE Plasmids

1. Transform pCAPE1 and pCAPE2-GOI from Subheading 3.5, **step 5**, into *Agrobacterium tumefaciens* GV3101, GV3850, or similar by electroporation [12]. After electroporation 1 mL YEP is added, the cells are transferred to a culture tube and incubated for 3 h at 28 °C at 100 rpm. Plate 50 and 200 μL on two plates with YEP agar supplemented with 50 μg/mL rifampicin and 50 μg/mL kanamycin and for GV3101 25 μg/mL gentamicin or 100 μg/mL carbenicillin or 50 μg/mL ampicillin for GV3850. Incubate for 2 days at 28 °C.
2. Check content of selected *Agrobacterium* colonies transformed with pCAPE2 constructs by colony PCR using specific primers for the GOI fragment.

3.7 Agroinoculation with pCAPE Plasmids

1. Day 1. For each construct, a starter culture of Agrobacteria is initiated from a single colony in 3 mL YEB or LB supplemented with 50 μg/μL kanamycin. Incubate overnight at 28 °C at 200 rpm. Use 700 μL of this culture to prepare a

glycerol stock by adding 300 μL 50% sterile glycerol. Store at −80 °C (*see* **Note 5**).

2. Day 2. For plant inoculation, agrobacteria are cultured in YEB or LB supplemented with 50 μg/μL kanamycin. Approximately 2.5 mL pCAPE1 and 2.5 mL pCAPE2 culture is needed for each plant (*see* **Note 6**). The cultures are initiated with 0.6 μL starter culture per mL for pCAPE1 and 0.3 μL starter culture per mL for pCAPE2. Incubate at 28 °C for 16–18 h at 200 rpm.
3. Harvest bacteria at OD550 = 1.2–1.5 by centrifugation at 3500 × *g* at room temperature.
4. Resuspend cells gently in infiltration medium at room temperature. Use 100 μL × OD550 per mL harvested culture (*see* **Note 7**).
5. Add acetosyringone to a final concentration of 100 μM (1 μL 100 mM per mL), mix, and incubate for 1–2 h at room temperature.
6. Mix pCAPE1 and pCAPE2 cultures 1:1 and infiltrate leaves of 1-month-old plants using a 1 mL syringe. Hold the syringe opening against the abaxial side of the leaf, and add sufficient pressure for the culture to enter the leaf without tearing it.
7. Observe pCAPE1/pCAPE2-MtPDS inoculated plants for the appearance of photobleaching (Fig. 2a) that can be expected 3 weeks after inoculation. Observe test plants for changes in phenotype (Fig. 2d, e), and conduct assays of relevance to your experiment.

4 Notes

1. Picking colonies from a fresh transformation often results in better plasmid DNA yields.
2. We have observed that some fragments induce more efficient silencing than others. Also, analysis of more than one VIGS fragment per target gene will provide supporting evidence that the phenotypes observed are actually caused by specific silencing of the target gene. To target two or more genes in a gene family, a fragment of each gene is inserted end to end in pCAPE2 using USER fusion [13].
3. The 476 bp *MtPDS* corresponds to nt 91,475–91,409, nt 91,187–90,966, nt 90,867–90,763, nt 85,318–85,274, and nt 85,036–84,991 from AC1447715.1
4. Sometimes PCR reactions perform better on linearized DNA. If you want to target a specific gene in a gene family, choose a silencing fragment in a variable region.

5. Agrobacteria from glycerol stocks are selected on plates with a full complement of antibiotics before they are used for starter cultures. For the first experiment is recommended to start a couple of pCAPE1 cultures from individual colonies and test the activity together with pCAPE2-MtPDS.
6. Fill only culture flasks with 1/5 volume medium to ensure good aeration.
7. Example of calculation: Agroculture of 100 mL with an OD550 = 1.5 is resuspended in 15 mL infiltration medium.

References

1. Grønlund M, Constantin G, Piednoir E et al (2008) Virus induced gene silencing in *Medicago truncatula* and Lathyrus odorata. Virus Res 135:345–349
2. Serwatowska J, Roque E, Gómez-Mena C et al (2014) Two euAGAMOUS genes control C-function in *Medicago truncatula*. PLoS One 9:e103770. https://doi.org/10.1371/journal.pone.0103770
3. Qian W, Yu C, Qin H et al (2007) Molecular and functional analysis of phosphomannomutase (PMM) from higher plants and genetic evidence for the involvement of PMM in ascorbic acid biosynthesis in Arabidopsis and Nicotiana benthamiana. Plant J 49:399–413
4. Constantin GD, Krath BN, MacFarlane SA, Nicolaisen M, Johansen IE, Lund OS (2004) Virus-induced gene silencing as a tool for functional genomics in a legume species. Plant J 40:622–631
5. Blein T, Pulido A, Vialette-Guiraud A (2008) A conserved molecular framework for compound leaf development. Science 322:1835–1839
6. Berbel A, Ferrándiz C, Hecht V et al (2012) VEGETATIVE1 is essential for development of the compound inflorescence in pea. Nat Commun 3:797. https://doi.org/10.1038/ncomms1801
7. Wang Z, Luo Y, Li X et al (2008) Genetic control of floral zygomorphy in pea (Pisum sativum L.). PNAS 105:10414–10419
8. Constantin G, Grønlund M, Johansen IE et al (2008) Virus induced gene silencing (VIGS) as a reverse genetic tool to study development of symbiotic root nodules. Mol Plant-Microbe Interact 21:720–727
9. Grønlund M, Olsen A, Johansen IE, Jakobsen I (2010) Protocol: using virus-induced gene silencing to study arbuscular mycorrhizal symbiosis in *Pisum sativum*. Plant Methods 6:28. https://doi.org/10.1186/1746-4811-6-28
10. Grønlund M (2015) VIGS for dissecting mechanisms involved in the symbiotic interaction of microbes with plants. In: Mysore KS, Senthil-Kumar M (eds) Plant gene silencing: methods and protocols, Methods in molecular biology, vol 1287. Springer, Heidelberg, pp 255–265
11. Kumagai MH, Donson J, della-Cioppa G, Harvey D, Hanley K, Grill LK (1995) Cytoplasmic inhibition of carotenoid biosynthesis with virus-derived RNA. Proc Natl Acad Sci U S A 92:1679–1683
12. Shen WJ, Forde GB (1989) Efficient transformation of agrobacterium spp. by high voltage electroporation. Nucleic Acids Res 25:8385
13. Nour-Eldin HH, Geu-Flores F, Halkier BA (2010) USER cloning and USER fusion: the ideal cloning techniques for small and big laboratories. In: Fett-Neto AG (ed) Plant secondary metabolism engineering: methods and applications, Methods in molecular biology, vol 643. Springer, Heidelberg, pp 185–200

Chapter 9

Stable Inactivation of MicroRNAs in *Medicago truncatula* Roots

Hélène Proust, Jérémie Bazin, Céline Sorin, Caroline Hartmann, Martin Crespi, and Christine Lelandais-Brière

Abstract

MicroRNAs are key regulators in the development processes or stress responses in plants. In the last decade, several conserved or non-conserved microRNAs have been identified in *Medicago truncatula*. Different strategies leading to the inactivation of microRNAs in plants have been described. Here, we propose a protocol for an effective inactivation of microRNAs using a STTM strategy in *M. truncatula* transgenic roots.

Key words MicroRNA inactivation, Mimicry, STTM strategy, Hairy root, Stem-loop reverse transcription, Plant

1 Introduction

MicroRNAs (miRNA) are 20–24-nucleotide small noncoding RNAs that mediate the regulation of many developmental processes and stress responses in plants through posttranscriptional gene-silencing mechanisms [1]. MiRNAs loaded into a RISC complex bind to their target RNAs via sequence complementary. This leads to the cleavage of the target between the 10th and 11th positions of the miRNA-target duplex or to the repression of its translation [1, 2]. Like in other species, deep sequencing of small RNA libraries in the model legume *Medicago truncatula* revealed thousands of conserved and non-conserved miRNA families differentially expressed during development or in response of different biotic or abiotic stresses [3–5]. However, in legume (Fabaceae) plants, the functions of most of these species- or legume-specific miRNAs remain largely unknown [6].

Plant miRNAs are coded by genes transcribed by the RNA POLYMERASE II as noncoding RNA precursors called the pri-miRs. These precursors present a double-stranded hairpin structure

Luis A. Cañas and José Pío Beltrán (eds.), *Functional genomics in* Medicago truncatula*: Methods and Protocols*, Methods in Molecular Biology, vol. 1822, https://doi.org/10.1007/978-1-4939-8633-0_9,

allowing their processing by DICER-LIKE ribonuclease III enzymes, in particular DCL1 [1, 7]. In plants, the precursor's length is of up to a few hundred nucleotides, and because of this relatively small size, mutations are difficult to obtain in miRNA genes. In addition, miRNA families are generally multigenic with several genes encoding the same mature miRNA or very closed variants [8]. Due to this redundancy, triple or quadruple mutants (or more) may be required to obtain a loss of function mutant. In addition, except in the region of the miRNA, the different members of one miRNA family generally have divergent sequences. Thus, classical strategies for gene function analysis such as the characterization of mutants or RNA interference are not appropriate to miRNAs. For years, functional analysis of miRNAs was restricted to overexpression of one of the corresponding miRNA gene. Another alternative was to compare the phenotypes obtained after ectopic expression of one miRNA target or a miRNA-resistant form of this target [9]. More recently, different technics have been developed to inactivate miRNAs, such as "miRNA sponge" in animals [10] or "target mimic (TM)" in plants [11]. In theory, these technics allow to inactivate either a miRNA specifically or all variants of one family. Target mimic is a technic based on the nonprotein coding gene *IPS1* (*induced by phosphate starvation*) from *Arabidopsis thaliana* which possesses a non-perfect complementary sequence of *MIR399* [11]. It has been shown that *IPS1* is not cleaved after the binding of miR399 due to mismatches at the expected miRNA cleavage site (between the 10th and 11th nt of the duplex). In consequence, the *IPS1/MIR399* binding is stabilized and leads to the sequestration of the miRNA. The exchange of the miR399 complementary sequence of *IPS1* with the complementary sequence of any microRNA of interest but with mismatches at positions 10 and 11 appears to be quite efficient to inactivate this miRNA in plants [12]. Recently, the "Short Tandem Target Mimic" (STTM) technic has been developed. In STTM constructs, modified binding sites of the corresponding miRNA are repeated two or three times separated by specific linkers [13, 14]. This technic has been used in different plant species and seems to be very efficient. In addition, selection of appropriate sequences may allow inactivating several variants of the same miRNA family [14–18]. In this chapter, we propose a protocol we used to inactivate miRNAs in *M. truncatula* transgenic roots using a modified STTM construct first described by Debernardi et al. [15].

2 Materials

2.1 Cloning

1. pMF2 binary vector (or another destination vector with a kanamycin selection cassette as well as a promoter and a terminator allowing expression of the MIMICRY artificial RNA) (*see* **Note 1**).

2. *EcoRI* and *KpnI* restriction enzymes and the corresponding buffer.
3. Gel and PCR Clean-up Column Kit.
4. T4 DNA ligase and its corresponding buffer.
5. *Escherichia coli* One Shot™ TOP10 competent cells.
6. Petri dishes containing LB medium supplemented with 50 mg/L kanamycin.
7. Plasmid Miniprep Kit.

2.2 Growth of *Agrobacterium rhizogenes*

1. Electrocompetent *A. rhizogenes* strain *Arqua1*.
2. Liquid LB medium.
3. Kanamycin and streptomycin stock solutions at 50 mg/mL.
4. Petri dishes containing LB medium supplemented with 50 mg/L streptomycin and 50 mg/L kanamycin.

2.3 *Medicago truncatula* Culture

1. *Medicago truncatula* genotype Jemalong A17 seeds.
2. Bayrochlore® solution: dissolve one tablet of Bayrochlore® into 400 mL of H_2O.
3. Concentrated sulfuric acid.
4. 125-by-125 mm square plates containing water supplemented with 15% (m/v) of Bacto-Agar.
5. 125-by-125 mm square plates containing Fahraeus medium [19] supplemented with 15% (m/v) of Bacto-Agar and 20 mg/L kanamycin.
6. 125-by-125 mm square plates containing Fahraeus [19] medium supplemented with 15% (m/v) of Bacto-Agar.
7. Cutted and sterilized brown germination paper (Anchor Paper Company).

2.4 RNA Extraction, Reverse Transcription, and Quantitative Real-Time PCR

1. Trizol (Tri) reagent.
2. Chloroform.
3. Isopropanol.
4. EtOH 75%.
5. SuperScript™ III Reverse Transcriptase Kit (Invitrogen).
6. LightCycler® 480 SYBR Green I Master Kit (Roche).
7. Appropriated RT and PCR primers (*see* later and Fig. 2).

3 Methods

3.1 Obtaining of the pMF2-STTM Construct

1. The STTM construct (Fig. 1) is designed with three modified binding sites of the miRNA to inactivate (*see* **Note 2**). Each binding site presents a bulge of three nucleotides including

Kpn1

```
GGTACCGCCGTAGCCGGCAGGTCTTCTCCCTCTAGAAANNNNNNNNNNGGANNNNNNNNNNNAGCTTCGGTTTTTCTCTTTGGANNNNNNNNNNGGANNNNNNNNNNNTTTTTCAATTT
TTTTGGTTGGANNNNNNNNNNGGANNNNNNNNNNNTTTTGATGGAAGATCTGCCGTAGCCGGCGAATTC
```

EcoR1

Fig. 1 Sequence of a STTM construct from Debernardi et al. [15]. Three modified miRNA binding sites (in light gray) are separated by spacers of 22 nt. The two restriction sites, *KpnI* and *EcoRI* (underlined), have been added at the extremities of the construct for subcloning

the 10th and 11th positions (*see* **Note 3**). All three modified binding sites are separated by a linker of 22 bp length.

2. To allow the further cloning into the binary vector pMF2, we inserted *EcoRI* and *KpnI* restriction sites flanking the STTM construct (Fig. 1).
3. This designed construct is ordered through GeneArt™ gene synthesis from Invitrogen. It is subcloned by GeneArt into a plasmid containing a selection cassette. This plasmid containing STTM construct is called pM-STTM.
4. Digest both pM-STTM and the binary destination vector pMF2 with *EcoRI* and *KpnI* in 20 μL final volume containing 500 ng of DNA, 1 μL of each enzyme and 2 μL of 10× FastDigest buffer. Digestion is performed 30 min at 37 °C.
5. Load the digestion products into a 1% agarose electrophoresis gel. The pM-STTM fragment should be approximatively 180 pb in length, and the linearized pMF2 should be approximatively 15, 28 kb in length.
6. Cut out the bands corresponding to pM-STTM fragment and linearized pMF2, and purify the DNA fragments using NucleoSpin® columns following the manufacturer's instructions.
7. Ligation of the STTM fragment into pMF2 is performed overnight at 16 °C in a 10 μL final volume containing 25 ng of linearized pMF2, 5 ng of pM-STTM digest fragment, 1 μL of 10× ligase buffer, and 1 μL of T4 DNA ligase (5 U/μL).
8. Then, the resulting pMF2-STTM vector is transformed into *Escherichia coli* for amplification using classical protocol of bacteria transformation and purified through a miniprep column following manufacturer's instructions.
9. Proceed to a Sanger's sequencing to verify the construct.
10. Mix 25 μL of electrocompetent *A. rhizogenes* strain *Arqua1* with 50 ng of the pMF2-STTM vector, and proceed to the electroporation at 2.5 V.
11. Recover the bacteria with 250 μL of LB medium, and place them to agitate for 2 h at 30 °C.

12. Plate bacteria on solid LB medium supplemented with 50 mg/L streptomycin and 50 mg/L kanamycin and grow overnight.

3.2 Transformation of *M. truncatula* roots with *A. rhizogenes* (See Note 4)

This protocol is adapted from Boisson-Dernier et al. [20].

1. *M. truncatula* seeds are scarified 3 min in concentrate acid sulfuric (*see* **Note 5**).
2. Rinse five times with sterile water.
3. Seeds are surface sterilized in Bayrochlore solution for 20 min with agitation
4. Rinse five times with sterile water to remove all the Bayrochlore.
5. Seeds are soaked in sterile water for 2 h with gentle horizontal agitation.
6. Seeds are ordered in plates containing water and 1.5% agar and kept in dark at 4 °C for 1 week.
7. Inoculate a single transformed colony of the pMF2-STTM *A. rhizogenes* strain *Arqua1* into 5 mL of liquid LB medium containing 50 mg/L streptomycin and 50 mg/L kanamycin, and incubate overnight at 30 °C.
8. Plate 1 mL of the overnight culture for 2 days on Agar-LB medium containing 50 mg/L streptomycin and 50 mg/L kanamycin, and incubate 2 days at 30 °C to get a continuous bacterial lawn.
9. After 1 week at 4 °C, plates with seeds are put at room temperature for few hours in dark.
10. The seedling radicle is sectioned approximatively 5 mm above the root tip. The section is then coating with *A. rhizogenes* bacteria.
11. Cutted seedlings are positioned on plates containing Fahraeus medium supplemented with 20 mg/L kanamycin with the inoculated section in contact with the medium.
12. Plates are placed vertically at 20 °C for 1 week and then transferred at 24 °C.
13. After 1 week at 24 °C, transgenic roots emerging from the inoculated section of the seedlings are distinguishable. Non-transformed roots emerging above the section should be removed.
14. Plantlets are transferred on fresh plates containing Fahraeus medium covered with sterile brown germination paper for at least 1 week and grown at 24 °C (*see* **Note 6**).

3.3 Validation of miRNA Inactivation

To validate the efficiency of the construct, a reverse transcription-PCR (RT-PCR) will be performed to determine that the STTM artificial RNA is truly overexpressed and that the miRNA-target

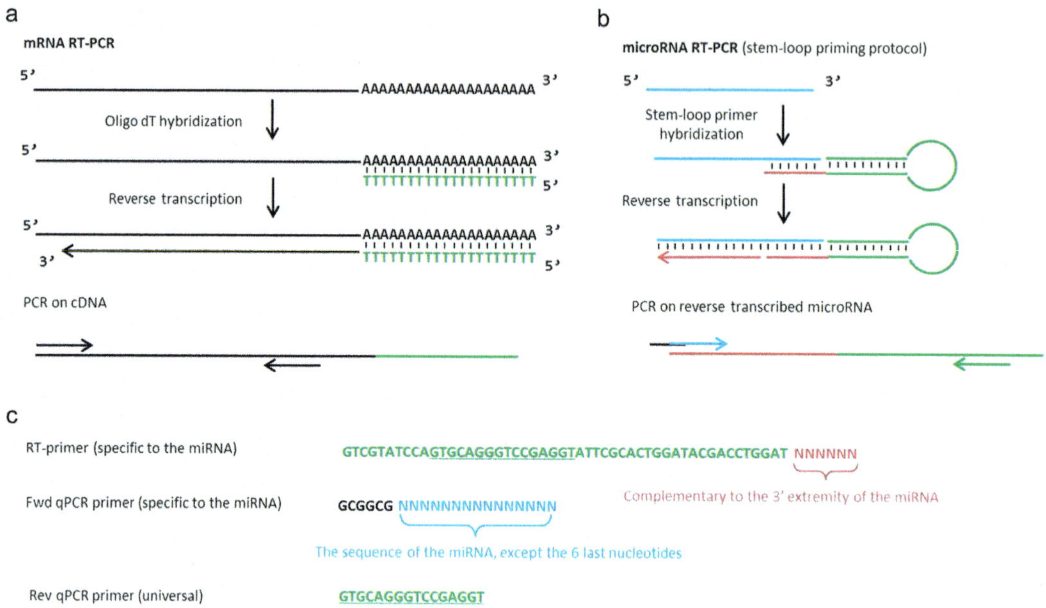

Fig. 2 Simultaneous amplification of mRNAs and microRNAs by RT-PCR. (**a**) Reverse transcription of mRNA. An oligo dT is hybridized on the polyA tail of the mRNA allowing the priming of the reverse transcription. Amplification of cDNA is performed using cDNA-specific primers. (**b**) MicroRNA reverse transcription using the stem-loop priming protocol, adapted from Varkonyi-Gasic et al. and Chen et al. [21, 25]. A stem-loop primer (in green) is hybridized to the mature microRNA (in blue) through six nucleotides (in red) complementary to the mature miRNA. The transcribed microRNA is amplified using a microRNA-specific forward primer (in blue) and a universal primer (in green) corresponding to the RT primer. (**c**) Design of microRNA RT-qPCR primers [25]

transcripts are overaccumulated which is a confirmation of the inactivation of the miRNA. Moreover, several authors reported that the IPS1 or STTM strategies often induce a decrease of accumulation levels of the sequestered miRNA [12, 14, 15, 17]. Thus, one additional argument for the efficiency of the STTM may be the quantification of the mature miRNA by RT-qPCR. Simultaneous reverse transcription of the mRNAs and the miRNAs is performed, thanks to a protocol adapted from the stem-loop RT-PCR pulsed protocol described by Varkonyi-Gasic et al. (Fig. 2, *see* **Note 7**) [21].

1. Grind frozen root tissues in liquid nitrogen (*see* **Note 8**).
2. Extract total RNA with Trizol reagent (Sigma), according to the manufacturer's recommendations except for one step. After precipitation with isopropanol, centrifuge 30 min at 20,000 × *g* at 4 °C. This longer centrifugation will allow keeping the small RNA fraction in the pellet (*see* **Note 9**).
3. DNase treatment is performed on 1.5 μg total RNAs with 1 U of DNase I according manufacturer's indications.
4. Prepare a first RT mix by adding the following components (corresponding to one reaction):

1 µg DNase-treated total RNA

1 µL of 50 µM dT_{17} oligonucleotide

1 µL of 10 mM dNTPs

1 µL of 1 µM miRNA RT primers (*see* **Note 10**)

Nuclease-free water to 14 µL

5. Incubate for 5 min at 65 °C.
6. Immediately incubate for 5 min on ice.
7. Prepare a second RT mix as the following for 1 reaction:

   1 µL of 0.1 M DTT

   4 µL of reverse transcriptase 5× buffer

   1 µL of SuperScript™ III Reverse Transcriptase (200 U.μL^{-1})
8. Add the 2nd RT mix to the 1st mix.
9. Incubate for 45 min at 50 °C and then 15 min at 55 °C.
10. Inactivate the enzyme with incubation at 70 °C for 15 min.
11. Prepare the qPCR mix adding the following components for a single reaction (*see* **Note 11**):

    3 µL of sterilized water

    5 µL of SYBR Green I Master Mix

    0.5 µL of 10 µM forward primer (cDNA or miRNA specific (*see* **Note 12**))

    0.5 µL of 10 µM reverse primer (cDNA specific or universal for miRNA amplification (*see* **Note 12**))

    1 µL of cDNA (*see* **Note 13**)
12. Quantitative real-time PCR is performed on LightCycler® 480 system (Roche) in the following conditions: 94 °C for 10 min followed by 50 cycles of 94 °C for 15 s, 60 °C for 15 s, and 72 °C for 15 s, followed by a dissociation step (*see* **Note 14**).
13. Quantitative PCR data are analyzed following the relative quantification method which compares the amplification of the gene of interest to at least one endogenous control gene constitutively expressed whatever the tissue or treatment (*see* **Note 15**). Transcript level of mRNA or miRNA is analyzed using the ΔΔCt method.

4 Notes

1. pMF2 contains a 35S CaMV promoter and a nopaline synthase (NOS) terminator separated by a multiple cloning site including *EcoRI* and *KpnI* restriction site and a kanamycin selection cassette [22]. Other binary destination vectors can of course be used. For example, the pK7WG2D, 1 vector allows

cloning the STTM construct between a 35S promoter and a 35S terminator via the Gateway technology (Invitrogen). Moreover, such binary vector possessing a GFP (or RFP, DsRed…) reporter gene can be useful to determine if transformation of all roots is truly efficient.

2. In this protocol, we used the STTM strategy. However, other strategies such as the IPS1 strategy are efficient too in *M. truncatula* [23].
3. The target cleavage occurs at a precise position between the 10th and the 11th nucleotide of the miRNA in the miRNA-target duplex [1]. The bulge formed by three nucleotides between 10th and 11th position in the binding site in STTM construct avoids the target cleavage and allows the trapping of the miRNA.
4. Be careful; sulfuric acid is highly corrosive and toxic. It should be used in a fume hood. Wear gloves and safety glasses.
5. *A. rhizogenes* mediated transformation allows obtaining composite plants with stable transgenic roots and untransformed aerial parts. This protocol is quick and useful to test the efficiency of the construct and to analyze different phenotypes in transgenic roots. Of course, stable transgenic lines of *M. truncatula* expressing the STTM construct can be obtained by *Agrobacterium tumefaciens*-mediated transformation [23].
6. Seedlings can be kept more than 1 week on the brown germination paper, depending on the experiment.
7. Because of the small size of mature miRNAs, specific protocols of northern blot or RT-PCR are required for their quantification. Polyadenylation and stem-loop priming are the two major technics used for miRNA RT-PCR experiments [21, 23–27]. Here, we used the stem-loop priming protocol combined with a standard RT reaction, allowing the simultaneous analysis of miRNAs and mRNAs.
8. Note that each root corresponds to one independent transformation event. Thus, it would be more correct to validate STTM efficiency in independent roots. However, it is often necessary to pool several roots to have enough material to extract.
9. To be sure to extract total RNAs containing the small RNA fraction, avoid the classic total RNA extraction kits, and use phenol-based, trizol-based protocols or dedicated miRNA extraction kits.
10. Reverse transcription for miRNAs requires a specific stem-loop RT primer for each miRNA tested. The design of these primers is described in Chen et al., 2005 and Varkonyi-Gasi et al., 2007 [21, 25]. Each RT primer comprises a universal

sequence forming a stem-loop and six nucleotides complementary to the 3′ extremity of the miRNA (Fig. 2b). At least ten different RT primers may be pooled in the RT reaction. Reverse transcription of mRNAs requires, here, an additional oligo dT primer which is complementary to the polyA extremity of the mRNA (Fig. 2a).

11. Because the target mimic strategies are based on the complementary of miRNA with a modified target binding site, the specificity of the inactivation should be considered. So, it could be necessary to verify the accumulation of other miRNA variants from the same family (i.e., mature miRNAs differing by one or two nucleotides).
12. For miRNA amplification, specific forward primers should be designed. Each primer contains a GC-rich region followed by the 15 first nucleotides of the miRNA. The reverse primer is the same for all the miRNAs. The primer sequence corresponds to 16 nucleotides complementary to the universal sequence of the RT miRNA primer (Fig. 2b).
13. It may be necessary to dilute the cDNA if the microRNA is known to be well accumulated.
14. Melting curve analysis permits to determine the specificity of the primers. It is particularly important to test miRNA amplification due to the putative possibility to amplify different miRNA variants from the same family. If you observe only one melting peak, this is likely that only one product so only one miRNA is amplified. If you observe more than one peak, this can be explained by the amplification of several products.
15. For miRNA relative expression, it is more accurate to use a small noncoding RNA as endogenous control gene. U6 snoRNA is a good candidate because it is considered as constitutively expressed in all tissues of the plant. However, miRNAs known to be not induced or repressed in the conditions tested are also suitable as endogenous controls.

References

1. Jones-Rhoades MW, Bartel DP, Bartel B (2006) MicroRNAs and their regulatory roles in plants. Annu Rev Plant Biol 57:19–53
2. Borges F, Martienssen RA (2015) The expanding world of small RNAs in plants. Nat Rev Mol Cell Biol 16(12):727–741
3. Lelandais-Brière C, Naya L, Sallet E et al (2009) Cloning and characterization of small RNAs from Medicago truncatula reveals four novel legume-specific microRNA families. Plant Cell 21(9):2780–2796. https://doi.org/10.1105/tpc.109.068130
4. Formey D, Sallet E, Lelandais-Brière C et al (2014) The small RNA diversity from Medicago truncatula roots under biotic interactions evidences the environmental plasticity of the miRNAome. Genome Biol 15(9):457. https://doi.org/10.1186/s13059-014-0457-4
5. Long RC, Li MN, Kang JM et al (2015) Small RNA deep sequencing identifies novel and salt-stress-regulated microRNAs from roots of Medicago sativa and Medicago truncatula. Physiol Plant 154(1):13–27. https://doi.org/10.1111/ppl.12266

6. Jagadeeswaran G, Zheng Y, Li YF et al (2009) Genome-wide Medicago truncatula small RNA analysis revealed novel microRNAs and isoforms differentially regulated in roots and nodules. New Phytol 184(1):85–98. https://doi.org/10.1111/j.1469-8137.2009.02915.x
7. Kurihara Y, Watanabe Y (2004) Arabidopsis micro-RNA biogenesis through Dicer-like 1 protein functions. Proc Natl Acad Sci U S A 101(34):12753–12758
8. Allen E, Xie Z, Gustafson AM et al (2004) Evolution of microRNA genes by inverted duplication of target gene sequences in Arabidopsis thaliana. Nat Genet 36(12):1282–1290
9. Tang G, Tang X (2013) Short tandem target mimic: a long journey to engineered molecular landmine for selective destruction/blockage of microRNAs in plants and animals. J Genet Genomics 40:291–296
10. Ebert MS, Neilson JR, Sharp PA (2007) MicroRNA sponges: competitive inhibitors of small RNAs in mammalian cells. Nat Methods 4:721–726
11. Franco-Zorilla JM, Valli A, Todesco M et al (2007) Target mimicry provides a new mechanism for regulation of microRNA activity. Nat Genet 39:1033–1037
12. Todesco M, Rubio-Somoza I, Paz-Ares J et al (2010) A collection of target mimics for comprehensive analysis of microRNA function in Arabidopsis thaliana. PLoS Genet 6:e1001031
13. Tang G, Yan J, Gu Y et al (2012) Construction of short tandem target mimic (STTM) to block the functions of plant and animal microRNAs. Methods 58:118–125
14. Yan J, Gu Y, Jia X et al (2012) Effective small RNA destruction by the expression of a short tandem target mimic in Arabidopsis. Plant Cell 24:415. https://doi.org/10.1105/tpc.111.094144
15. Debernardi JM, Rodriguez RE, Mecchia MA et al (2012) Functional specialization of the plant miR396 regulatory network through distinct MicroRNA–target interactions. PLoS Genet 8:e1002419
16. Bazin J, Khan GA, Combier JP et al (2013) miR396 affects mycorrhization and root meristem activity in the legume Medicago truncatula. Plant J 74(6):920–934. https://doi.org/10.1111/tpj.12178
17. Gu Z, Huang C, Li F et al (2014) A versatile system for functional analysis of genes and microRNAs in cotton. Plant Biotechnol J 12(5):638–649. https://doi.org/10.1111/pbi.12169 Epub 2014 Feb 12
18. Li F, Wang W, Zhao N et al (2015) Regulation of nicotine biosynthesis by an endogenous target mimicry of microRNA in tobacco. Plant Physiol 169:1062–1071
19. Fahraeus G (1957) The infection of clover root hairs by nodule bacteria studied by a simple glass slide technique. J Gen Microbiol 16(2): 374–381
20. Boisson-Dernier A, Chabaud M, Garcia F et al (2001) Agrobacterium rhizogene-transformed roots of Medicago truncatula for the study of nitrogen-fixing and endomycorrhizal symbiotic associations. Mol Plant Microbe Interact 14:695–700
21. Varkonyi-Gasic E, Wu R, Wood M et al (2007) Protocol: a highly sensitive RT-PCR method for detection and quantification of microRNAs. Plant Methods 3(1):12
22. Merchan F, de Lorenzo L, Gonzales-Rizzo S et al (2007) Identification of regulatory pathways involved in the reacquisition of root growth after salt stress in Medicago truncatula. Plant J 51:1–17
23. Hobecker KV, Reynoso MA, Bustos-Sanmamed P et al (2017) The MicroRNA390/TAS3 pathway mediates symbiotic nodulation and lateral root growth. Plant Physiol 174(4): 2469–2486
24. Song Y, Nolan KE, Rose RJ (2013) Stable transformation of Medicago truncatula cv Jemalong for gene analysis using Agrobacterium tumefaciens. Methods Mol Biol 1069:203–214. https://doi.org/10.1007/978-1-62703-613-9_14
25. Chen C, Ridzon AA, Broomer AJ et al (2005) Real-time quantification of microRNAs by stem-loop RT-PCR. Nucleic Acids Res 33:e179
26. Shi R, Chiang VL (2005) Facile means for quantifying microRNA expression by real-time PCR. BioTechniques 39(4):519–525
27. Varkonyi-Gasic E, Hellens RP (2011) Quantitative stem-loop RT-PCR for detection of microRNAs. Methods Mol Biol 744: 145–157. https://doi.org/10.1007/978-1-61779-123-9_10

Chapter 10

Non-isotopic RNA In Situ Hybridization for Functional Analyses in *Medicago truncatula*

Concepción Gómez-Mena and Edelín M. Roque

Abstract

Different strategies have been developed and implemented during the last decades aiming to decipher the function of particular genes. Among the different techniques, in situ hybridization of mRNA remains an essential experiment to fully understand gene function. Here, we describe a protocol for the in situ localization of gene transcripts in plants. It is optimized for use of paraffin-embedded tissues and DIG-labeled probes and has successfully applied to floral bud tissues from *Medicago truncatula*. Using this protocol, we have analyzed the expression of MADS-box transcription factors where some of them have been preserved as duplicates in the genome. When duplicated genes are analyzed, the tissue and cellular location of the transcripts is the only technique that accounts for small variations in the pattern of gene expression that occurred after duplication and diversification. The use of a well-standardized in situ hybridization protocol is vital for the systematic analysis of the function of genes in *Medicago truncatula*.

Key words In situ hybridization, mRNA, Transcript, Riboprobe, Digoxigenin, *M. truncatula*

1 Introduction

In situ hybridization (ISH) of mRNA is a molecular biology technique used to localize specific mRNAs in cells or tissues using complementary strands of genetic material [1, 2]. It enables researchers to know the timing, cellular localization, and cell type in which a gene is expressed [3]. As mRNAs constitute the genetic machinery for protein production, ISH is a crucial step for understanding the function of genes.

ISH follows the same principles of nucleic acid hybridization that the Northern blot assays. Both are based on the specific annealing of a labeled probe to complementary sequences of a target nucleic acid [4]. Northern blot analysis requires the isolation of mRNAs and their separation by size on a gel, blotting them on to nitrocellulose and probing with a radioactive single-stranded DNA fragment from a particular gene. However, in ISH the hybridization

Luis A. Cañas and José Pío Beltrán (eds.), *Functional genomics in* Medicago truncatula*: Methods and Protocols*, Methods in Molecular Biology, vol. 1822, https://doi.org/10.1007/978-1-4939-8633-0_10,

reaction between a labeled nucleotide probe and the complementary target RNA sequences takes place in a tissue slice.

A number of different types of probes are available for ISH. It can use cDNA, oligonucleotide, or RNA probes, the latter being considered the best. These probes can be labeled with either radioactive-, fluorescent-, or antigen-labeled bases. Depending on the probe used, autoradiography, fluorescence microscopy, and immunohistochemistry are used for visualization, respectively [3]. Radioisotope labeling is the more traditional approaches for labeling a probe. It is still widely used because of their high sensitivity in detection but has some disadvantages such as the biohazard, poor spatial resolution, and disposal of radioactive waste as well as being time-consuming [5]. In order to circumvent these problems, non-isotopic in situ hybridization methods have been developed by using alternative labeling. For this approach, compounds including biotin, fluorescein, digoxigenin, alkaline phosphatase, or bromodeoxyuridine are used and visualized by histochemistry or immunohistochemistry [6]. These methods are comparatively clean, rapid, and simple, providing equal sensitivity and a higher degree of cellular resolution than radiolabeled probes.

Since the first reported detection of nucleic acids by in situ hybridization in 1969 [7], a wide variety of complex and dissimilar protocols have been described. However, all procedures attempt to fix and render the tissue permeable to the nucleotide probe without loss of RNA, retain good morphology in the tissue section, select and design probes which can effectively penetrate tissue and hybridize specifically, and use appropriate controls to verify the specificity of reactions.

In our lab, we have studied the mRNA expression patterns of the *Medicago truncatula* MADS-box transcription factors within floral buds at different stages of floral development [8–11]. Here, we describe in detail the non-isotopic in situ hybridization protocol that we have routinely used for these purposes. This protocol describes the use of digoxigenin-labeled single-stranded RNA probes to detect expression of the genes of interest in paraffin-embedded tissue sections.

2 Materials

Most parts of the protocol are RNAse sensitive, and therefore all solutions and materials should be handled under the maximum clean conditions. It is important to prepare new and fresh solutions for this particular protocol and keep them separate. For preparing the solutions, we use analytical-grade reagents and sterile Milli-Q water avoiding the use of DEPC-treated water. Solutions are stored at room temperature (unless indicated otherwise).

When possible, glass and plastic containers should be used exclusively for the protocol. After being used, it is needed to rinse them thoroughly with distillated water and store them in a clean container.

2.1 Fixation and Embedding

1. Formaldehyde (37%).
2. Acetic acid.
3. Absolute ethanol.
4. Histoclear® (National Diagnostics) (*see* **Note 1**).
5. Paraffin chips (Paraplast Plus, Sigma).
6. Plastic tissue cassettes (Fig. 1a).

2.2 Block Preparation and Sectioning

1. Metal embedding molds (Fig. 1b).
2. Poly-L-lysine-coated adhesive microscope slides.

2.3 RNA Probe Synthesis

1. DIG RNA labeling mix (10 mM ATP, 10 mM CTP, 10 mM GTP, 6.5 mM UTP, 3.5 mM DIG-11-UTP; Roche).
2. DNAse I, RNAse-free.
3. RNAse inhibitor.

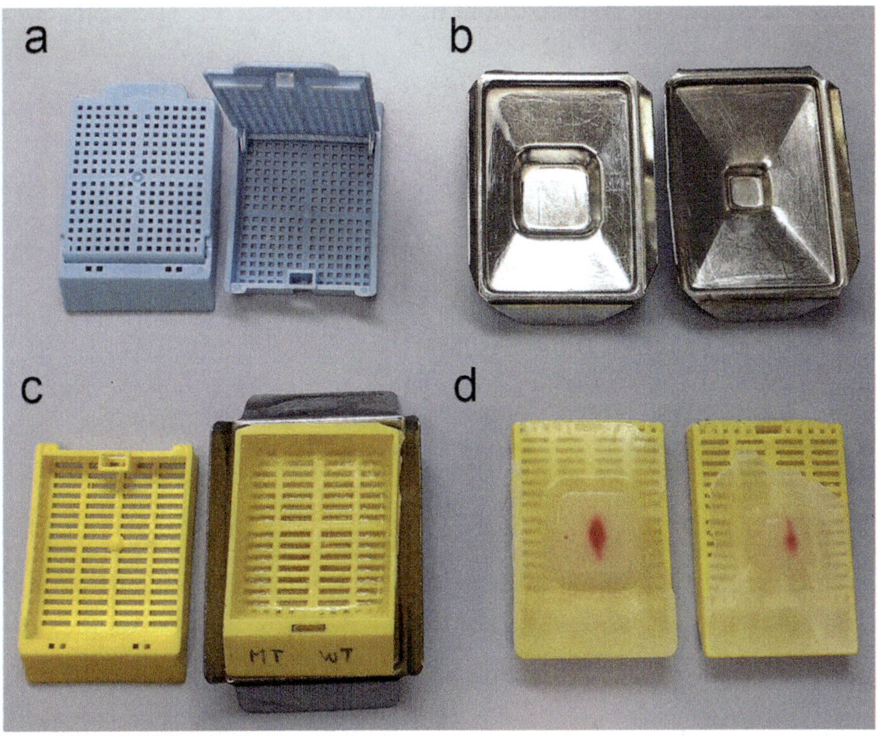

Fig. 1 Materials for tissue processing and embedding in paraffin blocks. (**a**) Plastic tissue cassettes for tissue processing. (**b**) Metal embedding molds for different sample sizes. (**c**) Rear plastic support (left) and paraffin block during solidification in the mold. (**d**) Paraffin-embedded samples after completion of the process stained with eosin

4. RNA polymerase (T7, SP6, or T3).
5. Carbonate buffer pH 10.2 (80 mM $NaHCO_3$ and 120 mM Na_2CO_3).

2.4 RNA Probe Quantification by Dot Blot

1. Positively charged nylon membrane.
2. DIG-labeled control RNA (100 ng/μL; Roche).

2.5 Pre-hybridization

1. Glass microscopy slide racks (for 10–20 slides).
2. Glass slide staining dishes (250 mL).
3. Proteinase K recombinant PCR grade (18.6 mg/mL; Roche).
4. Glycine.
5. 10× PBS: 1.3 M NaCl; 0.03 M Na_2HPO_4; 0.03 M NaH_2PO_4.

2.6 Hybridization and Washes

1. Plastic RNase-free coverslips.
2. Hybridization oven.
3. Square Petri dishes (120 × 120 × 17).
4. 20× SSC: saline sodium citrate; 3 M NaCl in 0.3 M sodium citrate; pH 7.0.
5. 20% SDS (sodium dodecyl sulfate).
6. Hybridization buffer: 6× SSC; 3% SDS; 50% formamide; 100 μg/mL tRNA.
7. Washing buffer: 50% formamide; 2× SSC.

2.7 Immuno-detection of DIG-Labeled Probes

1. Anti-DIG-AP Fab antibody (Roche).
2. 10× TBS (Tris-buffered saline): 1 M Tris-HCl PH 7.5; 4 M NaCl.
3. Blocking buffer: 0.5% blocking reagent (Roche) in TBS 1× (*see* **Note 2**).
4. BSA buffer: 1% bovine serum albumin (BSA); 0.3% Triton X-100 in 1× TBS.
5. Alkaline buffer: 100 mM Tris-HCl pH 9.5; 100 mM NaCl; 50 mM $MgCl_2$.
6. Nitro blue tetrazolium (NBT; 100 mg/mL).
7. 5-Bromo-4-chloro-3′-indolyphosphate (BCIP; 50 mg/mL).
8. Detection buffer: alkaline buffer with NBT/BCIP (150 μL of NBT and 150 μL of BCIP in 100 mL of buffer).

3 Methods

3.1 Tissue Fixation and Embedding in Paraffin

1. Prepare the fixative FAE (3.7% formaldehyde; 5% acetic acid; 50% ethanol; v/v) (*see* **Note 3**).

2. Harvest plant samples directly into vials containing the fixative and using a volume of five to ten times the volume of the samples. The size of the sample should not be bigger than 4 mm.
3. To help fixative penetration, apply vacuum for 15′, release vacuum slowly, and repeat.
4. Replace fixative solution for a fresh one, and keep the samples at room temperature for 3 h. After this time, replace fixative with 70% ethanol. At this point samples could be store at 4 °C for several months or continue with the protocol.
5. Transfer samples to plastic tissue cassettes labeled with pencil. Dehydrate samples by incubating in a series of ethanol solutions of increasing concentration (85, 96 and 100% ethanol) for 30 min each at room temperature. To improve sample visualization, the tissue can be stained with eosin by incubating the samples overnight in a solution 0.2% eosin in 100% ethanol at 4 °C.
6. On the next day, allow samples to temperate for 30 min before initiating the manual infiltration process (*see* **Note 4**). The ethanol needs to be slowly replaced by Histoclear by a series of incubation at room temperature:

   100% ethanol for 30 min

   100% ethanol for 30 min

   3:1 (v/v) ethanol: Histoclear for 1–2 h

   1:1 (v/v) ethanol: Histoclear for 1–2 h

   1:3 (v/v) ethanol: Histoclear for 1–2 h

   100% Histoclear for 1–2 h

   100% Histoclear for 1–2 h
7. To start paraffin infiltration, remove on a third of the Histoclear volume, and replace it with molten paraffin. Incubate overnight at 58 °C.
8. During the next 2 days, replace Histoclear/paraffin mix with liquid paraffin every 3–4 h.

3.2 Paraffin Block Preparation and Sectioning

1. Proceed with paraffin blocks preparation shortly after embedding; sample should not be kept in hot paraffin more than 2 days. The samples need to be individually located and oriented into the metallic molds before paraffin solidification (Fig. 1b–d). Paraffin will completely solidify after 30 min at room temperature, and blocks can be stored at 4 °C during years.

   In case metallic molds are not available, samples could be distributed and oriented into Petri dishes containing liquid paraffin (by using a hot plate underneath, paraffin will remain liquid). After cooling down, the paraffin can be cut into small pieces to individualize the samples.

2. Cut the paraffin-embedded tissue with a microtome using a clean blade to obtain 8 μm sections. Take strips of five to eight sections (depending of the size of the samples), and float them into a warm water bath (at 37 °C) until wax spreads. Pick up the strip from below using a fresh glass microscope slide, and transfer into a 40 °C hot plate to dry overnight (*see* **Note 5**).

3.3 RNA Probe Synthesis and DIG-Labeling

1. Cloning of gene fragments in a suitable vector.

   To generate the probe, a specific region from the gene of interest needs to be chosen. This is especially important when analyzing transcripts from closely related genes families. Gene fragments are amplified by PCR amplification using cDNA as a template and cloned into a vector containing an RNA polymerase promoter (T7, SP6, or T3).

   Fragments need to be cloned in the antisense orientation to the RNA polymerase promoter form the plasmid to generate the probe. As a negative control, plasmids with fragments cloned in the sense orientation are used.

   To obtain a template for the transcription reaction, linearize the plasmid with an appropriated restriction enzyme cutting downstream of the cloned fragment. Alternatively a DNA fragment containing both the RNA polymerase promoter and the cloned fragment could be PCR amplified and used for the synthesis of the probe. Templates should be gel purified and quantified before used.

2. Transcription and labeling of the RNA probe.

   Set the reaction for the synthesis of DIG-labeled RNA probes as follows:

   DNA template (1 μg of linearized vector or 200 ng of PCR fragment)
   2 μL DIG RNA labeling mix (Roche)
   2 μL (10×) RNA polymerase buffer
   1 μL RNase inhibitor
   2 μL RNA polymerase (T7, SP6, or T3)
   Water to 20 μL

   Incubate for 2 h at 37 °C. Add 2 μL de DNAse I, RNase-free, and incubate for 15 additional minutes. Stop reaction by adding 2 μL of 0.2 M EDTA (pH 8.0).

3. Precipitate the probe by adding 1 μL of tRNA (100 mg/mL), 1 volume of 7.5 M NH4Ac and 2.5 volumes of cold (−20°C) 100% ethanol followed by incubation at −70 °C >1 h or −20 °C overnight.

   Centrifuge for 10 min at 14,000 x *g* and 4 °C. Discard supernatant, and add 500 μL of 70% cold ethanol to wash pellet. Centrifuge for 5 min and remove ethanol. Dissolve pellet in 50 μL of Milli-Q sterile. Check the integrity of the amplified probe by running a 1.5% agarose gel.

4. Probe hydrolysis (*see* **Note 6**).

   Add 1 volume (50 μL) of 200 mM carbonate buffer to each probe, and incubate at 60 °C for the calculated time using the formula: $t = (L_i - L_f)/(K \times L_i \times L_f)$.

   t = time in minutes

   L_i = initial length in kb

   L_f = final length in kb (0.15 kb)

   K = 0.11 kb/min

   Stop reaction with 1/10 of 10% acetic acid, and precipitate the probe by adding 1/10 volume of 3 M NaAc (pH 4.8) and 2.5 volumes of cold ethanol. When pellet is completely dry, add 10 μL of Milli-Q sterile water and 90 μL of hybridization buffer before freezing at −20 °C.

   Check 1 μL of the reaction in a 1% agarose gel to confirm probe integrity and yield.

5. Dot blot of DIG-labeled probes.

   Prepare four serial dilutions (1/10; 1/100; 1/1000; 1/10000) for the RNA probes and the DIG-labeled control RNA (100 ng/μL). Blot 1 μL into a positively charged nylon membrane labeled with a pencil, and bind the RNA to the membrane by UV cross-linking. Incubate the membrane with the dots facing up in the following solutions and times:

   1× TBS for 5 min.

   Blocking buffer for 10 min.

   BSA buffer for 5 min.

   Anti-DIG-AP antibody diluted 1:3000 in BSA buffer for 20 min.

   Wash using BSA buffer for 5 min (repeat).

   Alkaline buffer for 2 min.

   Detection buffer for 12 min in the dark.

Rinse membrane in water to stop reaction. Compare the intensity of the spots with the control dilutions. Riboprobes are commonly used at a concentration of 0.1–0.5 ng/μL/kb.

3.4 Pre-hybridization

Check slides for tissue quality under a stereomicroscope, and mark with a pencil the slides for the sense and antisense probes. We typically use a glass slide rack that permits to allocate 20 slides in zigzag disposition. The rack fits into a glass container of 250 mL. Prepare all the solutions in advance before starting the procedure, and use a single container for every solution if possible. All incubations are performed at room temperature except for the proteinase K treatment that needs to be warmed in advance at 37 °C.

1. Deparaffinate and rehydrate by immersion of the rack into the following solutions and times:

   Histoclear for 10 min

   Histoclear for 10 min

   100% ethanol for 2 min

   100% ethanol for 2 min

   95% ethanol for 2 min

   70% ethanol for 2 min

   50% ethanol for 2 min

   30% ethanol for 2 min.

   Milli-Q H_2O for 2 min

2. Acid hydrolysis:

   0.2 M HCl for 20 min

   Milli-Q H_2O for 5 min

   2× SCC for 5 min

   Milli-Q H_2O for 5 min

3. Proteinase K treatment:

   1 μg/mL proteinase K (in 100 mM Tris; 50 mM EDTA pH 8.0) at 37 °C for 18 min

   PBS for 2 min

   2 mg/mL glycine in PBS for 2 min

   PBS for 2 min

   PBS for 2 min

4. Fixation:

   4% formaldehyde in PBS for 10 min

   PBS for 5 min

   PBS for 5 min

5. Rehydrate:

   Milli-Q H_2O for 2 min

   Milli-Q H_2O for 2 min

   30% ethanol for 2 min

   50% ethanol for 2 min

   70% ethanol for 2 min

   95% ethanol for 2 min

100% ethanol for 2 min

100% ethanol for 2 min

Leave slides to dry completely before adding the hybridization mix.

3.5 Hybridization and Washes

1. Probe preparation

   Calculate the amount of hybridization buffer needed according to the number of slides to be used (60 μL per slide). Add the required amount of probes for the both the antisense and sense (negative control) according to the dot-blot results. For every probe, the ideal amount to use needs to be tested empirically, but as estimation, 5–15 μL of probe for 1000 μL of hybridization buffer are sufficient.

   Denature probe by incubating at 80 °C during 2 min. Chill on ice, and transfer to a thermo block at 50 °C.

2. Apply 60 μL to every slide, and cover with a plastic coverslip avoiding bubbles. Transfer slides into a humidity chamber, and seal it with tape to avoid evaporation (*see* **Note 7**).
3. Incubate overnight in the oven at 50 °C (*see* **Note 8**).
4. Transfer slides to a rack, and incubate briefly in washing solutions until all coverslips fall off. Be very careful not damaging the tissue by pulling the coverslips.
5. Wash twice in washing solution (50% formamide; 2× SCC) for 90 min at 50 °C.

3.6 Immunological Detection

Transfer slides facing up to a big square Petri dish containing 1× TBS. A volume of 50 mL is sufficient to completely cover the slides.

1. Immunolocalization

   Incubate slides in the following solutions and times with horizontal gentle agitation (before adding a new solution try to remove the previous completely):

   1× TBS buffer for 5 min.

   Blocking buffer for 1 h.

   BSA buffer for 30 min.

   Anti-DIG-AP antibody diluted 1:3000 in BSA buffer for 90 min.

   Wash using BSA buffer for 30 min (three times).

2. Detection

   Transfer slides to a fresh square Petri dish containing alkaline buffer for 5 min. Incubate in detection buffer (containing NBT/BCIP substrates) for 24–72 h in the dark.

3. Visualization of the signal

   Stop reaction by immersion of the slides in Milli-Q H_2O. In the absence of background, pictures can be taken before permanent mounting adding some drops of water. Usually short incubation times (1–2 min) in ethanol series are needed to wash off the background. Proceed as follows:

   Milli-Q H_2O for 1–2 min

   70% ethanol for 1–2 min

   95% ethanol for 1–2 min

   100% ethanol for 1–2 min

   Histoclear for 2 min

   Histoclear for 2 min.

Mount the slides using a few drops of a toluene-based mounting media, and let the media to harden overnight. Slides can be stored for years at room temperature. Check the presence of background or unspecific signal in the control slides hybridized with the sense probe. Visualize the signal using a microscope with differential interference contrast (DIC) optical. The signal appears as a purple or blue precipitate located on the cells and tissues where the specific gene is expressed (Fig. 2).

Results shown in Fig. 2 correspond to expression analysis of duplicated floral MADS-box genes (*AGAMOUS*-like and *APETALA3*-like genes) in floral apices of *Medicago truncatula* [9, 10]. Both *MtAGa* and *MtAGb* transcripts began to accumulate at stage 2 of flower development. However, *MtAGb* accumulates in the center of the floral primordia, while *MtAGa* signal was detected throughout the floral meristem (Fig. 2a, b). Meanwhile, duplicated *Medicago AP3*-like genes show a non-overlapping and complementary expression pattern of both genes in petals and stamens. *MtTM6* was expressed predominantly in the outer cell layers of both floral organs and *MtNMH7* in the inner cell layers of petals and stamens (Fig. 2c–f). Expression analyses of these duplicated genes using in situ hybridization were essential to stablish their specific functions during flower development in this species.

4 Notes

1. Histoclear® is a safe substitute of xylene (toxic by inhalation and contact) when preparing histological sections.
2. This solution must be prepared in advance, the blocking reagent needs to be heated at 65 °C to dissolve, and then the solution needs to cool down to room temperature.
3. PFA (4% paraformaldehyde in PBS) is an alternative fixative also recommended for in situ hybridization. In this case, perform the fixation at 4 °C overnight.

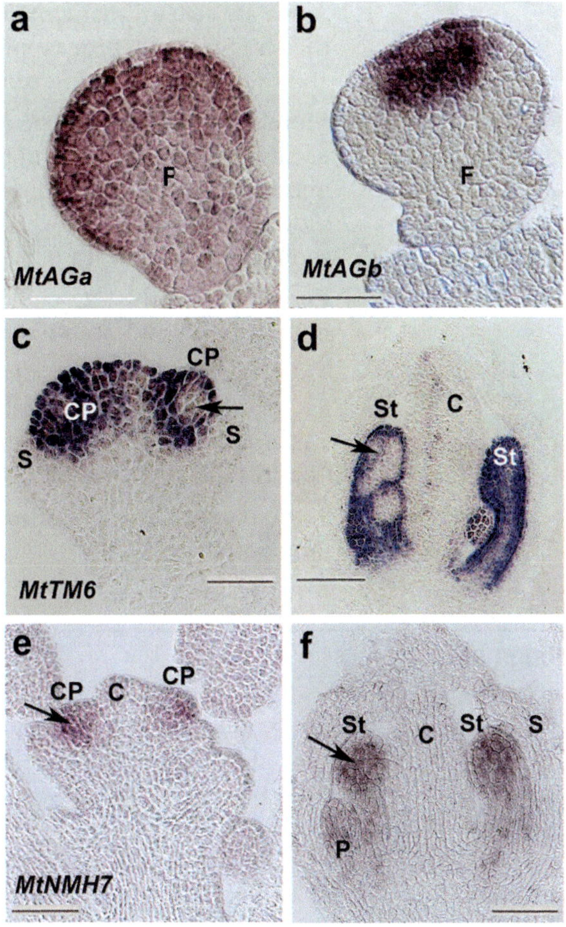

Fig. 2 RNA in situ hybridization on floral sections using digoxigenin-labeled probes applied to the functional analysis of duplicated *M. truncatula* floral homeotic genes. (**a**) and (**b**) Differential expression of duplicated *MtAGAMOUS* genes (*MtAGa* and *MtAGb*) at floral stage 2. *MtAGa* transcripts are localized in the whole floral meristem, while transcripts of its paralogous gene *MtAGb* are localized in the center of the floral meristem. (**c**) to (**f**) Differential expression of the duplicated *APETALA3*-like genes (*MtNMH7* and *MtTM6)* in *M. truncatula* flower buds. At stage 4, *MtTM6* mRNA is strongly expressed in the most external cell layers (arrow) of the common primordia and in the central region of the floral meristem (**c**). At stage 6, *MtTM6* mRNA is only detected in the outer cell layers (arrow) of petals and stamens (**d**). At stage 4, *MtNMH7* transcripts are located in the inner cell layers (arrow) of the common primordia (**e**). At stage 6, *MtNMH7* expression is found in the inner cell layers (arrow) of the developing petals and stamens. Expression was not observed in the outer cell layers of these organs (**f**). *F* floral meristem, *CP* common primordia, *S* sepal, *P* petal, *St* stamen, *C* carpel. Bars represent 50 μm

4. Alternatively automatic tissue processors can be used allowing the simultaneous processing of a high number of specimens and reducing processing time to less than 24 h.
5. More than one strip can be transferred to a single slide, but try not overfilling the slide and that strips do not overlap. It is very important to use poly-L-lysine-coated adhesive microscope slides; otherwise, tissue will be lost during hybridization. Slides should be marked with pencil.
6. For probes on the size range of 300–400 bp, hydrolyzation is not necessary and does not improve hybridization results.
7. To prepare the humidity chamber, use a square Petri dish (or any flat plastic box) with moistened filter paper at the bottom. Clean ELISA plates could be located on top of the paper to obtain a flat surface to put the slides on.
8. Some temperature optimization can be required being in the range of 48–53 °C.

Acknowledgments

This work was supported by a grant from the Spanish Ministry of Economy and Competitiveness MINECO (BIO2016-75485-R).

References

1. Polak JM and McGee JOD (1990). In: *In situ* hybridization: Principles and Practice. Oxford University Press, New York
2. Nakamura RM (1990) Overview and principles of *in situ* hybridization. Clin Biochem 23: 255–259
3. Jensen E (2014) Technical review: *In Situ* hybridization. Anat Rec 297:1349–1353
4. Baumgart E, Schad A, Grabenbauer M (2001) In: Beesley JE (ed) Immunocytochemistry and in situ hybridization in the biomedical sciences. Birkhäuser Boston, Boston, MA, pp 108–137
5. Tsukamoto T, Kusakabe M, Saga Y (1991) *In situ* hybridization with non-radioactive digoxigenin-11-UTP-labeled cRNA probes: localization of developmentally regulated mouse tenascin mRNAs. Int J Dev Biol 35:25–32
6. Jin L, Lloyd RV (1997) *In situ* hybridization: methods and applications. J Clin Lab Anal 11:2–9
7. Gall JG, Pardue ML (1969) Formation and detection of RNA-DNA hybrid molecules in cytological preparations. Proc Natl Acad Sci U S A 63:378–383
8. Benlloch R, Roque E, Ferrandiz C, Cosson V, Caballero T, Penmetsa RV, Beltran JP, Cañas LA, Ratet P, Madueño F (2009) Analysis of B function in legumes: PISTILLATA proteins do not require the PI motif for floral organ development in *Medicago truncatula*. Plant J 60:102–111
9. Roque E, Serwatowska J, Cruz Rochina M, Wen J, Mysore KS, Yenush L, Beltran JP, Cañas LA (2013) Functional specialization of duplicated *AP3*-like genes in *Medicago truncatula*. Plant J 73:663–675
10. Serwatowska J, Roque E, Gomez-Mena C, Constantin GD, Wen J, Mysore KS, Lund OS, Johansen E, Beltran JP, Cañas LA (2014) Two *euAGAMOUS* genes control C-function in *Medicago truncatula*. PLoS One 9:e103770
11. Roque E, Fares MA, Yenush L, Rochina MC, Wen J, Mysore KS, Gomez-Mena C, Beltran JP, Cañas LA (2016) Evolution by gene duplication of *Medicago truncatula PISTILLATA*-like transcription factors. J Exp Bot 67:1805–1817

Chapter 11

In Situ Hybridization Method for Localization of mRNA Molecules in Medicago Tissue Sections

Olga Kulikova, Carolien Franken, and Ton Bisseling

Abstract

Here we describe an in situ hybridization (ISH) method using Invitrogen™ ViewRNA™ ISH Tissue Assay (ThermoFisher Scientific) optimized for *Medicago* root and nodules sections. The method is based on branched (b)DNA signal amplification technology originally developed for use in microplate format and further adapted for detection of (m)RNAs in mammalian tissue sections. Signal amplification is achieved via a series of sequential hybridizations of linking sequences which are anchored to complementary sequences present on specific oligonucleotide probes. The typical (m)RNA probe set contains ~20 synthetic adjacent oligonucleotide pairs. Each probe is composed of a 20bp primary sequence designed to target sequence of interest and a secondary extended sequence serving as a template for hybridization of a preamplifier oligonucleotide. The preamplifier forms a stable hybrid only if it hybridizes to two adjacent probes. By this principle, background is reduced. Other regions on the preamplifier are designed to hybridize to multiple bDNA amplifier molecules that create a branched structure. Finally, alkaline phosphatase (AP)-labeled oligonucleotides, which are complementary to bDNA amplifier sequences, bind to the bDNA molecule by hybridization. By adding Fast Red substrate, red punctuated precipitates are formed that can be detected by light bright and/or fluorescent microscope. ThermoFisher Scientific (https://www.thermofisher.com/nl/en/home.html) designs and synthesizes probe sets for a gene of interest and Invitrogen™ ViewRNA™ ISH Tissue Assay kits include all components required for pretreatment of plant tissues, hybridization and signal amplification.

Key words ViewRNA™ ISH Tissue Assays, Branched DNA, RNA in situ probe sets, Preamplifier, Amplifier, Alkaline phosphatase (AP)

1 Introduction

For *Medicago*, several mutant libraries and numerous transcriptome data are available. Mining of these has led to the identification of many genes with functions in various processes. To unravel the biological roles of these genes, the knowledge of their expression patterns at a cellular level is essential.

Several methods are routinely used for the examination of the expression pattern of a gene. Promoter-reporter constructs are frequently used for this, but this approach suffers from several flaws.

Luis A. Cañas and José Pío Beltrán (eds.), *Functional genomics in* Medicago truncatula*: Methods and Protocols*, Methods in Molecular Biology, vol. 1822, https://doi.org/10.1007/978-1-4939-8633-0_11,

For some genes, it is difficult to identify all regulatory elements required for full complementation of knockout mutants. Besides, knockout mutations of many genes often do not result in a clear phenotype under laboratory conditions, and as a result it is hard to identify functional promoters. Further, an independent validation to ensure that the transgene expression pattern mimics that of the endogenous gene is typically required.

In situ hybridization (ISH) is the most reliable technique to localize (m)RNA in whole-mount preparations and tissue sections and does not require insight in the regulatory sequences controlling the expression of a gene. Highly efficient whole-mount ISH procedures are routinely used for *Arabidopsis* seedlings [1]. However, *Medicago* roots and nodules are too thick to obtain expression patterns with sufficient resolution with whole-mount procedures. Therefore, for *Medicago* root and nodule, thin sections are used for ISH. To detect mRNAs, the probe can be labeled by radioactive or nonradioactive methods. The radioactive ISH technique is sensitive but has difficulties in obtaining precise localization of the signal. For example, phosphorus (^{32}P) emits highly energetic beta particles which increase sensitivity, reduce exposure times (days), but hinder accurate signal localization. Probes labeled with ^{3}H-UTP have the opposite property, low sensitivity and high resolution [2–4]. Tritium emits weakly energetic beta particles which increase resolution of signal as the particles do not travel so far but at the expense of greatly increased exposure times (weeks). ^{35}S-UTP has often been used to label probes, resulting in rather high detection efficiency, but it reduces the cellular resolution due to the high energy of its radiation. In addition, there are some special requirements for the laboratories working with radioactivity. Therefore, nonradioactive ISH is in general preferable. Such methods rely on detecting labeled antisense RNA probes hybridized with endogenous mRNA molecules either directly (e.g., with fluorescence) or with antibody-based detection strategies. In the last years, a number of new nonradioactive ISHs offering excellent sensitivity have been developed.

Here, we outline one of such protocols for ISH on *Medicago* tissue sections using Invitrogen™ ViewRNA™ ISH Tissue Assay (ThermoFisher Scientific). This is a very powerful technology to detect any RNA of interest with single-cell resolution. This assay is based on highly specific probe sets and branched DNA (bDNA) signal amplification technology [5]. The bDNA signal amplification technology utilizes nonradioactive synthetic oligonucleotide probes and has been used extensively in micro-well format to detect specific nucleic acid sequences in clinical applications [6]. A typical probe set consists of ~20 pairs of oligonucleotide probes that hybridize to specific regions across the target mRNA sequence (Fig. 1).

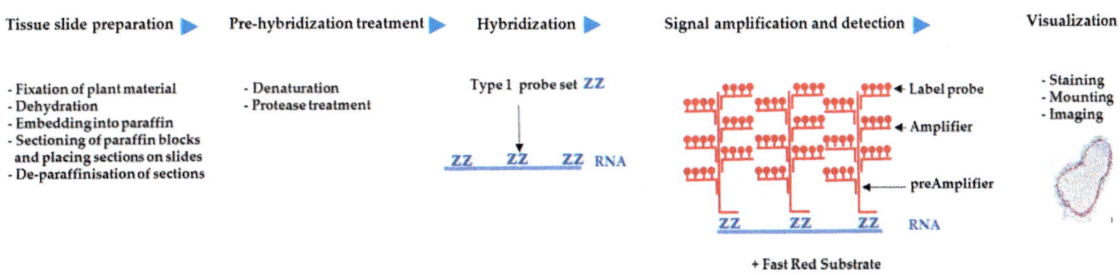

Fig. 1 ViewRNA ISH Tissue 1-Plex assay workflow

Each probe is composed of a region of 20 nucleotides that is complementary to the RNA of interest, a short linker region, and one half of a split preamplifier binding site. Only the hybridization of both members of a pair probe side by side on the target sequence allows the binding of a preamplifier molecule [6, 7] (Fig. 1). This markedly reduces background signal. The bound preamplifier molecules allow the binding of amplifier molecules, which in turn allow binding of label molecules conjugated to alkaline phosphatase (AP). The sequential hybridizations build up a branched structure in order to increase the number of binding sites for the AP-conjugated oligonucleotide probe. Finally, the addition of Fast Red substrate produces red precipitates (dots), and each dot represents a single RNA molecule that is intact and properly unmasked. This precipitate (=hybridization signal) has a dual nature: chromogenic and fluorescent. Therefore, it can be detected by both bright-field (counterstaining tissue sections with hematoxylene) and fluorescent (counterstaining with DAPI) microscopes.

As two different substrates are available for AP, Fast Red and Fast Blue, the ViewRNA ISH Tissue Assays can be used to detect one (Invitrogen ViewRNA™ ISH Tissue1-Plex Assay) or two target mRNAs simultaneously (Invitrogen ViewRNA™ ISH Tissue2-Plex Assay). In the latter case, two different types of probe sets need to be made, type 1 and type 6. Each probe set, after hybridization, is detected by specific preamplifier, amplifier, and Label Probe 1-AP or Label Probe 6-AP. Two sequential rounds of amplification and detection need to be done in this case because Label Probe 1 and Label Probe 6 are conjugated to the same AP and enzyme conjugates are unable to differentiate between Fast Red and Fast Blue. The first round is to amplify and detect the type 6 probe set, after which Label 6-AP must be quenched with AP stop buffer. Second round includes amplification and detection of probe set type 1.

Both protocols were successfully used for RNA ISH on *Medicago* nodule sections [8, 9]. In this chapter, the protocol for detection of single mRNAs in *Medicago* tissue sections is described. It is modified from the original (https://cdn.panomics.com/downloads/17400RevF%20140822_ViewRNA%20ISH%20

Tissue%201-Plex.pdf) manual. This RNA ISH method on plant tissue sections includes a fixation of plant material and embedding it into paraffin, sectioning and immobilization of tissue on slides, de-paraffinization and pretreatment of tissue, hybridization, signal amplification and detection. The workflow of this protocol is shown in Fig. 1.

2 Materials

The Invitrogen ViewRNA ISH Tissue 1-Plex Assay (ThermoFisher Scientific) consists of two kits, each sold separately and available in two sizes: (1) ViewRNA ISH Tissue 1-Plex Assay Kit (QVT0050–24 assays, QVT0051–96 assays) and (2) ViewRNA Chromogenic Signal Amplification Kit (QVT0200–24 assays, QVT0201–96 assays).

2.1 ViewRNA ISH Tissue 1-Plex Assay Kit Components (Storage Conditions Are Indicated on Each Container)

1. 100× pretreatment solution.
2. Protease QF enzyme.
3. Probe Set Diluent QF.
4. Amplifier Diluent QF.
5. Label Probe Diluent QF.
6. Wash Buffer Component 1.
7. Wash Buffer Component 2.

2.2 ViewRNA Chromogenic Signal Amplification Kit Components (Storage Conditions Are Indicated on Each Container)

1. PreAmp1 QF	DNA in aqueous buffered solution
2. Amp1 QF	DNA in aqueous buffered solution
3. AP Enhancer Solution	Aqueous buffered solution
4. Fast Red Tablets	Red substrate for the detection of alkaline phosphatase activity
5. Naphthol Buffer	Buffer required for preparation of Fast Red substrate
6. Label Probe-AP	Alkaline phosphatase-conjugated oligonucleotide in aqueous buffered solution

Some ViewRNA type 1 probe sets for *Medicago* are already available and can be found and ordered from a catalog online https://www.thermofisher.com/nl/en/home/life-science/cancer-research/cancer-rna-expression/viewrna-tissue-probe-sets.html, and new target-specific probe sets can be designed and synthesized by request at ThermoFisher Scientific.

2.3 Required Equipment

1. ThermoBrite System (110/120 V) and ThermoBrite humidity strips (Leica Biosystems).
2. Hot plate at 42 °C and water bath at 92–95 °C.

3. Shaker.
4. Table-top microtube centrifuge.
5. Dry incubator (oven) capable of maintaining 40 ± 1°C and 58–60°C.
6. Microtome.
7. Microscope (transmitted light and fluorescence).
8. Dissecting microscope.

2.4 Required Materials

1. Staining dish with cover.
2. Removable glass slide racks.
3. Handle for glass slide racks.
4. A stainless steel rack with a fixed handle.
5. Hydrophobic barrier pen.
6. 20 mL glass vials.
7. Slides positively charged (ThermoFisher Scientific).
8. Cover slips 24 × 50 mm.
9. Paraformaldehyde.
10. Glutaraldehyde (25%, EM grade).
11. Paraplast X-TRA.
12. Mounting Citifluor medium.
13. 100% ethanol.
14. Sterile MQ water.
15. 10× PBS.
16. Gill's Hematoxylin I.
17. Histo-Clear II.
18. 37% formaldehyde.
19. 30% ammonium hydroxide.
20. DAPI (4', 6-diamidino-2-phenylindole)

3 Methods

3.1 Tissue Slide Preparation

3.1.1 Fixation

1. The fixative (4% paraformaldehyde mixed with 0.25% glutaraldehyde in 1× PBS (pH 7.2–7.4) has to be made fresh. 600 mg of paraformaldehyde in 7.5 mL of MQ in a tube, incubate at 60°C till it is dissolved. *Paraformaldehyde is toxic; this solution should therefore be prepared in a fume hood. Paraformaldehyde is stored at 4 °C and should be bought new every 6–9 months.* Add 40 mL of 1 N NaOH. *Cool in ice before adding* 150 μL of 25% of glutaraldehyde. Add 1.5 mL of 10× PBS, check pH which should be around 7.2–7.4, and then add MQ till 15 mL.

2. Immediately place dissected tissues (e.g., roots, nodule primordia or nodules) in freshly prepared fixative in glass scintillation vials on ice. *Volume of fixative should be at least 10 higher than specimen volume.*
3. Apply a vacuum on tissues for 15–30 min to improve infiltration and to degas the tissues. Watch plant material for "boiling" of fixative. If this happens plant material might get out of the fixative; turn off the pump, and let the vial stay under vacuum for 15–30 min (if necessary knock plant material back into fixative). Repeat this step until the tissues sink after release of vacuum. Then keep plant material in fixative overnight at 4 °C.

3.1.2 Dehydration

1. Remove fixative, and pass plant material through the following dehydration steps at room temperature (RT), each step 30 min: 1× PBS, MQ, then the following ethanol series 10%, 30%, 50%, 70%, 90%, and 96% ethanol +0.1% eosin B that stains the cell wall making the samples more visible during embedding and sectioning, and finally 100% ethanol for 3 × 1 h (plant material can be left overnight in 100% ethanol). Ethanol series are made of 100% ethanol and sterile MQ water. *It is recommended to use a large excess of each solution!*
2. Plant material can be stored in 70% ethanol or 100% ethanol at −20 °C for months without significant loss of RNA.

3.1.3 Infiltration and Embedding in Paraffin

1. Remove ethanol, and replace subsequently with Histo-Clear II: 100% ethanol mixture—1:3, 1:1, 3:1, and finally 3 times 100% Histo-Clear II. Incubate for 1 h for each step at RT.
2. Replace Histo-Clear once more with 10–15 mL of fresh Histo-Clear and add four to five Paraplast chips, and incubate overnight on a shaker at RT. Melt paraffin in glass beaker covered with aluminum foil in oven at 58–60 °C overnight.
3. On the next day, add more Paraplast chips (~½ volume of the paraffin-Histo-Clear II solution), and transfer the vials to the oven at 58–60 °C.
4. After 5–6 h, replace the paraffin-Histo-Clear mixture with pure molten paraffin, and incubate overnight.
5. On the next day, replace paraffin with a new portion of melted paraffin, and incubate samples at 58–60 °C for 2 days. During these 2 days, paraffin has to be replaced at least five to six times.
6. Set up a hot plate at 60 °C. Pre-warm an aluminum foil embedding mold at 60 °C on this plate. Gently stir the samples, and pour into the mold. If required add molten paraffin. Position plant material with a needle. Plant samples should be oriented correctly for sectioning. At the end, switch off the plate. When

the paraffin is solidified, the mold can be stored at 4 °C in square Petri dishes (120 × 120 mm) sealed with Parafilm. We have stored paraffin molds up to 1 year without significant loss of RNA.

3.1.4 Preparation of Slides

1. Cut a block of paraffin from the mold so that the cutting surface has a trapezoid shape. The sample is orientated so that cutting of the trapezoid goes from its long side to its short side. This will produce a ribbon containing adjacent sections of the sample.
2. Melt some paraffin, and cover the surface of a block holder. Allow this to set than place a drop of molten paraffin in the middle of the block holder and gently press on the paraffin block containing the sample. This is best done by gently holding the block in a pair of forceps. Use a heated spatula to melt paraffin around the edge of the block to produce a good seal.
3. Place the block onto the microtome such that the longer of the two parallel faces is at the bottom. Carefully bring the block forward to the blade, and make sure that the surface of the block is parallel to the blade.
4. Cut 5 ± 1 μm thick sections by using a microtome (with disposable knife).
5. Collect paraffin ribbon (shining side down) in a paper box.
6. Choose the desired sections with a dissecting microscope and stretch them in a drop of freshly autoclaved MQ in a center of glass slides on hot plate at 42 °C. Area of ~20 × 30 mm can be occupied by sections. Routinely, we place five to eight sections from three to four different blocks on one slide (Fig. 2a).
7. Remove excess of water by tilling slides gently, and keep them at room temperature vertically in a rack overnight.
8. Place ThermoBrite humidity strips in cylinder with MQ water.
9. On the next day, select the best slides, and label them with a pencil.
10. Set the ThermoBrite hybridization system to 60 ± 1 °C. Keep the lid open, and bake slides for 60 min to immobilize tissue sections.

After this procedure, slides can be kept in a box at −20 °C for months without significant loss of RNA; avoid multiple freezing-thawing steps.

3.1.5 Removal of Paraffin

1. Pour 200 mL of Histo-Clear II into a glass staining dish, and insert an empty slide rack.
2. Set the hybridization system to 80°C, keep the lid open.
3. Heat the slides for 3 min at 80 °C to melt the paraffin.

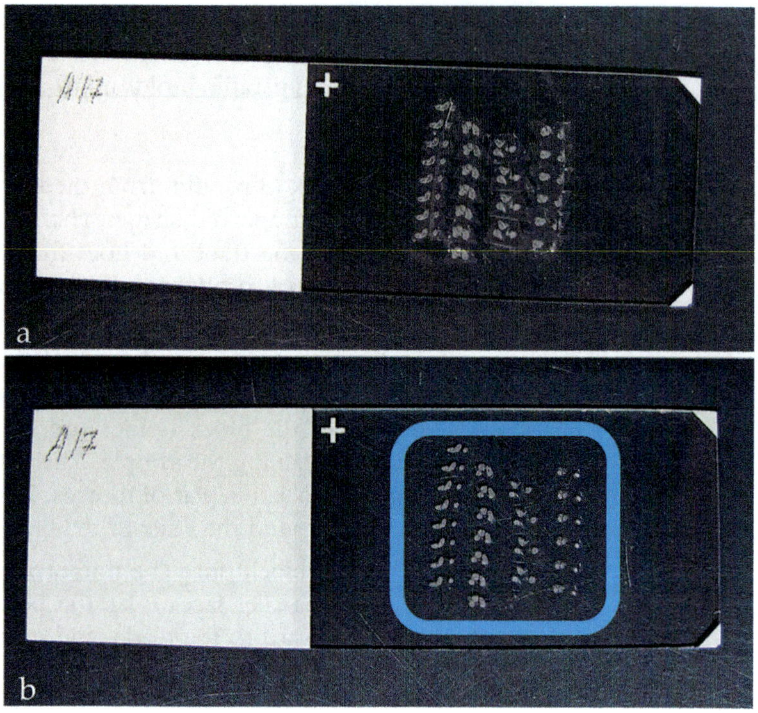

Fig. 2 Positioning of nodule sections on a slide. Sections must occupy not more than 20 × 30 mm in the center and not to be too close to the edges of a slide (**a**). In this way, a hydrophobic barrier around sections can be created after deparaffinization (**b**)

4. Immediately insert the warm slides into the Histo-Clear II, and agitate frequently by moving the rack up and down for 5 min at RT. Work in a fume hood.
5. Discard the used Histo-Clear II, and refill the dish with another 200 mL of fresh Histo-Clear II.
6. Agitate frequently by moving the rack up and down for another 5 min at RT.
7. Remove the slide rack from the Histo-Clear II, and wash the slides twice, each time with 200 mL of 100% ethanol for 5 min with frequent agitation.
8. Remove the slides from the rack, and place them face up on a paper towel to air dry at RT for 5 min. These slides can be stored in a slide box at RT for up to 2 weeks or at −20 °C for up to 1 year (*avoid multiple freezing-thawing steps*). Create a hydrophobic barrier around sections (Fig. 2b). Allow barrier to dry at RT for 20–30 min.

3.2 Prehybridization Treatment

1. Cover the beaker containing the 500 mL of 1× pretreatment solution (5 mL 100× pretreatment solution and 495 mL MQ in 1 L beaker) with aluminum foil, insert a thermometer in

and place the beaker on an open flame, and heat the solution to a temperature of 96 °C.

2. Load the slides into the stainless steel rack with a fixed handle.
3. Place the beaker in water bath at 92–95 °C. Using a pair of forceps, submerge the slide rack into the heated 1× pretreatment solution, *maintain the temperature of the solution at 90–95 °C for 5 min.*
4. After pretreatment, remove the slide rack with forceps from beaker, and submerge it into a staining dish containing 200 mL of MQ water, and wash for 1 min with frequent agitation.
5. Repeat the wash one more time with 200 mL of fresh MQ water.
6. Transfer the slide rack to a staining dish containing 1× PBS.

Do not let the tissue sections dry out from this point onward!

7. Insert wed stripes in ThermoBrite hybridization system.
8. Prepare the working protease solution diluting 4 μL protease QF in 396 μL 1× PBS (1:100) per slide; calculate reagents according to the number of slides to be run and briefly vortex to mix.
9. Remove each slide, and flick it to remove excess 1× PBS. Without completely drying out the sections, tap the slides on the edge, and then wipe the backside with a soft paper tissue.
10. Place the slides one by one face up on a flat platform, and immediately add 400 μL of the working protease solution onto the tissue sections.
11. Transfer the slides to the hybridization system, and incubate at 40 °C for 15 min.
12. Pour 200 mL of 1× PBS into a staining dish, and insert an empty slide rack into the dish.
13. After the incubation, decant the working protease solution from the slides, insert the slides into the rack, and *wash gently* but thoroughly by moving the rack up and down for 1 min.
14. Repeat the wash one more time with another 200 mL of fresh 1× PBS.
15. Transfer the slide rack to a staining dish containing 200 mL of 4% formaldehyde (178 mL 1× PBS + 22 mL 37% formaldehyde), and fix for 5 min at RT in a fume hood.
16. Wash the slides twice, each time with 200 mL of fresh 1× PBS for 1 min with frequent agitation.

3.3 Hybridization

Performing "hybridization, amplification and detection" we follow directly the protocol of ThermoFisher Scientific, only for some steps, time of incubation is extended.

1. Prepare the working probe set solution including 392 μL of Probe Set Diluent QF (pre-warmed to 40 °C) and 8 μL of ViewRNA type 1 probe set (1:50 dilution, if gene is not very highly expressed, use 1:40 dilution or even 1:30).
2. Remove each slide, and flick it to remove excess 1× PBS. Without completely drying out the sections, tap the slides on the edge, and then wipe the backside with a tissue paper.
3. Place the slides face up on a flat platform, and immediately add 400 μL of working probe set solution to each sample.
4. Transfer the slides to the hybridization system, and incubate at 40 °C for 3 h.
5. In the meantime, prepare 3 L of Wash Buffer and 200 mL Storage Buffer. Wash Buffer is made by mixing 27 mL Wash Comp 1 in 1 L MQ water, then add 7.5 mL Wash Comp 2, and add MQ water till 3 L. Storage Buffer consists of 60 mL Wash Comp 2 and 140 mL MQ water.
6. Insert an empty slide rack into a microscope slide staining dish containing 200 mL of Wash Buffer.
7. After incubation, decant the working probe set solution from the slides, and insert them into the slide rack.
8. Wash the slides three times, each time with 200 mL of fresh Wash Buffer at RT for 2 min with constant and vigorous agitation.
9. Store slides in a staining dish containing 200 mL of Storage Buffer at RT overnight. Cover the dish to prevent evaporation.
10. Insert the ThermoBrite humidity strips in cylinder with MQ water overnight.

3.4 Signal Amplification and Detection

1. Next day pre-warm Amplifier Diluent QF and Label Probe Diluent QF to 40 °C.
2. Briefly spin down the Label Probe 1-AP and place on ice. Bring Fast Red Tablets, Naphthol Buffer, and AP Enhancer Solution to RT.
3. Remove the rack with slides from Storage Buffer, and place it in staining dish with Wash Buffer.
4. Wash slides two times, each time with 200 mL of fresh Wash Buffer at RT for 2 min with constant and vigorous agitation.
5. Dilute PreAmp1 QF 1:100 in pre-warmed Amplifier Diluent QF, and briefly vortex to mix. Scale reagents according to the number of slides to be run.
6. Remove slides one by one from Wash Buffer, tap the slide on its edge, and then wipe the backside with a paper tissue. Place the slides face up on a platform, and immediately add 400 μL of working PreAmp1 solution to each slide.

7. Transfer slides to the hybridization system, and incubate at 40°C for 30 min.
8. Insert an empty slide rack into a clear staining dish containing 200 mL of Wash Buffer.
9. After incubation, decant the working PreAmp1 solution from the slides, and insert them into the slide rack.
10. Wash the slides three times, each time with 200 mL of fresh Wash Buffer at RT for 2 min with constant and vigorous agitation.
11. Prepare the working Amp1 solution diluting 4 mL of Amp1 QF in pre-warmed Amplifier Diluent QF (1:100) per slide, and briefly vortex to mix.
12. Remove each slide, and flick to remove the Wash Buffer. Without completely drying out the sections, tap the slide on its edge, and then wipe the backside with a paper tissue. Place the slides face up on a flat platform, and immediately add 400 μL of working Amp1 solution to each slide.
13. Transfer slides to the hybridization system, and incubate at 40 °C for 20 min.
14. Insert an empty slide rack into a staining dish containing 200 mL of Wash Buffer.
15. After incubation, decant the working Amp1 solution from the slides, and insert them into the slide rack.
16. Wash the slides three times, each time with 200 mL of fresh Wash Buffer at RT for 2 min with constant and vigorous agitation.
17. Briefly vortex and spin down Label Probe-AP before using.
18. Dilute Label Probe-AP 1:1000 in pre-warmed Label Probe Diluent QF, and briefly vortex to mix. Scale reagents according to the number of slides to be run.
19. Remove each slide, and flick to remove the Wash Buffer. Tap the slide on its edge, and then wipe the backside with a paper tissue. Place the slides face up on a flat platform, and immediately add 400 μL of working Label Probe-AP solution to each slide.
20. Transfer the slides to the hybridization system, and incubate at 40 °C for 20 min.
21. Insert an empty slide rack into a clear staining dish containing 200 mL of Wash Buffer.
22. After incubation, decant the working Label Probe-AP solution from the slides, and insert them into the slide rack.
23. Wash the slides three times, each time with 200 mL of fresh Wash Buffer at RT for 3 min with constant and vigorous agitation.

24. Remove each slide, and flick it to remove the Wash Buffer. Tap the slide on its edge, and then wipe the backside with a paper tissue. Place the slides face up on a platform.
25. Immediately add 400 μL of the AP Enhancer Solution to each tissue section, and incubate at RT for 5 min while preparing the Fast Red substrate.
26. Add 5 mL of Naphthol Buffer and one Fast Red Tablet to a 15 mL conical tube. Vortex to completely dissolve the tablet. Protect from light by wrapping the tube in aluminum foil.
27. Pour off the AP Enhancer Solution, and flick the slide *twice* to completely remove any excess AP Enhancer Solution. Tap the slide on its edge, and then wipe the backside with a paper tissue.
28. Immediately add 400 μL of Fast Red substrate onto each slide.
29. Transfer the slides to the hybridization system, and incubate at 40 °C for 40 min.
30. Insert an empty slide rack into a clear staining dish containing 200 mL of 1× PBS.

    After incubation, pour off the Fast Red substrate from the slides, and insert them into the slide rack.
31. Rinse off the excess Fast Red substrate from the slides by moving the rack up and down for 1 min.
32. Transfer the slide rack to the staining dish containing the 200 mL of Gill's hematoxylin and stain for 10 s at RT.
33. Wash the slides three times, each time with 200 mL of fresh MQ water for 1 min by moving the slide rack up and down.
34. Pour off the MQ water, refill with 200 mL of 0.01% ammonium hydroxide in MQ, and incubate the slides for 10 s.
35. Wash the slides once more in 200 mL of fresh MQ water by moving the rack up and down for 1 min.
36. Remove the slides from the slide rack, and flick to remove the excess MQ water. Tap the slide on its edge, and then wipe the backside with a paper tissue. Place slides vertically in a rack. Keep slides in the dark. Slides should be completely dry before mounting.

3.5 Mounting and Imaging

1. Add three drops of Citifluor mixed with 0.5 μg/mL DAPI to each slide, and cover it with a cover slip 24 × 50 mm. *Do not use mounting solutions containing alcohol.*
2. Image the results using a bright-field and/or fluorescence microscope. Store slides in a fridge. The hybridized mRNAs are detected as red dots. Example of an in situ result is presented in Fig. 3.

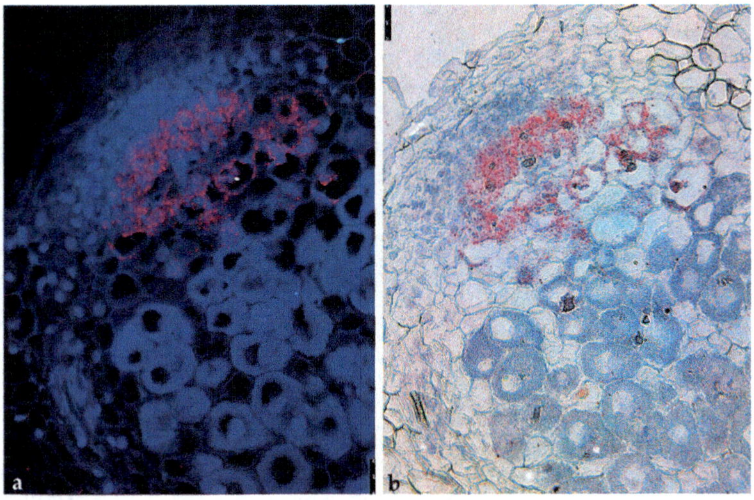

Fig. 3 Result of ViewRNA ISH on *Medicago* nodule section with *Mt*ENOD12 probe set. The hybridized mRNAs are detected as red fluorescence dots under fluorescent microscope using Cy3/TRITC filter set (nodule section is stained with DAPI) (**a**) and as red dots under bright-field microscope (nodule section is stained also with Gill's Hematoxylin I) (**b**). Hybridization signals are detected in the infection zone of a nodule. Scale = 25 μm

4 Notes

All possible troubleshooting and recommended actions are described in details in the manufacturer's user manual of ThermoFisher Scientific. From our own experience, we would like to indicate the following issues:

1. Time for fixation of plant material should not exceed 24 h.
2. Get rid of all air bubbles from tissue during fixation.
3. Gradually embed plant tissue into paraffin.
4. Temperature during incubation in pretreatment solution should be exactly in a range of 90–95 °C.
5. Keep correct order during amplification steps.
6. Post-fixation of tissue sections in 4% paraformaldehyde after protease treatment should not take more than 5 min.
7. When recommended to wash slides with constant and vigorous agitation, do so.
8. Besides, ThermoFisher Scientific includes in the assay two slides with 5 ± 1 μm thick rat kidney sections attached to positively charged slides and type 1 rat *SPP1* probe set (VC1-10477-06). Performing ISH for the first time includes one positive (rat *SPP1* probe set) and one negative (omit probe sets) slide to control the procedure starting from the hybridization.

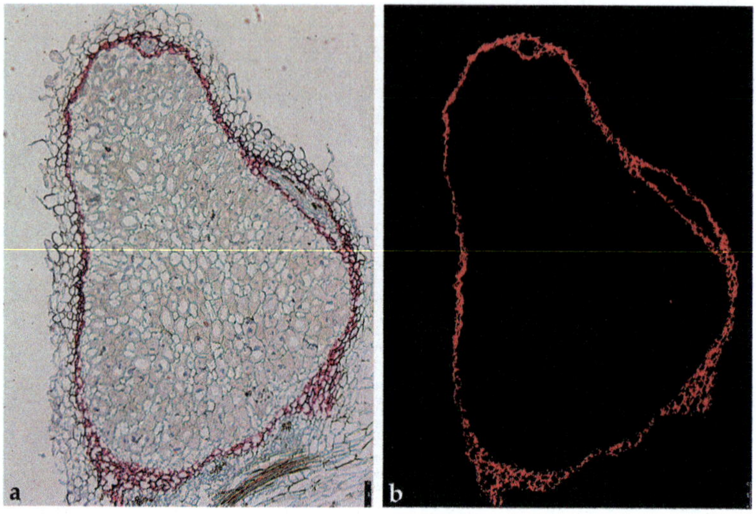

Fig. 4 Result of ViewRNA ISH on *Medicago* nodule section with *MtENOD2* probe set. Hybridization signals are detected in nodule parenchyma. Images have been taken under bright-field microscope (**a**) and fluorescence microscope (**b**). Scale = 25 μm

For these control slides, use a 10 min pretreatment boiling time and a 20 min protease incubation time. *SPP1* is highly expressed in rat kidney, so the positive control slide should show a uniform and high signal. The negative control slide should exhibit a clean background of less than one dot per ten cells. We would also recommend in the first ISH experiment to include one extra slide with *Medicago* nodule sections to detect a highly expressed *Medicago* gene with well-known pattern as a control. *MtENOD2* (catalog number VF1–14233-01, ThermoFisher Scientific) would be the excellent choice. *MtENOD2* is a gene highly expressed in nodule parenchyma (10, Fig. 4). Then, the entire ISH procedure including fixation, embedding in paraffin, slide preparation, and slide pretreatment will be under control.

References

1. Hejátko J, Blilou I, Brewer P, Friml J, Scheres B, Benkova E (2006) *In situ* hybridization technique for mRNA detection in whole mount *Arabidopsis* samples. Nat Protoc 1:1939–1946. https://doi.org/10.1038/nprot.2006.333
2. Cassidy A, Jones J (2014) Developments in *in situ* hybridisation. Methods 70(1):39–45
3. Scheres B, van Engelen F, van der Knaap E, van de Wiel C, van Kammen A, Bisseling T (1990) Sequential induction of Nodulin gene expression in the developing pea nodule. Plant Cell 2:687–700
4. Scheres B, McKhann HI, Zalensky A, Lobler M, Bisseling T, Hirsch AM (1992) The *PsENOD12* gene is expressed at two different sites in Afghanistan pea pseudonodules induced by auxin transport inhibitors. Plant Physiol 100:1649–16554
5. Player AN, Shen L-P, Kenny D, Antao VP, Kolberg JA (2001) Single-copy gene detection using branched DNA (bDNA) *in situ*

Hybridization. J Histochem Cytochem 49(5):603–611
6. Tsongalis GJ (2015) Branched DNA technology in molecular diagnostics. Am J Clin Pathol 126(3):448–453
7. Wang F, Flanagan J, Su N, Wang L-C, Bui S, Nielson A, Wu X, Vo H-T, Ma X-J, Luo Y (2012) RNAscope: A novel *in Situ* RNA analysis platform for formalin-fixed, paraffin-embedded tissues. J Mol Diagn 14(1):22–29
8. Roux B, Rodde N, Jardinaud M-F, Timmers T, Sauviac L, Cottret L, Carrere S, Sallet E, Courcelle E, Moreau S, Debelle F, Capela D, de Carvalho F, Niebel A, Gouzy J, Bruand C, Gamas P (2014) An integrated analysis of plant and bacterial gene expression in symbiotic root nodules using laser-capture microdissection coupled to RNA sequencing. Plant J 77:817–837
9. Franssen HJ, Xiao TT, Kulikova O, Wan X, Bisseling T, Scheres B, Heidstra R (2015) Root developmental programs shape the *Medicago truncatula* nodule meristem. Development 142:2941–2950
10. van de Wiel C, Scheres B, Franssen H, Lierop MJ, Lammeren A, van Kammen A (1990) The early nodulin transcript ENOD2 is located in the nodule parenchyma (inner cortex) of pea and soybean root nodules. EMBO J 9(1):1–7

Chapter 12

Editing the *Medicago truncatula* Genome: Targeted Mutagenesis Using the CRISPR-Cas9 Reagent

Shaun J. Curtin

Abstract

Medicago truncatula is an annual plant used for studying legume biology, in particular symbioses with nitrogen-fixing rhizobia and arbuscular mycorrhizal fungi. Efforts to decipher the genetic basis of these ecologically and economically important traits are a major goal of plant and crop biology. *M. truncatula* is an excellent model system for this purpose, as it has several publicly available sequenced genomes, has a rapid seed-to-seed generation time, and is highly transformable. Various mutagenesis platforms such as Tnt1 retrotransposons and RNAi knockdown have been used successfully in forward and reverse genetic studies to identify and functionally characterize candidate genes. The CRISPR/Cas9 reagent is the most recent mutagenesis platform and is highly effective at generating site-directed double-stranded breaks (DSB) in *M. truncatula*. This protocol will demonstrate the construction of reagents using two genome engineering platforms that have successfully generated mutant plants in *M. truncatula*, *M. sativa*, and soybean systems. The reagents are easy to assemble, can be quickly retrofitted to test novel regulatory sequences for improved efficiency, and can be used for more advanced genome engineering strategies such as gene insertion or gene replacement.

Key words *Medicago*, CRISPR/Cas9, Gene knockout, Large site-directed deletions

1 Introduction

Site-directed mutagenesis technologies include reagents that induce *in planta* double-strand breaks (DSBs) at specific genetic loci, resulting in random sequence modifications. The DSB is rapidly repaired by the host's error-prone nonhomologous end-joining (NHEJ) repair pathway that oftentimes leads to nucleotide insertion/deletions (indels) and gene disruption [1].

The CRISPR/Cas9 is the most recent genome engineering reagent used to generate targeted DSBs in plants. It is a two-component system consisting of the Cas9 protein with DNA cleavage capability and a sequence determinant single-stranded guide RNA (gRNA) [2, 3]. It differs from previous platforms, such as the zinc finger nuclease (ZFN) [4] and TAL effector nuclease (TALEN)

Luis A. Cañas and José Pío Beltrán (eds.), *Functional genomics in* Medicago truncatula*: Methods and Protocols*, Methods in Molecular Biology, vol. 1822, https://doi.org/10.1007/978-1-4939-8633-0_12,

[5], in that it can be reprogrammed to target specific DNA by simply altering the 20 nucleotides located at 5′ end of the gRNA molecule, referred to as the gRNA spacer. It is this sequence that anneals to the target DNA adjacent to a NGG protospacer adjacent motif (PAM) according to Watson-Crick base pairing.

The gRNA molecules are expressed from a RNA polymerase III promoter such as the U6 or 7sL transcription unit and are introduced into the cell along with a Cas9 [6]. This is an efficient approach to induce single-gene knockouts, and its construction is described in Subheading 3.1. Alternatively, gRNAs can be expressed using a constitutive RNA polymerase II promoter co-expressed with Cas9 and an RNA-cleaving enzymes such as Csy4 [7]. This strategy allows for "mutliplexing," whereby a single reagent can express multiple gRNAs. The Csy4 enzyme cleaves the multiplex array transcript and releases gRNAs at equimolar levels. Multiple gene knockouts and large deletions have been observed in single transgenic events using this reagent, and its construction is described in Subheading 3.2.

To date, CRISPR/Cas9 has been used to edit the genomes of a wide range of organisms including human cells, *Caenorhabditis elegans*, zebrafish, *Drosophila*, mice, and rat, as well as many plant species, including several crops: *Arabidopsis*, tobacco, soybean, *Medicago*, tomato, rice, wheat, maize, and sorghum [6–18]. This protocol details the design and construction of CRISPR/Cas9 vectors for single- and multigene knockout using two versatile genome engineering platforms. While the focus of this protocol is on CRISPR/Cas9 mutagenesis of legumes, an extensive collection of reagents and protocols optimized for other plant species can be viewed at (http://cfans-pmorrell.oit.umn.edu/CRISPR_Multiplex/). For more advanced genome engineering strategies such as gene insertion and gene replacement, a recently published protocol is highly recommended [19].

2 Materials

2.1 Materials for Genome Engineering Platforms

2.1.1 Molecular Biology Reagents

1. Vectors from the CRISPR/Cas9 Gateway from platform [18, 20] and the Golden Gate multipurpose plant genome engineering kit [7] are available from Addgene, Cambridge, MA, USA (http://www.addgene.org/).
2. T4 polynucleotide kinase + T4 DNA ligase buffer (contains ATP).
3. LR Clonase II (Life Technologies).
4. Proofreading DNA polymerase and buffer such as Phusion® High-Fidelity DNA Polymerase.
5. Taq DNA polymerase and buffer.

6. AarI restriction enzyme and AarI oligonucleotide.
7. Esp3I restriction enzyme.
8. SapI restriction enzyme.
9. BsaI restriction enzyme.
10. T7 DNA ligase and 2× T7 ligase buffer.
11. T4 DNA ligase and 10× T4 DNA ligase buffer.

2.1.2 Stock Solutions and Media

1. LB medium: 5 g/L yeast extract, 20 g/L tryptone, 10 g/L sodium chloride.
2. 100 mg/mL carbenicillin stock.
3. 100 mg/mL kanamycin stock.
4. 100 mg/mL spectinomycin stock.
5. LB plates/liquid media with 100 mg/L carbenicillin (ampicillin), 100 mg/L kanamycin, or 100 mg/L spectinomycin.
6. 40 mg/mL X-gal dissolved in *N'*,*N*-dimethylformamide.

2.1.3 Other Supplies

1. dNTPs.
2. Gel electrophoresis equipment.
3. QIAprep Spin Miniprep Kit (Qiagen).
4. QIAquick Gel Extraction Kit (Qiagen).
5. Plasmid map viewing such as Geneious™ or Vector NTI™.
6. Sequence analysis (MEGA).

2.1.4 Cell Strains

1. *E. coli* DH5α chemically competent cells.
2. *E. coli* ccdB resistant strain chemically competent cells such as DB3.1.

2.2 Gateway Platform

1. Order pSC218GG (#104779), pNB184 (#104786), and pAH595 (#104784) vectors for the Gateway platform [18, 20] from Addgene, Cambridge, MA, USA (http://www.addgene.org/) (Fig. 1a).
2. Complementary oligos containing the gRNA target sequence.
3. T4 polynucleotide kinase + T4 DNA ligase buffer (contains ATP).
4. T4 DNA ligase and 10× T4 DNA ligase buffer.
5. Gel electrophoresis equipment.
6. DH5α chemically competent cells.
7. LB medium.
8. 100 mg/mL kanamycin or spectinomycin stock.
9. 40 mg/mL X-gal dissolved in either dimethyl sulfoxide or *N'*,*N*-dimethylformamide.

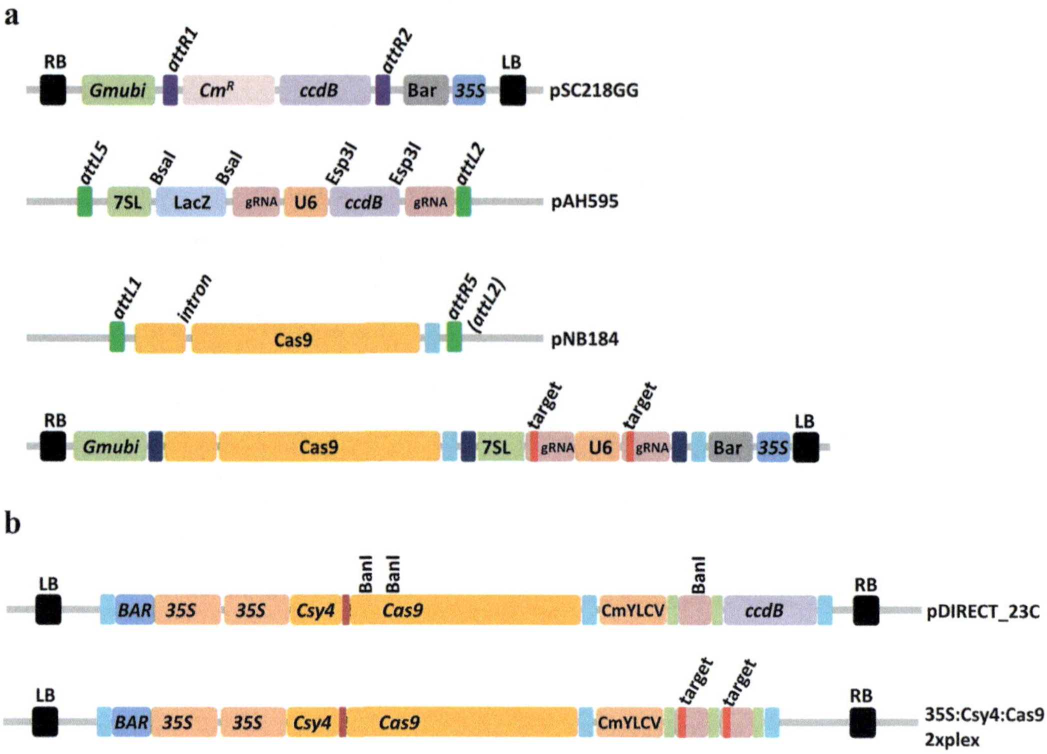

Fig. 1 The Gateway and Golden Gate reagent platforms for *M. truncatula* mutagenesis. (**a**) Schematic representations of the Gateway platform components, the empty destination vector (pSC218GG), and the entry clone (pAH595) used to express target gRNAs. The "target2" is first cloned into the Esp3I sites followed by cloning of "target1" into the BsaI sites. The second entry vector contains the Cas9 expression cassette with a synthetic intron. The two entry vectors are assembled into the destination vector by MultiSite Gateway® Pro 2-Fragment LR recombination reaction. (**b**) The direct cloning vectors used for *Medicago* mutagenesis are from the Voytas Lab Plant Genome Engineering Toolkit. These vectors are designed to speed up the cloning process as the gRNAs can be directly cloned into a transformation backbone that harbors a Cas9 expression cassette and a selectable marker of choice. The empty pDIRECT_23C vector has a Csy4:P2a:Cas9 fusion cassette and 35S:Bar selectable marker and can clone up six target gRNAs

10. LB liquid medium with 100 mg/L kanamycin or spectinomycin.
11. QIAprep Spin Miniprep Kit (Qiagen).
12. Primers used for Gateway targets.

    Pho2ab-F2-(Esp3I-target2) GATTGATTAGAACCTAAAG GAACCGG

    Pho2ab-R2 AAAACCGGTTCCTTTAGGTTCTAATC

    Pho2ab-F1-(BsaI-target1) GATCGACAGTCCCTACATT TGAAGAG

    Pho2ab-R1 AAAACTCTTCAAATGTAGGGACTGTC
13. Primers for sequencing:

AtU6 F1: 5′- AGAACAATAGTATTTCTTATATAGG (pAH595 vector)

At7sL F1: 5′- CTATAATGGGACTCAAAATAAGG (pAH595 vector)

Gmubi F1c: 5′- CGTTGATTCTATATCGTTTTTCG (pSC218GG vector)

OCS R2: 5′-CGAACGGATAAACCTTTTCACG (pSC218GG vector)

2.3 Golden Gate Platform

1. The pDIRECT_23C (#91140) vector efficiently generates multigene knockouts or chromosome deletions in *M. truncatula* and is a component of the multipurpose plant genome engineering kit [7]—available from Addgene, Cambridge, MA, USA (http://www.addgene.org/) (Fig. 1b).
2. Primers are designed using the online tools available at http://cfans-pmorrell.oit.umn.edu/CRISPR_Multiplex/.
3. Phusion® High-Fidelity DNA Polymerase and buffer.
4. dNTPs.
5. Restriction enzyme BanI.
6. Gel electrophoresis equipment.
7. QIAquick Gel Extraction Kit (Qiagen).
8. Restriction enzyme Esp3I
9. Restriction enzyme SapI.
10. T7 DNA ligase and 2× T7 DNA ligase buffer.
11. DH5α chemically competent cells.
12. LB medium.
13. 100 mg/mL kanamycin.
14. LB liquid medium with 100 mg/L kanamycin.
15. Taq DNA polymerase and buffer.
16. QIAprep Spin Miniprep Kit (Qiagen).
17. Primers used for Golden Gate targets (From output in Subheading 3.2, **step 5**).

    PCR reaction #1

    oCmYLCV

    TGCTCTTCGCGCTGGCAGACATACTGTCCCAC

    CSY_Pho2ab-target1

    TCGTCTCCTGTAGGGACTGTCTGCCTATACGGCAGTGAAC

    PCR reaction #2

    REP_Pho2ab-target1

TCGTCTCATACATTTGAAGAGTTTTAGAGCTAGAAATAGC

CSY_Pho2ab-target2

TCGTCTCCTTAGGTTCTAATCTGCCTATACGGCAGTGAAC

PCR reaction #3

REP_Pho2ab-target2

TCGTCTCACTAAAGGAACCGGTTTTAGAGCTAGAAATAGC

CSY_term

TGCTCTTCTGACCTGCCTATACGGCAGTGAAC

18. Primers for colony PCR and sequencing:

    TC320: 5′-CTAGAAGTAGTCAAGGCGGC

    M13F: 5′-GTAAAACGACGGCCAGT

    TC430: GTTGGATCTCTTCTGCAGCA

3 Methods

3.1 The Design and Construction of the Gateway™ Reagent for Single-Gene Knockouts

This protocol describes the construction the Gateway™ CRISPR/Cas9 reagent that uses the *Glycine max* ubiquitin (Gmubi) promoter to drive constitutive expression of Cas9. The platform consists of two entry vectors and a destination binary vector that are assembled by MultiSite Gateway® Pro 2-Fragment LR recombination reaction. The first entry vector is a promoter-less *Arabidopsis* optimized Cas9 with a synthetic intron (*att*L1-Cas9-Intron-*att*R5), and the second entry vector has the two gRNA targets each expressed by the *Arabidopsis* regulatory promoters 7sL and U6, respectively (*att*L5-At7sL:target1:gRNA and AtU6:target2gRNA-*att*L2). The destination vector is comprised of a Gmubi promoter with recombination sites compatible with the two entry vectors (Gmubi:*att*R1:Gateway cassette:*att*R2) (Fig. 1a). The system is interchangeable, and the destination vector can be exchanged for the AtUBQ10 or *rol*D promoters for improved germline or callus expression (available from Addgene). To date, the Gmubi expressed Cas9 reagent has successfully generated multiple whole plant mutants in *M. truncatula* and soybean (Fig. 2) [18, 20].

1. Purify plasmid DNA from 10 mL LB culture for pSC218GG and 5 mL LB cultures for pNB184 (*see* **Note 1**).
2. Select targets in the first and second exon using the $N_{20}NGG$ template. Alternatively, online tools such as CRISPR-P

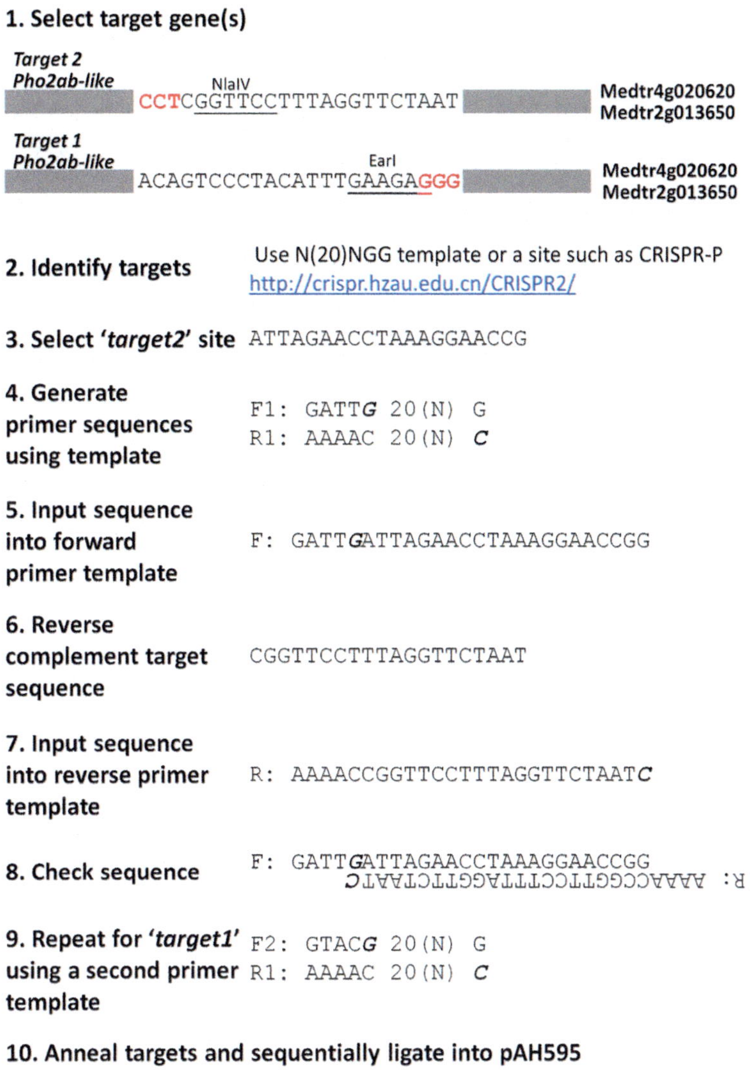

Fig. 2 A construction guide for the Gateway platform. Two targets in the *Medicago Pho2a-like* and *Pho2b-like* genes are used as examples for constructing the Gateway reagent

(http://crispr.hzau.edu.cn/CRISPR2/) [21] are useful for target site selection, prediction of on-target activity, identification of potential off-targets, and restriction enzyme site selection for screening T_0 plants (*see* **Note 2**).

>Pho2ab-target1

ACAGTCCCTACATTTGAAGA

>Pho2ab-target2

ATTAGAACCTAAAGGAACCG

3. Design two complementary oligonucleotides for "target1" and "target2" using the primer templates as shown in Fig. 2. The specific 4 bp overhang should match the appropriate promoter, either At7sL (GTAC) or AtU6 (GATT). See Subheading 2.2, **item 12** for primer sequences.
4. Phosphorylate the oligonucleotides. Prepare the following reaction:
   - 3 μL 10× T4 DNA ligase buffer
   - 3 μL 100 μM forward gRNA oligonucleotide
   - 3 μL 100 μM reverse gRNA oligonucleotide
   - 2 μL T4 polynucleotide kinase
   - 19 μL H_2O

   Incubate at 37 °C/1 h.
5. Anneal the phosphorylated oligonucleotides. Place the reaction from **step 2** in a PCR machine and run the following program: 100 °C/2 min + allow to slowly cool to room temperature.
6. Dilute the reaction by 25 times (1 μL phosphorylated oligo mixture +24 μL H_2O).
7. First insert the "target2" annealed oligonucleotides into the Esp3I sites of pAH595 via Golden Gate cloning. This reaction will include:
   - 2 μL 10× T4 DNA ligase buffer
   - 50 ng of selected pAH595 plasmid
   - 1 μL of 25× diluted phosphorylated and annealed "target2" oligo mixture
   - 0.5 μL Esp3I
   - 1 μL T4 DNA ligase
   - H_2O up to 20 μL
8. Place the reaction in a PCR machine and run the following program: 37 °C/5 min + 16 °C/10 min + 37 °C/15 min + 80 °C/5 min
9. Transform 5 μL of the reaction into *E. coli*. The reaction is highly efficient since the Esp3I digestion replaces a ccdB selection cassette with the "target2" annealed oligonucleotides. It is important to use DH5α or similar *E. coli* strain sensitive to the presence of ccdB gene and plate on LB + 100 mg/L spectinomycin. Incubate at 37 °C overnight.
10. The following day inoculate 2–3 colonies into 5 mL of LB supplemented with 100 mg/L spectinomycin. Incubate and shake overnight at 220 rpm, 37 °C.

11. In the following day, purify plasmids from the LB cultures and insert the "target1" annealed oligonucleotides into the BsaI sites of pAH595/target1 via Golden Gate cloning. This reaction will include:
    - 2 μL 10× T4 DNA ligase buffer
    - 50 ng of selected pAH595/target1 plasmid
    - 1 μL of 25× diluted phosphorylated and annealed "target1" oligo mixture
    - 0.5 μL BsaI
    - 1 μL T4 DNA ligase
    - H_2O up to 20 μL
12. Place the reaction in a PCR machine and run the following program: 37 °C/5 min + 16 °C/10 min + 37 °C/15 min + 80 °C/5 min
13. Transform 5 μL of the reaction into *E. coli*, use DH5α or similar *E. coli* strain sensitive to the presence of ccdB gene, and plate on LB + 100 mg/L spectinomycin with X-gal and IPTG. Incubate at 37 °C overnight.
14. The following day inoculate 2–3 white colonies into 5 mL of LB supplemented with 100 mg/L spectinomycin. Incubate and shake overnight at 220 rpm, 37 °C.
15. The following day purify plasmids from the LB cultures and verify by sequencing using the primer AtU6 F1 and At7sL F1.
16. Perform MultiSite Gateway® Pro 2-Fragment LR recombination reaction to complete reagent. Set up the following LR Clonase II reaction:
    - 50–150 ng pNB184 (Cas9-Intron) entry clone
    - 50–150 ng pAH595/target1/target2 entry clone (sequence confirmed)
    - 150 ng pSC218GG destination vector
    - TE buffer pH 8.0, up to 8 μL
    - 2 μL LR Clonase II
17. Incubate overnight at room temperature and transform the reaction into *E. coli* (DH5α or similar, but sensitive to the presence of ccdB gene) and plate on LB + 100 mg/L kanamycin.
18. The following day inoculate 1–3 colonies into 10 mL of LB supplemented with 100 mg/L kanamycin.
19. The following day purify plasmids from the LB cultures and sequence confirm using AtU6 F1, At7sL F1, Gmubi F1c, OCS R2 primers prior to *Agrobacterium* transformation.

3.2 Design and Construction of Reagents for Multigene Knockouts and Chromosome Deletions

This protocol describes cloning of multiple target gRNA arrays (2–6 gRNAs) into the pDIRECT_23C binary vector. This vector houses the 35S:Csy4:P2A:Cas9L: CmYLCV:Csy4:gRNA:Csy4 and 35S:bar cassettes. The pDIRECT_23C vector is used as both the cloning vector and the template for PCR amplification of individual target gRNA components. Targets are cloned by Golden Gate assembly of amplicons containing gRNA spacer sequences and Csy4 binding sites using SapI and Esp3I restriction enzymes. SapI opens the vector backbone, and Esp3I creates complementary overhangs on each amplicon to allow for ligation in the appropriate order and orientation. New targets are created by generating primer sequences using a dedicated online tool. When the target gRNA array transcript is expressed along with the Csy4 ribonuclease, individual gRNA units are cleaved and released into the cell along with Cas9. Once sequence confirmed, assembled vectors are ready for *Agrobacterium* and plant transformation (Fig. 1b).

1. Select the vector by opening the link to the Webtools for the Voytas Lab Plant Genome Engineering Toolkit (http://cfans-pmorrell.oit.umn.edu/CRISPR_Multiplex/vector.php) and clicking on the "Vector Selection" or "List All Vectors" link. To select a "Direct Cloning Backbone", follow **step 1** in Subheading 3.1, but select "35S:Csy4-P2A-AtCas9 + CmYLCV:gRNAs with Csy4 spacers" in **Step 4** of the selection process (*see* **Note 3**).

   Vector Selection Summary

   Step 1: Direct Cloning Backbone

   Step 2: T-DNA (pCAMBIA based, kanamycin for bacterial selection)

   Step 3: 2 × 35S:bar

   Step 4: 35S:Csy4-P2A-AtCas9 + CmYLCV:gRNAs with Csy4 spacers

   Vector Code: pDIRECT_23C [Download] (GenBank Flat File)

2. Order selected plasmids from Addgene, and purify plasmid DNA from 5 mL LB cultures using the QIAprep Spin Miniprep Kit (*see* **Note 4**).
3. Select targets for the gene(s) of interest (GOI) (*see* **Note 5**) or two targets flanking the gene's genomic locus to be deleted. The N_{20}NGG template can be used to select targets with no constraints on the first nucleotide. Online tools such as CRISPR-P (http://cbi.hzau.edu.cn/cgi-bin/CRISPR) [21] are excellent target site selection resources. These sites can predict on- and off-target activity and identify restriction sites for screening T_0 plants.

4. Create a list of target DNA sequences in FASTA format, using the first 20 nucleotides (N_{20}) of each target excluding PAM sequence, and save as a ".txt" file. See the example for a 2xplex reagent below. In this example, the two *Medicago* targets *Pho2a* and *Pho2b* genes are used [18].

   >Pho2ab-target1

   ACAGTCCCTACATTTGAAGA

   >Pho2ab-target2

   ATTAGAACCTAAAGGAACCG

5. Open the "Primer Design and Map Construction" tool at (http://cfans-pmorrell.oit.umn.edu/CRISPR_Multiplex/assembly.php). Use the "Browse..." button to select the "Pho2ab" text file, and select the target vector "pDIRECT23_C," promoter "CmYLCV," restriction enzyme "Esp3I," and splicing system "Csy4," and click "Submit." An example of the output generated by the program can be found in Subheading 2.3, **item 17**.
6. Order primers from the output and cut and paste or download vector sequences into a vector map software for reference.
7. Prepare PCR template for amplification of the first gRNA unit containing the promoter by digesting the target vector with the restriction enzyme BanI (*see* **Note 6**). Mix about ~2 μg of the pDIRECT_23C plasmid DNA with 2 μL of CutSmart™ buffer and 1 μL of BanI enzyme, and add ddH_2O to 20 μL. Incubate the reaction at 37 °C for 1 h, and run on an agarose gel.
8. Extract the promoter and gRNA scaffold-containing fragment from the gel using the QIAquick Gel Extraction Kit. For pDIRECT_23C, the fragment size will 4637-bp (*see* **Note 7**).
9. Amplify the gRNA units by Phusion DNA polymerase (NEB). Set up a PCR reaction for each primer pair designed by the Primer Design and Map Construction tool in **step 5**. Use the BanI fragment isolated in **step 8** as the template in the PCR reaction 1 and the undigested vector in all other PCR reactions:
   - 10 μL 5× Phusion HF buffer
   - 1 μL 10 mM dNTPs
   - 5.0 μL 10 mM forward and reverse primer pair
   - 5–20 ng template
   - 0.5 μL Phusion DNA polymerase (NEB) (*see* **Note 8**)
   - H_2O up to 50 μL
10. Place the PCR reaction in a thermocycler and run the following program:

98 °C/1 min + 30× (98 °C/10 s + 60 °C/15 s + 72 °C/1 5 s) + 72 °C/2 min + 4 °C hold

11. Confirm successful amplification by running 5 μL of each PCR product on a gel. The amplicon sizes should be ~500 bp for the CmYLCV promoter and ~150 bp for target gRNAs.
12. Dilute each PCR product ten times with dH_2O (*see* **Note 9**). Do not purify the PCR products.
13. Assemble the promoter and gRNA amplicons into pDIRECT_23C. Prepare the following reaction:
    - 10 μL 2× T7 DNA ligase buffer.
    - 50 ng of pDIRECT_23C.
    - 0.5 μL of each 10× diluted PCR product.
    - 0.5 μL SapI.
    - 0.5 μL Esp3I.
    - 1 μL T7 DNA ligase.
    - H_2O up to 20 μL.
    - Mix the reaction by pipetting up and down several times.
14. Place reaction in a thermocycler and run the following program: 10× (37 °C/5 min + 25 °C/10 min) + 4 °C hold (*see* **Note 10**).
15. Transform 5 μL of the Golden Gate reaction into *E. coli* (DH5α or similar, but sensitive to the presence of ccdB gene), and plate on LB + 100 mg/L kanamycin, and incubate at 37 °C overnight.
16. The following day confirm correct assembly of the reagent by screening 8–10 colonies (*see* **Note 11**). Use a Taq polymerase for colony PCR assay with TC320 and M13F primers.
17. Separate the PCR reactions by electrophoresis. The amplicons from correctly assembled clones will ladder according to the number of target gRNAs. For the 2xplex example vector, the most abundant amplicon size should be 576-bp.
18. Inoculate 1–2 PCR positive clones into 5 mL of LB supplemented with 100 mg/L of kanamycin antibiotic. Incubate and agitate overnight at 220 rpm, 37 °C.
19. The following day, purify plasmids from the LB cultures and sequence confirm using TC320, TC430, and M13F primers prior to *Agrobacterium* transformation.

4 Notes

1. pSC218GG is a low copy number plasmid that requires a 10 mL Miniprep.

2. It is recommended to identify targets that have a restriction site that can be disrupted by mutagenesis. This is useful for screening of T_0 mutant plants.
3. The preferred multi-gRNA expression system uses the Csy4 ribonuclease to release individual gRNAs units from the transcript. This system offers the highest efficiency in *Medicago*; however, there are two other systems available, using tRNA or ribozyme-based processing.
4. Prepping the plasmids with kits from other suppliers may reduce efficiency of the reaction.
5. The majority of the blunt DNA breaks created by CRISPR/Cas9 are repaired precisely. To prevent precise restoration of the original sequence, it is recommended to target each gene for deletion with two (or more) gRNAs. This appears to increase the efficiency of mutagenesis and can simplify T_0 mutant screening.
6. Due to the presence of two Csy4 repeats in each template vector, two products can be amplified when amplifying the first gRNA unit since the reverse primer binds to the repeated sequence. This is prevented by using BanI digested plasmid as the template. BanI cleaves the gRNA repeat sequence and separates the two Csy4 repeats.
7. The purified fragment can be stored at −20 °C and used for multiple reactions.
8. The protocol has good results using Phusion DNA polymerase. Taq polymerase should be avoided for this step.
9. The use of non-diluted products decreases cloning efficiency.
10. Do not heat inactivate T7 ligation reaction as it decreases transformation efficiency due to the presence of PEG in the reaction buffer.
11. The cloning efficiency decreases with the number of gRNA spacers cloned and may be as low as 50–60% with six gRNAs. Therefore we recommend screening 8–10 colonies. The cloning efficiency is much higher when cloning 2–3 gRNA spacers, therefore screening 2–4 colonies is sufficient.

Acknowledgments

Thanks to members of the Dan Voytas laboratory, especially Nicholas Baltes, Paul Atkins, and Tomas Čermák, for the development and optimization of these reagents. Thanks to Robert Stupar for the critical reading of the manuscript and helpful comments.

References

1. Voytas DF (2013) Plant genome engineering with sequence-specific nucleases. Annu Rev Plant Biol 64:327–350. https://doi.org/10.1146/annurev-arplant-042811-105552
2. Jinek M, Chylinski K, Fonfara I et al (2012) A programmable dual-RNA-guided DNA endonuclease in adaptive bacterial immunity. Science 337:816–821. https://doi.org/10.1126/science.1225829
3. Cong L, Ran F, Cox D et al (2013) Multiplex genome engineering using CRISPR/Cas systems. Science 339:819–823. https://doi.org/10.1038/nbt1319
4. Carroll D (2011) Genome engineering with zinc-finger nucleases. Genetics 188:773–782. https://doi.org/10.1534/genetics.111.131433
5. Christian M, Cermak T, Doyle EL et al (2010) TAL effector nucleases create targeted DNA double-strand breaks. Genetics 186:757–761. https://doi.org/10.1534/genetics.110.120717
6. Shan Q, Wang Y, Li J et al (2013) Targeted genome modification of crop plants using a CRISPR-Cas system. Nat Biotechnol 31:686–688. https://doi.org/10.1038/nbt.2650
7. Cermak T, Curtin SJ, Gil-Humanes J et al (2017) A multipurpose toolkit to enable advanced genome engineering in plants. Plant Cell 29:1196–1217. https://doi.org/10.1105/tpc.16.00922
8. Li J-F, Aach J, Norville JE et al (2013) Multiplex and homologous recombination-mediated plant genome editing via guide RNA/Cas9. Nat Biotechnol 31:688–691. https://doi.org/10.1038/nbt.2654
9. Nekrasov V, Staskawicz B, Weigel D et al (2013) Targeted mutagenesis in the model plant Nicotiana benthamiana using Cas9 RNA-guided endonuclease. Nat Biotechnol 31:691–693. https://doi.org/10.1038/nbt.2655
10. Liang Z, Zhang K, Chen K, Gao C (2014) Targeted mutagenesis in Zea mays using TALENs and the CRISPR/Cas system. J Genet Genomics 41:63–68. https://doi.org/10.1016/j.jgg.2013.12.001
11. Cho SW, Kim S, Kim Y et al (2014) Analysis of off-target effects of CRISPR Cas-derived RNA-guided endonucleases and nickases. Genome Res 24:132–141. https://doi.org/10.1101/gr.162339.113
12. Hwang WY, Fu Y, Reyon D et al (2013) Efficient genome editing in zebrafish using a CRISPR-Cas system. Nat Biotechnol 31:227–229. https://doi.org/10.1038/nbt.2501
13. Kim H, Kim SS, Ryu J et al (2016) A simple , flexible and high-throughput cloning system for plant genome editing via CRISPR-Cas system. J Integr Plant Biol 58:705–712. https://doi.org/10.1111/jipb.12474
14. Chang N, Sun C, Gao L et al (2013) Genome editing with RNA-guided Cas9 nuclease in zebrafish embryos. Cell Res 23:465–472. https://doi.org/10.1038/cr.2013.45
15. Jiang W, Zhou H, Bi H et al (2013) Demonstration of CRISPR/Cas9/sgRNA-mediated targeted gene modification in Arabidopsis, tobacco, sorghum and rice. Nucleic Acids Res 41:e188. https://doi.org/10.1093/nar/gkt780
16. Xie K, Zhang J, Yang Y (2014) Genome-wide prediction of highly specific guide RNA spacers for the CRISPR-Cas9 mediated genome editing in model plants and major crops. Mol Plant 7:923–926. https://doi.org/10.1093/mp/ssu009
17. Miao J, Guo D, Zhang J et al (2013) Targeted mutagenesis in rice using CRISPR-Cas system. Cell Res 23:1233–1236. https://doi.org/10.1038/cr.2013.123
18. Curtin SJ, Tiffin P, Guhlin J et al (2017) Validating genome-wide association candidates controlling quantitative variation in nodulation. Plant Physiol 173:921–931. https://doi.org/10.1104/pp.16.01923
19. Cermak T, Curtin SJ (2017) Design and assembly of CRISPR/Cas9 reagents for gene knockout, targeted insertion, and replacement in wheat. Methods Mol Biol 1679:187–212. https://doi.org/10.1007/978-1-4939-7337-8_12
20. Curtin SJ, Xiong Y, Michno J-M et al (2017) CRISPR/Cas9 and TALENs generate heritable mutations for genes involved in small RNA processing of Glycine max and Medicago truncatula. Plant Biotechnol J. https://doi.org/10.1111/pbi.12857
21. Lei Y, Lu L, Liu H-Y et al (2017) CRISPR-P: a web tool for synthetic single-guide RNA design of CRISPR-system in plants. Mol Plant 7:1494–1496. https://doi.org/10.1093/mp/ssu044

Chapter 13

Functional Genomics and Seed Development in *Medicago truncatula*: An Overview

Christine Le Signor, Vanessa Vernoud, Mélanie Noguero, Karine Gallardo, and Richard D. Thompson

Abstract

The study of seed development in the model species *Medicago truncatula* has made a significant contribution to our understanding of this process in crop legumes. Thanks to the availability of comprehensive proteomics and transcriptomics databases, coupled with exhaustive mutant collections, the roles of several regulatory genes in development and maturation are beginning to be deciphered and functionally validated. Advances in next-generation sequencing and the availability of a genomic sequence have made feasible high-density SNP genotyping, allowing the identification of markers tightly linked to traits of agronomic interest. A further major advance is to be expected from the integration of omics resources in functional network construction, which has been used recently to identify "hub" genes central to important traits.

Key words Desiccation tolerance, Longevity, Globulin, GWAS, Network, Dormancy, Flavonoid, AFL

1 Introduction

1.1 *Medicago truncatula*: Merits and Limits as a Model for Legume Seeds

Medicago truncatula was selected as a model plant for the legumes, primarily because of being a diploid, self-fertile species with a short life cycle, its relative facility of genetic transformation, and its suitability for studying the rhizobial symbiosis [1]. About the same time, a second legume model, *Lotus japonicus*, with similar features, was proposed [2], and both have been extensively used since. The development of the *Medicago truncatula* seed and pod show some notable morphological differences to those of major crop legumes, which at first sight may raise doubts about its suitability as a model for legume seed biology. However, its coding sequences are remarkably similar to pea and other close relatives, as the publication of the genomic sequence underlined [3], and the superficial physical differences occult close similarity in terms of gene expression. Crop legumes such as pea have undergone selection during domestication and plant breeding that has altered the

Luis A. Cañas and José Pío Beltrán (eds.), *Functional genomics in* Medicago truncatula*: Methods and Protocols*, Methods in Molecular Biology, vol. 1822, https://doi.org/10.1007/978-1-4939-8633-0_13,

timing of seed development, the size of the seed, and quantitative aspects of its composition, dormancy, and longevity compared to the closest wild relatives and forage legumes such as *M. truncatula* and *L. japonicus*. The basic physiological events remain, however, largely paralleled in cultivated legume seeds and *M. truncatula* seeds, and a detailed analysis of seed composition [4] confirmed this overall similarity.

1.2 Basic Morphology and Composition of M. truncatula Seeds in Comparison with Crop Legumes

Medicago seeds are reniform, 2.5–4.5 mm × 1.3–2.5 mm, and develop within a coiled pod (three to six turns) that forms a barrel-like structure, with the exposed face covered with spines, often curved or hooked. At maturity, the pod is hard walled and the seeds are difficult to release. The mature seed is covered by an amber to brown seed coat that contributes a variable degree of dormancy depending on the genotype [5]. Mature *Medicago truncatula* seeds have a composition that is broadly similar to that of pea except for the substitution of the major stored form of carbon, which is starch in pea, by lipids in *Medicago* seeds. In addition, as *M. truncatula* has not been selected for seed consumption, the seeds remained small and contain a high proportion of cell wall material in comparison with those of major pulse crops. It is of particular interest that the protein families represented are however the same in *M. truncatula* and major crop legume seeds, their patterns of accumulation are quite similar, and their amino acid compositions are practically identical, thus validating its choice as a model for studies on nutritional improvement in these crops.

1.3 M. truncatula Seed Development: The Physiological Stages

M. truncatula is self-fertile and seed set occurs before full anther emergence from the corolla. Seed development takes around 30 days to full maturity depending on ambient conditions. *M. truncatula* is an exalbuminous species, the precocious proliferation of the endosperm giving way to its progressive degradation, which supplies nutrients to the developing embryo. An initial phase of rapid cell divisions in the embryo up to around 8 DAP is followed by a seed-filling phase 9–25 DAP, during which period desiccation tolerance is also acquired, with few further cell divisions. Thus the final seed size is largely determined by cotyledon cell divisions, as reported for other legumes [6]. A final maturation phase during which time the seed progressively dehydrates determines the acquisition of dormancy and the longevity of the seed.

A comprehensive cytological study of *M. truncatula* seed development has not to our knowledge been carried out, but the development of the seed coat was analyzed at the cellular level [7]. Seed coat gene expression clustered into four clusters during seed development similarly to whole seed transcriptome clustering. After cell division ceases, (4–6 DAP), seed coat cells elongate and endoreduplication occurs. During seed filling (12 DAP onward) the innermost seed coat cell layer, the endothelium, displayed features typical of a high metabolic activity implying a role in solute transfer to the developing embryo. This is supported by features such as the temporary appearance of starch granules. Seed coat maturation is characterized by formation of a dense and rigid macrosclereid plus osteosclereid cell layer, with a cutinized external surface, and the possible presence of suberin at the hilum [7].

2 Approaches Used to Dissect Seed Development

As part of its exploitation as a legume model species, *M. truncatula* seed development has been investigated using omics and post-genomics techniques, summarized here in chronological order of their application.

2.1 Proteomics

M. truncatula seed development has been dissected over the past decade using a variety of –omics approaches. Protein fractionation by two-dimensional gel electrophoresis permitted the identification of several hundred polypeptides [8, 9], and more recently, the use of shotgun mass spectrometry techniques has increased this count to thousands and permitted the quantification of individual proteins at the same time, which was hitherto a limitation [10]. These have revealed the distinct roles played by the component tissues, the seed coat, endosperm and embryo, with discrete protein and transcript profiles, and many examples of tissue-specific specialization of metabolism. Thanks to an immature seed culture system [11], the seed-filling phase can be followed in vitro, allowing a study of this process in the absence of the influence of the mother plant. These studies have also served to confirm the important roles played by the seed coat and endosperm in supporting embryo development, particularly during embryogenesis.

An independent approach to identifying seed regulatory factors was taken by purifying nuclei and isolating and characterizing seed nuclear proteins [12]. Although few transcription factors were identified in this way, the study revealed a series of nuclear proteins involved in RNA maturation and epigenetic modifications during seed development.

2.2 Transcriptomics

Global analyses of transcripts were facilitated by the development of microarrays [13, 14]. In order to classify transcripts, the kinetics of RNA abundancy during seed development were followed and clustering analyses carried out on the resulting profiles. The results discriminated several patterns of gene expression and were significantly more complicated than a simple sigmoidal curve. As part of a larger project, transcript abundancies from developing seed samples were compared to those from many other tissues and under varying conditions, in a global resource, the "Medicago Gene Expression Atlas" [15]. Gene expression profiles obtained using the Affymetrix Medicago Gene Chip® were made accessible for the majority of *M. truncatula* genes covering all its major organ systems (roots, nodules, stems, petioles, leaves, vegetative buds, flowers, seeds, and seed pods), with detailed developmental time series for nodules and seeds. These data have been supplemented with transcriptome data from plants subjected to various kinds of abiotic and biotic stresses and data from specific cell and tissue types. The Medicago Gene Atlas thus gives access to a global picture of expression for a given gene, accessible through a web server [16].

A comparison of transcript profiles and the profiles of the proteins encoded (e.g., in 10) has revealed that protein accumulation frequently occurs several days after that of the corresponding mRNAs, implying a major component of posttranscriptional control during seed development.

Concerning the regulation of transcription, a "selective transcriptomics" approach was applied to monitor the expression during seed development of all the mRNAs putatively encoding transcription factors (TFs), using a library of primers made to predicted TF sequences from the genome of *M. truncatula* [17]. A clustering analysis grouped them according to the phase of development during which they act, and in a few cases functional studies on selected factors have since been done. Putative orthologs were identified to the best characterized TFs implicated in *Arabidopsis* seed development, suggesting that the basic seed regulatory framework has been conserved within dicotyledons. Another study focusing on regulation of early seed development [18] analyzed the timing and precise localization by in situ hybridization of selected TF transcripts in ovules from fertilization onward, providing an insight into very early events in embryogenesis. Terrasson et al. [19] in contrast focused on gene expression during the desiccation phase, characterizing what they termed the "desiccome," associated with the preparation for dormancy and longevity of the mature seed. The transcriptome associated with the acquisition of desiccation tolerance (DT) in seed development was compared with that produced during the reestablishment of DT upon polyethylene glycol treatment of seedlings, to identify genes specifically associated with this character.

Medicago truncatula also possesses a complement of miRNAs and smRNAs [20], but there are currently few if any reports of their regulatory roles during gene expression in seeds. It is likely, however, that the miRNAs from conserved families, which are expressed in seeds (our unpublished observations), play roles analogous to those seen in other species notably in *Arabidopsis* [21].

3 Exploitation of Genomic Sequence Information

Genomic sequence information has been regrouped for six legume species (*Medicago truncatula*, *Glycine max* (soybean), *Lotus japonicus*, *Phaseolus vulgaris* (common bean), *Cicer arietinum* (chickpea), and *Cajanus cajan* (pigeon pea) and two outgroup reference species, *A. thaliana* and *Poplar trichocarpa*, at the LegumeIP site [22]. On this site, a variety of resources are available, including gene expression atlases for four model legumes (*M. truncatula*, *G. max*, *L. japonicus*, and *C. arietinum*). These datasets are organized based on experimental ontology (EO) and plant ontology (PO) allowing data retrieval for selected analyses. Within dataset, phylogenetic analyses are also possible).

With the availability of the genomic sequence and deep transcriptome sequencing, higher-density versions of Medicago-expressed sequence microarrays were developed using Affymetrix [23] and NimbleGen [24] technologies.

4 Integration of Omics Data by the Visualization of Gene Regulatory Networks (GRNs)

In order to progress in understanding seed development from –omics studies, the findings have to be integrated into a functional framework. Verdier et al. [23] constructed GRNs using datasets of expression from Affymetrix microarray hybridization of transcriptomes at 11 stages during seed development. Differentially expressed probe sets were selected, and Pearson correlation coefficients (PCC) and principal component analysis (PCA) were used to first identify the main processes occurring at each developmental stage. Then, using the gene co-expression network, analysis module of ArrayMining (ArrayMining web resource (http://www.arraymining.net)), a topological weighted gene co-expression network, was assembled and visualized using Cytoscape. A web-based application for GRN analysis using legume gene expression data (http://legumegrn.noble.org) has also been produced [25] that permits GRN predictions and hosts gene expression datasets from *M. truncatula*, *L. japonicus*, and *G. max*.

Righetti and colleagues [24] have used a more comprehensive transcriptome data resource obtained by varying temperature, light intensity, osmotic stress, and soil water potential, to generate 104 seed transcriptomes analyzed using NimbleGen arrays and made available on the Medicago eFP browser (http://bar.utoronto.ca/efp_medicago/cgi-bin/efpWeb.cgi?dataSource=medicago_seed). This approach allowed the generation of MatNet, a regulatory co-expression network of seed maturation.

5 Methods for the Validation of Gene Function

5.1 Reverse Genetics

Medicago truncatula was selected as a plant model on the basis of its transformability, and a variety of methods have been described for stable transformation [26] and for transformation of root cultures using *A. rhizogenes* [27]. A report that *M. truncatula* was transformable by the simple floral dip method used for *A. thaliana* has however not been confirmed, and the frequency of transformation and the time taken for regeneration remain limiting compared to *Arabidopsis*. Furthermore there is a high genotype-dependent variation in transformation efficiency by *A. tumefaciens*. In practice, for phenotypes not easily tested in genetically transformed *Agrobacterium rhizogenes* root cultures, reverse genetics on mutant populations has been more widely used for validation of gene function. The mutant resources available include TILLING on EMS mutants [28, 29] or on fast neutron-generated deletions [30] and screens of a Tnt1 retrotransposon insertion library [31–34]. Platforms for recovering mutants have been set up [35–37]. The activation of an endogenous *M. truncatula* retrotransposon, Mere-1, has generated a further source of variation [38] usable as a reverse genetics tool.

5.2 Genome-Wide Association Studies (GWAS)

Medicago truncatula is a model for investigating legume genetics, including the genetics and evolution of the legume-Rhizobia symbiosis, and is beginning to be used in GWAS approaches to identify key loci. Whole-genome sequence data has been used to identify and characterize sequence polymorphisms and linkage disequilibrium (LD) in a diverse collection of 26 *M. truncatula* accessions. *M. truncatula* harbors both higher diversity and less LD than soybean (*Glycine max*) and exhibits patterns of LD and recombination similar to *Arabidopsis thaliana*. More than 3 million SNPs were detected, comprising a valuable resource for genome-wide association mapping of legume genes [39].

In a translational genomics approach between *M. truncatula* and pea, the position of seed protein PQLs, controlling one or more storage proteins, was compared [40]. Genes specifically expressed in *M. truncatula* seed within PQLs conserved or non-conserved between the two species were retained. Interestingly, a functional classification according to AgriGO revealed an overrepresentation of transcription factors. To narrow down the genes potentially responsible for protein composition variations, a GWAS approach was used in *M. truncatula* using the HapMap resource [41]. This study revealed sequence polymorphisms associated with variations in storage protein quantities analyzed by 1D or 2D gel electrophoresis. Numerous polymorphic loci were identified for the first time as gene candidates for regulating synthesis, transport, or maturation of legume storage proteins, and a model for regula-

Table 1
TF gene candidates from PQL (protein quantity loci) and GWAS approaches for regulating seed protein composition

PQLs and type of regulation **PQLs conserved with Pea*	ID *M. truncatula* (Mt4.0)	Pea Orthologue (Pscam#)	Definition (Interpro)	Expression in developing seeds of *M. truncatula* (dap) 10	12	16	20	24	36
*Vic and Conv (Trans), Leg (cis)	Medtr7g104480	017634	ABI5 bZIP	*21*	*24*	*299*	*1339*	*2421*	*2441*
*Leg and Vic (Trans)	Medtr4g097950	023491	bHLH18-like	*11*	*12*	*33*	*852*	*1931*	*2613*
Leg and Vic (Trans)	Medtr8g020490	050441	Myb	*581*	*565*	*782*	*1128*	*978*	*179*
Vic (Trans)	Medtr8g043960	038409	bZIP	*346*	*785*	*2475*	*2829*	*2484*	*218*

Expression data taken from Benedito et al. [15]

tion involving these genes was proposed [40]. As an example of these results, by comparing PQL and GWAS results, four TFs were selected as being likely to regulate expression of storage protein genes (Table 1). The genes lie within globulin PQL intervals and are specifically expressed in the filling seed. A mutant allele of one of these, *abi5*, reduced by 30% the accumulation of sulfur AA-poor vicilin storage proteins.

6 Analyses of Mutants Affecting the Developing Seed

6.1 Mutant Resources Available

As genetic transformation of *M. truncatula*, although quite feasible, is not trivial, rather than the analysis of transgenic plants, mutants have been mainly obtained from EMS mutagenesis (TILLING) collections of point mutants or transposon insertion (TnT1) resources [28, 29, 34]. A complementary resource, consisting of deletions, was created from a population of fast neutron-mutagenized plants, the De-TILLING collection [30]. All of these resources can be screened for mutations in genes of interest by PCR on DNA pools of several individuals. Furthermore, the TnT1 insertion sites have been partially sequenced, which allows access to mutant alleles in a gene of interest without screening the library and purifying via deconvolution the line concerned. The TnT1 collection has also been subjected to several forward screens. For example, Serrani-Yarce et al. [42] screened for reduced seed coat pigmentation and identified ten Tnt1 mutants. However, most of the seed-expressed genes studied to date have been selected on the basis of their homology with *Arabidopsis* genes of known function. In the section that follows, examples are given of the use of mutants to elucidate traits of agronomic importance. A more extensive review of *M. truncatula* functional studies can be found in [43].

6.2 Embryogenesis

Regulatory factors intervening during embryogenesis include important determinants of final seed size. Ge et al. [44] identified a novel transcription factor, *MtBIG SEEDS1*, of the TIFY family, which when deleted, resulted in enlarged lateral organ size, including seeds in *M. truncatula* and in soybean (Glycine max). *MtBS1* encodes a repressor normally repressing expression of the growth regulating factors GRF and GIF1 to limit growth rates. When mutated, increased cell divisions occur, resulting in organ enlargement. In contrast, two genes have been identified whose mutation reduces *M. truncatula* seed size. Both are expressed specifically in the endosperm but also affect embryo development. *DASH*, (*D*OF *a*cting in *s*eed embryogenesis and *h*ormone accumulation), an endosperm-specific transcription factor of the DOF family, is expressed during embryogenesis [45], (Fig. 1). In *dash* mutants at the globular stage, the endosperm degenerates, and embryo development is arrested, resulting in a severely reduced seed size which is at the limit of viability. Auxin addition partly reverts the phenotype and allows the production of viable seeds. The phenotype is restricted to the seeds, vegetative plant parts being unaffected. The DASH factor is assumed to control auxin perception, distribution, and/or metabolism in the endosperm because of complementation by auxin addition and as pod auxin levels are 50-fold higher in the mutant than in WT seeds.

The second endosperm-specific gene identified to play a role in the control of seed size encodes a subtilase, *MtSBT1.1* [46]. The gene was identified within a seed weight QTL interval spanning 18 seed-expressed genes. Of these, *SBT1.1* was one of only two to be expressed during embryogenesis. Mutant alleles at the locus derived from TILLING were associated with reduced seed weight. A comparison of cell size and number in WT and *sbt1.1* seed sections showed that while cell cross-sectional area was unchanged in the mutant, the number of cells was reduced. It was thus concluded that *SBT1.1* promotes cell divisions, possibly by cleaving a peptide hormone precursor to liberate an active growth factor, similar to the action of *CLE8*, which encodes a peptide ligand controlling embryo and suspensor proliferation in *Arabidopsis* [47].

An AP2/ERF transcription factor termed the MEDICAGO TRUNCATULA SOMATIC EMBRYO RELATED FACTOR 1 (MtSERF1) and the putative ortholog of *WUSCHEL*, *MtWUS*, were shown to be important for somatic embryogenesis (SE) in *M. truncatula* [48]. RNAi of these genes caused strong inhibition of

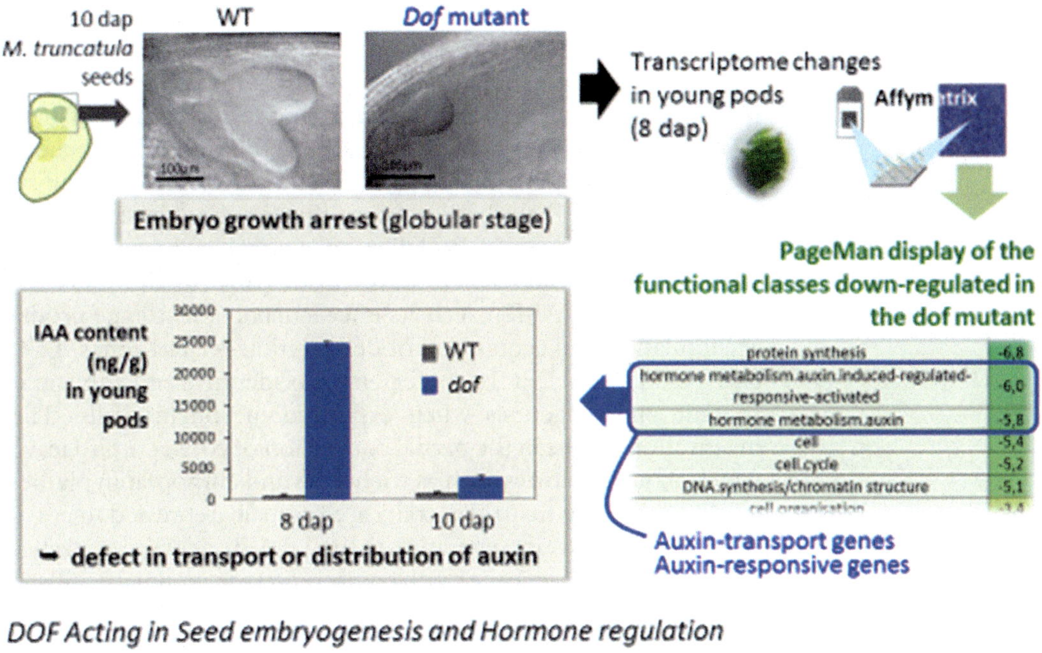

Fig. 1 Strategy used for characterizing DASH, (Dof acting in seed embryogenesis and hormone accumulation), an endosperm-specific TF required for embryo development. Comparison of WT and dash pod transcriptomes at a stage preceding embryo arrest revealed enrichment for auxin-related genes. Auxin measurements showed the mutant pods to accumulate high levels of auxin, leading to the conclusion that auxin transport or perception is affected

SE. *MtSERF1* and *MtWUS* are also expressed in zygotic embryos, where *MtSERF1* is proposed to enable a connection between stress and development. MtSERF1 may be induced by MtWUS, based on the presence of MtWUS binding sites on its promoter and the co-localization of their sites of expression.

6.3 Control of Seed Filling

The phase of seed filling in *M. truncatula* occurs after embryogenic cell divisions. This phase largely determines mature seed composition. Seed filling has been studied at the protein and transcript levels, and genes encoding potentially regulatory transcription factors were identified by analogy with *Arabidopsis*, where a group of "master regulators" has been functionally characterized [49]. The master regulators include LEAFY COTYLEDON1 (LEC1), a HAP3 subunit of the CCAAT-binding protein, which does not appear to bind DNA directly [50] but interacts as part of a complex, the LAFL complex, which also includes three related B3 domain transcription factors, ABSCISIC ACID INSENSITIVE3 (ABI3), FUSCA3 (FUS3), and LEAFY COTYLEDON2 (LEC2). The LAFL transcription factors LEC1, LEC2, and FUS3 are expressed both in the developing embryo and the endosperm, while ABI3 is embryo specific. While good candidates for LEC1,

ABI3, and FUS3 orthologs have been found in *M. truncatula*, with similar expression patterns, in contrast to *Arabidopsis*, no LEC2-like sequence was found [17]. In fact, LEC2 homologues are missing from several plant families, suggesting its role can be assumed by another TF [51]. In *Arabidopsis*, the LAFL transcription factors LEC2, FUS3, and ABI3 are partly functionally redundant and recognize the same DNA binding sites; they associate in similar complexes with LEC1 or LEC1-like to control gene targets [50]. However, FUS3 and LEC2 differ from ABI3 in the traits controlled. Whereas ABI3 exerts its effects mainly on storage product accumulation and acquisition of dehydration-related genes, LEC2 and FUS3, with LEC1, confer embryo identity, and can induce somatic embryogenesis when expressed in somatic cells. Their inactivation results in the partial conversion of cotyledons to leaves, which acquire features such as trichomes and anthocyanin pigmentation, indicating that these traits are normally repressed in cotyledons. In contrast, in vegetative tissues, LAFL genes are normally repressed by the action of polycomb repressor complexes −1 and −2 (PRC1 and 2), and more specific repression and activation mechanisms operate on different LAFLs (Table 2) to prevent storage product accumulation in these cell types [49].

AFL transcription factors LEC2, FUS3, and ABI3 bind at RY-boxes on target promoters to regulate production of target storage proteins, oleosome proteins (oleosins), and other downstream-acting transcription factors including WRINKLED1 (WRI1), MYB118, and AGL15. WRI1 notably controls accumulation of oil in the embryo. MYB118 is expressed exclusively in the endosperm and represses LEC2 expression and storage protein deposition there. AGL15 encodes a MADS box factor positively controlled by LEC2. The presence of G-boxes adjacent to RY targets of LAFL factors implies a possible interaction with bZIP transcription factors, and interactions between ABI3 and LEC1 and specific bZIP transcription factors, potentiating their action, have been shown [50, 52]. The WD-40 repeat TF TRANSPARENT TESTA GLABRA-1 (TTG1) negatively regulates the accumulation of embryo storage proteins and the seed-filling stage more generally; it is itself repressed by FUS3 [53].

The LAFL factors have been shown in *Arabidopsis* to act in part by controlling plant hormone homeostasis. Thus, during *Arabidopsis* embryogenesis, LEC2 and FUS3 repress AtGA3ox2, which encodes an enzyme of gibberellin biosynthesis [54]. LEC2 induces YUC2 and YUC4 expression which encodes Yucca-type flavin monooxygenases implicated in auxin biosynthesis [55]. Similar mechanisms are likely to occur in *Medicago* seeds, but as a LEC2 ortholog has not been identified, it may be substituted by another factor.

Table 2
Epigenetic regulation of LAFL gene expression

TF	Repressor				PTGS			
Locus concerned	PRC1/PRC2	PICKLE/PRK2	CHR5	SCL15	VAL	miR166	miR156	ASIL1/ASIL2 HDA6/SIL1
Site of action	Vegetative tissues	Vegetative tissues	Vegetative tissues	Vegetative tissues (vascular)	Seedlings	Vegetative tissues	Vegetative tissues	Embryo during early embryogenesis
Genes affected: LEC1	–	–		–	–			
LEC2	–	–			–	–	–	–
FUS3	–	–	+		–	–	–	–
ABI3	–	–	+	–				
Reference								

The table summarizes epigenetic controls reported for *Arabidopsis* seed development and reviewed in Fatihi et al. [49]

6.4 Regulation of Nutritional Supply to the Developing Seed: Example of Sulfate

In order to study the response of *M. truncatula* to sulfur starvation, an original approach was taken, comparing the strategies of nutrient allocation with seed yield and quality measurements. The study showed that sulfur starvation at the vegetative stage, i.e., appearance of tertiary ramifications, affects foliage and seed production and reduces carbon allocated to the seeds. The seeds also have low germinative vigor and low sugar and RFO oligosaccharide contents needed for elongation of the embryo axis during imbibition. Similar results were seen for oilseed rape [56] showing the importance of maternal sulfur nutrition for physiological seed quality (Coll. J-C Avice, UMR EVA Caen). The relation between seed composition and seed vigor was also studied in *M. truncatula*, highlighting the negative impact of low sucrose conversion into RFO during seed development on seed vigor [57]. Sulfur starvation shows differential effects on accumulation of *M. truncatula* storage proteins depending on their S-amino acid contents (Fig. 2). The relatively sulfur-rich albumins are greatly reduced, and of the globulins, the legumins are reduced, whereas the sulfur-poor vicilins and convicilins increase in proportion. The effects are greater the earlier the starvation is imposed.

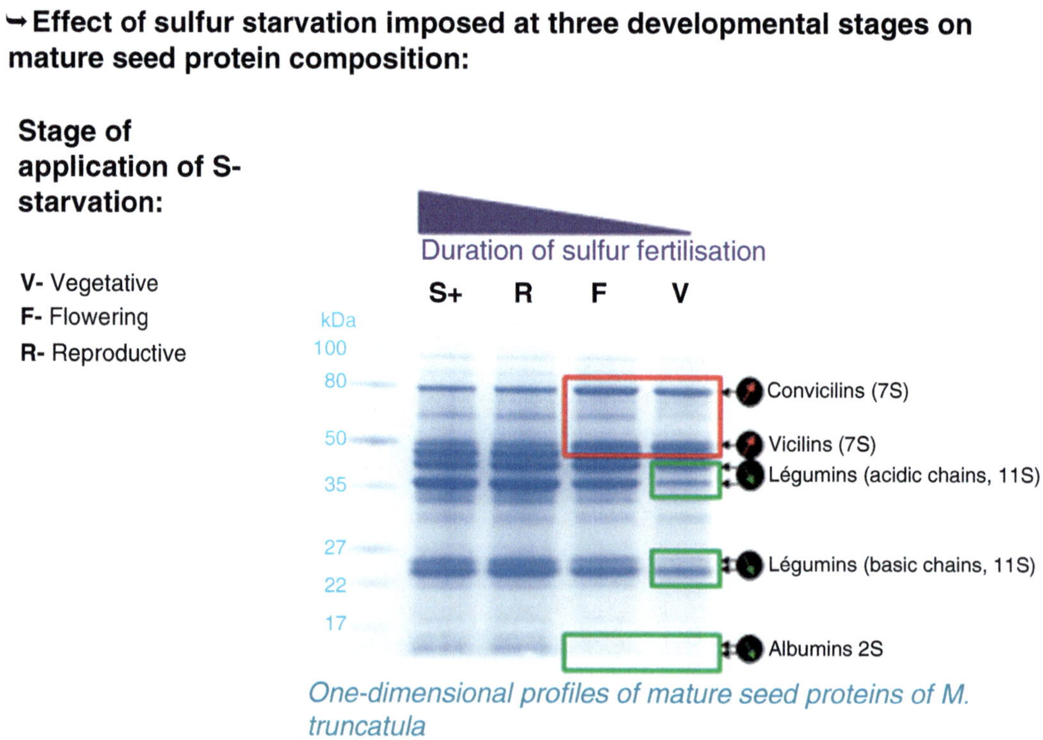

Fig. 2 Effect of sulfur starvation (S-) on *M. truncatula* mature seed composition. Sulfur deprivation applied at three stages, reproductive (R), flowering (F), and vegetative (V). The sulfur-poor convicilins and vicilins accumulate preferentially (boxed in red), whereas the relatively sulfur-rich legumins and albumins accumulate less (boxed in green)

6.5 Late Maturation

During the maturation of the seed, several traits vital for seed vigor and plant dispersal are acquired including ability to germinate, desiccation tolerance, longevity, and dormancy. These processes may partly overlap the seed-filling stage or begin afterward, depending on the species. For *M. truncatula*, capacity to germinate and desiccation tolerance are acquired during the seed-filling phase, whereas markers for late seed maturation (loss of chlorophyll, accumulation of raffinose family oligosaccharides) are expressed after seed filling has ceased [58]. In contrast, in *Arabidopsis*, these markers overlap the seed-filling stage [24]. Another factor distinguishing desiccation tolerance and longevity in *Medicago* is that, whereas the acquisition of desiccation tolerance is independent of environmental conditions and is described as a "robust" trait, the acquisition of longevity displays "plasticity" and is very dependent on the environment of the maturing seed.

The most abundant class of proteins synthesized during the late maturation stage in most dicot seeds including *Medicago* is that of the late embryogenesis accumulating (LEA) proteins. LEA proteins present in *Medicago* belong to several distinct sequence-related classes, which share overall physical properties. They are amphiphilic and thought to protect cellular structures during dehydration and/or contribute to longevity, although hard evidence is still scarce, possibly because of their redundancy. Although initially implicated in acquisition of desiccation tolerance, they for the most part accumulate well after this event during late maturation in *Medicago*. There is a significant delay between transcript appearance and protein accumulation indicating a major posttranscriptional control of LEA expression, the mechanism of which is unclear.

An advance in understanding the developmental regulation of *Medicago* seed formation has been obtained by analyzing the relationships between genes based on their co-expression [23]. A topological weighted gene co-expression network was assembled and visualized using Cytoscape [59] (Fig. 3 of Verdier et al. 2013). The projection separates clearly genes expressed at different developmental stages. Superimposed on the network, discrete clusters of genes associated with desiccation tolerance (DT), LEA proteins, abscission, and longevity are evident. The representation also revealed that the proportion of seed-specific genes associated with each character varied greatly, being high for LEA and DT and much lower for abscission and longevity. The acquisition of DT and dormancy module was enriched in genes with GO terms related to responses to stresses (e.g., response to desiccation, heat, temperature, abiotic stimulus) and in GSH metabolic process. They included many LEA genes, HSPs or small heat shock proteins (sHSPs; DNAJ), universal stress proteins, aldose reductases, and abscisic acid (ABA)-induced genes. Interestingly, the cluster containing most of the genes associated with longevity had these located in a separate sub-cluster from those increasing around the point of abscission, suggesting two discrete biological processes are involved; this was also supported by the different identities of the genes in the two sub-clusters. LEA protein accumulation is partly controlled by LAFL transcription factors and also subject to regulation by transcription factors acting specifically in late maturation such as ABI4, ABI5, EEL, and DOG1, although each TF controls a different spectrum of LEAs [60], and further candidate controlling transcription factors are still to be identified [23].

Sequences in the longevity module are enriched in GO terms for translation, RNP complex biogenesis, and ribosome biogenesis, suggesting a role in the posttranscriptional control of LEA proteins. Transcripts of genes potentially involved in translational regulation increase 10 d later than LEA transcripts and in advance of the accumulation of the LEA protein levels.

To further investigate transcriptional regulation, a co-expression gene regulatory network was constructed by combining the seed-specific probe sets and the seed-specific TF probe sets [23]. In this representation, clusters corresponding to a DT/LEA module (which was separable into the two components) and a longevity module were produced, each incorporating several transcription factors. The DT and LEA clusters included previously characterized transcription factors (ABI3, 4, 5), but in all identified clusters several new transcription factors were found, e.g., an uncharacterized MtAP2/EREPB TF in the LEA module, that now require functional validation. The putative *Medicago* ortholog of the *Arabidopsis* seed-specific HSFA9 [61] is intriguingly located at the interface between DT, LEA, and longevity clusters, suggesting a role for HSPs in all three processes. Gonzalez-Morales [62] analyzed co-expression networks of *Arabidopsis* seed-expressed transcription factors that failed to be activated in desiccation-intolerant mutants, i.e., related to DT, and identified two subsets of transcription factors: transcription factorsSeed-subnetDT1 which included mainly AP2/ERF family members but also bZIP12, MADS (AGL67), and C2H2 (DOGL4) and transcription factorsSeed-subnetDT2, which included NACs including ATAF1, AP2-type, C3H Zn finger, AT-rich and Zn-binding protein family (PLATZ1 and PLATZ2) transcription factors, ABI5, and EIL5. Of these, ATAF1, AGL67, and PLATZ1 and PLATZ2 all showed effects on desiccation tolerance, and their orthologues in *Medicago* may also be good candidates for a role in desiccation tolerance.

A group of genes that were overrepresented in the *Medicago* "acquisition of DT and dormancy" module are related to the biosynthesis of reduced glutathione (GSH), such as a gamma-glutamyl-Cys ligase, which catalyzes the first and rate-limiting step, and a gamma-glutamyl-Cys synthetase [63]. A requirement for GSH in seed maturation has been shown for *Arabidopsis* [64], and it was implied in desiccation tolerance in *Sporobolus* [65].

By adding data points for *Medicago* seed development cultivated under a range of environmental conditions, a gene co-expression network of seed maturation was generated, termed MatNet [24], on which developmental phases are easily distinguished. From MatNet, a gene module related to longevity was extracted, by selecting a set of transcripts whose expression profiles were highly correlated to P50 (the time required for 50% seed mortality at 35 °C, 75% relative humidity). Likewise, a desiccation tolerance module was identified, corresponding to genes which increased in abundance between desiccation-intolerant and desiccation-tolerant stages. *Arabidopsis* interaction data in Genemania.org was mined to identify homologues to genes in the *Medicago* longevity module. Most of these were found to be similarly connected in *Arabidopsis* and in *Medicago*. For a subset of genes, a functional validation of their implication in longevity was demonstrated in *Arabidopsis*.

The longevity module also contained many genes with GO terms associated with biotic stress. Functional analyses in *Arabidopsis* of two transcription factors in this module confirmed their roles in longevity and linked them in addition to a role in resistance to necrotrophic pathogens [24].

6.5.1 Flavonoid Composition of the Seed Coat

An aspect of *M. truncatula* seed composition that has received much attention is the flavonoid (anthocyanin and condensed tannin or proanthocyanidin (PA)) content of the seed, these components being concentrated in the seed coat and permitting a visual screen for mutant seeds. The regulation of PA accumulation in the *Medicago* seed has been studied with the long-term objective of increasing PA deposition in vegetative tissues to regulate digestion and prevent pasture bloat, which constrains the use of *M. truncatula* and certain other legumes as forage crops. From studies in other plants, it was known that a major role in control of PA accumulation is played by a ternary complex associating *M*yb, *b*HLH, and *W*D-40 transcription factors, the MBW complex [66]. In a forward genetics screen of *M. truncatula* Tnt1 insertion mutants [42], ten Tnt1 mutants with clearly reduced seed coat pigmentation were identified, and many of these have been functionally characterized. Table 3 lists genes involved in flavonoid accumulation in *M. truncatula* identified through mutant analysis. In addition to transcription factors, genes involved in flavonoid glycosylation and intracellular flavonoid transport are involved.

6.6 Molecular Genetic Dissection of Seed Responses to Abiotic and Biotic Stresses

M. truncatula has been studied extensively for its adaptation to drought and salt as these are encountered in its natural habitat, and genetic variability in the physiological responses occurs [75–78]. A study of the transcriptome and metabolome responses of vegetative plant parts to a progressive drought stress, and the effects of

Table 3
Loci affected in *M. truncatula* seed coat anthocyanin or proanthocyanidin accumulation

Gene class	Name	Source	Pathway affected	References	Notes
Regulator of transcription	MtLAP1	Overexpression of Myb gene related to AthTT2	Anthocyanin accumulation	[67]	
	MtPAR	Tnt1 insertion	PA accumulation	[68]	
	MtMYB5	Tnt1 insertion	PA accumulation	[69]	
	MtMYB14	Tnt1 insertion	PA accumulation	[69]	
	MtTT8	Tnt1 insertion	Anthocyanin and PA accumulation	[70]	
	MtMYB2	Homology with anthocyanin repressor AthMYBL2	Repressor of anthocyanin and PA accumulation	[71]	
Glycosyl transferase	UGT78G1	Induced by MtLAP1, Tnt1 insertion	Anthocyanin accumulation	[67]	
	UGT72L1	Tnt1 insertion	PA accumulation	[72]	
Transporter	MATE1	Tnt1 insertion	Vacuolar transport of PA precursor	[73]	Transport specificities established in yeast
	MATE2	Homology to AthTT12, Tnt1 insertion	Anthocyanin vacuolar transport	[74]	Transport specificities established in yeast

MtLAP1, legume anthocyanin production1; *MtPAR*, proAnthocyanidin regulator; *MtMYB5, MtMYB14*, Myb-class TFs; *UGT78G1*, UDP-glucosyl transferase78G1; *UGT72L1*, UDP-glucosyl transferase72L1; *MATE1*, multidrug and toxic compound extrusion1; *MATE2*, multidrug and toxic compound extrusion2

rewatering, has been reported [79]. The responses of shoots and roots were very different. Both plant organs showed distinct short-term and long-term responses, with the short-term responses reflecting drought avoidance (as exhibited by non-tolerant species) whereas the longer-term responses reflected progressive adaptation. Among the metabolites accumulating were osmoprotectants including myoinositol, proline, several amino acids, and sugars, which probably contribute to drought avoidance. Few studies to date in *Medicago* have focused on the effects of abiotic or biotic stresses specifically on seed development and composition, and the extent to which results for shoots or roots are valid for seeds is open to question. Interestingly, many genes previously considered to be seed specific were found to be expressed in the transition from mild to severe drought stress in *Medicago* shoots and roots, suggesting their recruitment during the acquisition of drought tolerance by the angiosperm seed.

Studies of seed vigor suggest that this may give an indication of resistance to, or tolerance of, environmental stresses. Vandecasteele et al. [57] investigated whether the composition or amount of soluble sugars (sucrose and RFO) was one of the genetic determinants of seed vigor in *M. truncatula* using two recombinant inbred line (RIL) populations. The correlation and co-location of Suc/RFO ratio with germination and radicle growth QTLs suggested that an increased Suc/RFO ratio in seeds might negatively affect seed vigor.

Several physiological parallels exist between responses to abiotic and biotic stresses and steps in late seed maturation. Desiccation-tolerant seeds and vegetative plant tissues adapted to water deficit are functional parallels and share many identical or similar gene products. As an example, seeds express constitutively several proteins associated with abiotic stress responses, including heat shock proteins (HSPs). sHSPs have a chaperone function that may contribute to seed longevity. HSP expression is controlled by heat shock factors and in seeds partly by a seed-specific HEAT SHOCK TRANSCRIPTION FACTOR9 (HSFA9). When sunflower HSFA9 was overexpressed in tobacco, seeds were more heat resistant, and when the gene was inactivated, seeds were more susceptible to controlled deterioration [80]. Interestingly, this gene was located at the interface of the desiccation tolerance and longevity gene expression modules in *Medicago*, suggesting its involvement in both of these aspects of seed maturation [24].

Concerning biotic stresses, seeds are protected by constitutive expression of plant defense compounds, many of which accumulate in the seed's outer cell layers. These cell layers are modified by secondary plant products (condensed tannins, anthocyanins, waxes, suberin, lignin, etc.) which have roles in defense against pests and pathogens and also assure physical integrity of the seed. A gene regulatory network (GRN) analysis of gene expression in developing *M. truncatula* and *Arabidopsis* seeds [24] revealed the existence in both species of a gene module that included both longevity- and defense-related genes, implying coordination of their expression. When *M. truncatula* seeds were challenged by an incompatible pathogen (*Xanthomonas campestris pv. campestris*), transcriptional reprogramming occurred. The resulting defense response induced synthesis of pathogen-related protein transcripts and was accompanied by a downregulation of genes involved in seed maturation in an apparent trade-off of resources [81].

References

1. Barker D, Bianchi S, Blondon F, Dattée Y, Duc G, Essad S et al (1990) *Medicago truncatula*, a model plant for studying the molecular genetics of the Rhizobium-legume symbiosis. Plant Mol Biol Rep 8:40–49. https://doi.org/10.1007/BF02668879
2. Handberg K, Stougaard J (1992) *Lotus japonicus*, an autogamous, diploid legume species for classical and molecular-genetics. Plant J 2:487–496. https://doi.org/10.1111/j.1365-313X.1992.00487.x
3. Young ND, Debellé F, Oldroyd GED, Geurts R, Cannon SB, Udvardi MK et al (2011) The Medicago genome provides insight into the evolution of rhizobial symbioses. Nature 480:520–524. https://doi.org/10.1038/nature10625
4. Djemel N, Guedon D, Lechevalier A, Salon C, Miquel M, Prosperi JM et al (2005) Development and composition of the seeds of nine genotypes of the *Medicago truncatula* species complex. Plant Physiol Biochem 43:557–566. https://doi.org/10.1016/j.plaphy.2005.04.005
5. Lesins K, Lesins I (1979) Genus Medicago: a taxogenetic study. W. Junk, The Hague
6. Munier-Jolain NG, Ney B (1998) Seed growth rate in grain legumes II. Seed growth rate depends on cotyledon cell number. J Exp Bot 49:1971–1976. https://doi.org/10.1093/jxb/49.329.1971
7. Verdier J, Dessaint F, Schneider C, Abirached-Darmency M (2013a) A combined histology and transcriptome analysis unravels novel questions on *Medicago truncatula* seed coat. J Exp Bot 64:459–470. https://doi.org/10.1093/jxb/ers304
8. Gallardo K, Le Signor C, Vandekerckhove J, Thompson RD, Burstin J (2003) Proteomics of *Medicago truncatula* seed development establishes the time frame of diverse metabolic processes related to reserve accumulation. Plant Physiol 133:664–682. https://doi.org/10.1104/pp.103.025254
9. Watson BS, Asirvatham VS, Wang L, Sumner LW (2003) Mapping the proteome of barrel medic (*Medicago truncatula*). Plant Physiol 131:1104–1123. https://doi.org/10.1104/pp.102.019034
10. Gallardo K, Firnhaber C, Zuber H, Hericher D, Belghazi M, Henry C et al (2007) A combined proteome and transcriptome analysis of developing *Medicago truncatula* seeds: evidence for metabolic specialization of maternal and filial tissues. Mol Cell Proteomics 6:2165–2179. https://doi.org/10.1074/mcp.M700171-MCP200
11. Gallardo K, Kurt C et al (2006) In vitro culture of immature *M. truncatula* grains under conditions permitting embryo development comparable to that observed in vivo. Plant Sci 170(6):1052–1058
12. Repetto O, Rogniaux H et al (2008) Exploring the nuclear proteome of *Medicago truncatula* at the switch towards seed filling. Plant J 56(3):398–410
13. Kuster H, Hohnjec N et al (2004) Construction and validation of cDNA-based Mt6k-RIT macro- and microarrays to explore root endosymbioses in the model legume *Medicago truncatula*. J Biotechnol 108(2):95–113
14. Firnhaber C, Puhler A et al (2005) EST sequencing and time course microarray hybridizations identify more than 700 Medicago truncatula genes with developmental expression regulation in flowers and pods. Planta 222(2):269–283
15. Benedito VA, Torres-Jerez I, Murray JD, Andriankaja A, Allen S, Kakar K et al (2008) A gene expression atlas of the model legume *Medicago truncatula*. Plant J 55:504–513. https://doi.org/10.1111/j.1365-313X.2008.03519.x
16. He J, Benedito VA, Wang M, Murray JD, Zhao PX, Tang Y et al (2009) The *Medicago truncatula* gene expression atlas web server. BMC Bioinform 10:441. https://doi.org/10.1186/1471-2105-10-441
17. Verdier J, Kakar K, Gallardo K, Le Signor C, Aubert G, Schlereth A et al (2008) Gene expression profiling of M-truncatula transcription factors identifies putative regulators of grain legume seed filling. Plant Mol Biol 67:567–580. https://doi.org/10.1007/s11103-008-9320-x
18. Kurdyukov S, Song Y, Sheahan MB, Rose RJ (2014) Transcriptional regulation of early embryo development in the model legume *Medicago truncatula*. Plant Cell Rep 33:349–362. https://doi.org/10.1007/s00299-013-1535-x
19. Terrasson E, Buitink J, Righetti K, Ly VB, Pelletier S, Zinsmeister J et al (2013) An emerging picture of the seed desiccome: confirmed regulators and newcomers identified using transcriptome comparison. Front Plant Sci 4:497. https://doi.org/10.3389/fpls.2013.00497
20. Bustos-Sanmamed P, Bazin J et al (2013) Small RNA pathways and diversity in model legumes: lessons from genomics. Front Plant Sci 4:236

21. Sunkar R, Zhu JK (2004) Novel and stress-regulated microRNAs and other small RNAs from Arabidopsis. Plant Cell 16(8):2001–2019
22. Li J, Dai X, Liu T, Zhao PX (2012) LegumeIP: an integrative database for comparative genomics and transcriptomics of model legumes. Nucleic Acids Res 40:D1221–D1229. https://doi.org/10.1093/nar/gkr939
23. Verdier J, Lalanne D, Pelletier S, Torres-Jerez I, Righetti K, Bandyopadhyay K et al (2013) A regulatory network based approach dissects late maturation processes related to the acquisition of desiccation tolerance and longevity of *Medicago truncatula* seeds. Plant Physiol 163:757–774. https://doi.org/10.1104/pp.113.222380
24. Righetti K, Vu JL et al (2015) Inference of longevity-related genes from a robust coexpression network of seed maturation identifies regulators linking seed storability to biotic defense-related pathways. Plant Cell 27(10):2692–2708
25. Wang M, Verdier J, Benedito VA, Tang Y, Murray JD, Ge Y et al (2013) LegumeGRN: a gene regulatory network prediction server for functional and comparative studies. PLoS One 8:e67434. https://doi.org/10.1371/journal.pone.0067434
26. Thomas MR, Rose RJ, Nolan KE (1992) Genetic transformation of Medicago truncatula using Agrobacterium with genetically modified Ri and disarmed Ti plasmids. Plant Cell Rep 11:113–117. https://doi.org/10.1007/BF00232161
27. Boisson-Dernier A, Chabaud M, Garcia F, Bécard G, Rosenberg C, Barker DG (2001) Agrobacterium rhizogenes-transformed roots of *Medicago truncatula* for the study of nitrogen-fixing and endomycorrhizal symbiotic associations. Mol Plant-Microbe Interact 14:695–700. https://doi.org/10.1094/MPMI.2001.14.6.695
28. Le Signor C, Savois V, Aubert G, Verdier J, Nicolas M, Pagny G et al (2009) Optimizing TILLING populations for reverse genetics in *Medicago truncatula*. Plant Biotechnol J 7:430–441. https://doi.org/10.1111/j.1467-7652.2009.00410.x
29. Penmetsa RV, Cook DR (2000) Production and characterization of diverse developmental mutants of Medicago truncatula. Plant Physiol 123:1387–1398. https://doi.org/10.1104/pp.123.4.1387
30. Rogers C, Wen J, Chen R, Oldroyd G (2009) Deletion-based reverse genetics in Medicago truncatula. Plant Physiol 151:1077–1086. https://doi.org/10.1104/pp.109.142919
31. Cheng X, Wang M, Lee H, Tadege M, Ratet P, Udvardi M et al (2014) An efficient reverse genetics platform in the model legume Medicago truncatula. New Phytol 201:1065–1076. https://doi.org/10.1111/nph.12575
32. Tadege M, Ratet P, Mysore KS (2005) Insertional mutagenesis: a Swiss Army knife for functional genomics of *Medicago truncatula*. Trends Plant Sci 10:229–235. https://doi.org/10.1016/j.tplants.2005.03.009
33. D'Erfurth I, Cosson V, Eschstruth A, Lucas H, Kondorosi A, Ratet P (2003) Efficient transposition of the Tnt1 tobacco retrotransposon in the model legume Medicago truncatula. Plant J 34:95–106. https://doi.org/10.1046/j.1365-313X.2003.01701.x
34. Jiang C, Chen C, Huang Z, Liu R, Verdier J (2015) ITIS, a bioinformatics tool for accurate identification of transposon insertion sites using next-generation sequencing data. BMC Bioinform 16(72):72. https://doi.org/10.1186/s12859-015-0507-2
35. Porceddu A, Panara F, Calderini O, Molinari L, Taviani P, Lanfaloni L et al (2008) An Italian functional genomic resource for *Medicago truncatula*. BMC Res Notes 1:129. https://doi.org/10.1186/1756-0500-1-129
36. Carelli M, Calderini O et al (2013) Reverse genetics in Medicago truncatula using a TILLING mutant collection. Methods Mol Biol 1069:101–118
37. Iantcheva A, Vassileva V et al (2009) Development of functional genomic platform for model legume *Medicago truncatula* in Bulgaria. Biotechnol Biotechnol Equip 23(4):1440–1443
38. Rakocevic A, Mondy S, Tirichine L, Cosson V, Brocard L, Iantcheva A et al (2009) MERE1, a low-copy-number copia-type retroelement in *Medicago truncatula* active during tissue culture. Plant Physiol 151:1250–1263. https://doi.org/10.1104/pp.109.138024
39. Branca A, Paape TD et al (2011) Whole-genome nucleotide diversity, recombination, and linkage disequilibrium in the model legume Medicago truncatula. Proc Natl Acad Sci U S A 108(42):E864–E870
40. Le Signor C, Aimé D et al (2017) Genome-wide association studies with proteomics data reveal genes important for synthesis, transport and packaging of globulins in legume seeds. New Phytol 214:1597. https://doi.org/10.1111/nph.14500
41. Kang Y, Sakiroglu M, Krom N, Stanton-Geddes J, Wang M, Lee Y-C et al (2015) Genome-wide association of drought-related and biomass traits with HapMap SNPs in Medicago

truncatula. Plant Cell Environ 38:1997–2011. https://doi.org/10.1111/pce.12520
42. Serrani-Yarce JC, Lee H-K, Tadege M, Ratet P, Mysore KS (2013) Forward genetics screening of Medicago truncatula Tntl insertion lines. Methods Mol Biol. Legume genomics 1069:93–100
43. Kang Y, Li MY et al (2016) A snapshot of functional genetic studies in *Medicago truncatula*. Front Plant Sci 7:1175
44. Ge L, Yu J, Wang H, Luth D, Bai G, Wang K, Chen R (2016) Increasing seed size and quality by manipulating BIG SEEDS1 in legume species. Proc Natl Acad Sci U S A 113:12414–12419
45. Noguero M, Le Signor C, Vernoud V, Bandyopadhyay K, Sanchez M, Fu C et al (2015) DASH transcription factor impacts Medicago truncatula seed size by its action on embryo morphogenesis and auxin homeostasis. Plant J 81:453–466. https://doi.org/10.1111/tpj.12742
46. D'Erfurth I, Le Signor C, Aubert G, Sanchez M, Vernoud V, Darchy B et al (2012) A role for an endosperm-localized subtilase in the control of seed size in legumes. New Phytol 196:738–751. https://doi.org/10.1111/j.1469-8137.2012.04296.x
47. Fiume E, Fletcher JC (2012) Regulation of Arabidopsis embryo and endosperm development by the polypeptide signaling molecule CLE8. Plant Cell 24(3):1000–1012
48. Mantiri FR, Kurdyukov S et al (2008) The transcription factor MtSERF1 of the ERF subfamily identified by transcriptional profiling is required for somatic embryogenesis induced by auxin plus Cytokinin in Medicago truncatula. Plant Physiol 146(4):1622–1636
49. Fatihi A, Boulard C et al (2016) Deciphering and modifying LAFL transcriptional regulatory network in seed for improving yield and quality of storage compounds. Plant Sci 250:198–204
50. Baud S, Kelemen Z et al (2016) Deciphering the molecular mechanisms underpinning the transcriptional control of gene expression by master transcriptional regulators in Arabidopsis seed. Plant Physiol 171(2):1099–1112
51. Carbonero P, Iglesias-Fernandez R et al (2017) The AFL subfamily of B3 transcription factors: evolution and function in angiosperm seeds. J Exp Bot 68(4):871–880
52. Lara P, Oñate-Sánchez L et al (2003) Synergistic activation of seed storage protein gene expression in Arabidopsis by ABI3 and two bZIPs related to OPAQUE2. J Biol Chem 278(23):21003–21011
53. Chen S-K, Kurdyukov S et al (2009) The association of homeobox gene expression with stem cell formation and morphogenesis in cultured Medicago truncatula. Planta 230(4):827–840
54. Curaba J, Moritz T et al (2004) AtGA3ox2, a key gene responsible for bioactive gibberellin biosynthesis, is regulated during embryogenesis by LEAFY COTYLEDON2 and FUSCA3 in Arabidopsis. Plant Physiol 136(3):3660–3669
55. Stone SL, Braybrook SA et al (2008) Arabidopsis LEAFY COTYLEDON2 induces maturation traits and auxin activity: implications for somatic embryogenesis. Proc Natl Acad Sci U S A 105(8):3151–3156
56. D'Hooghe P, Dubousset L, Gallardo K, Kopriva S, Avice J-C, Trouverie J (2014) Evidence for proteomic and metabolic adaptations associated to alterations of seed yield and quality in Sulphur-limited Brassica napus L. Mol Cell Proteomics 13:1165–1183
57. Vandecasteele C, Teulat-Merah B et al (2011) Quantitative trait loci analysis reveals a correlation between the ratio of sucrose/raffinose family oligosaccharides and seed vigour in Medicago truncatula. Plant Cell Environ 34(9):1473–1487
58. Leprince O, Pellizzaro A et al (2017) Late seed maturation: drying without dying. J Exp Bot 68(4):827–841
59. Smoot ME, Ono K et al (2011) Cytoscape 2.8: new features for data integration and network visualization. Bioinformatics 27(3):431–432
60. Bies-Etheve N, Gaubier-Comella P et al (2008) Inventory, evolution and expression profiling diversity of the LEA (late embryogenesis abundant) protein gene family in Arabidopsis thaliana. Plant Mol Biol 67(1–2):107–124
61. Kotak S, Vierling E, Bäumlein H, von Koskull-Döring P (2007) A novel transcriptional cascade regulating expression of heat stress proteins during seed development of Arabidopsis. Plant Cell 19:182–195
62. González-Morales SI, Chávez-Montes RA et al (2016) Regulatory network analysis reveals novel regulators of seed desiccation tolerance in Arabidopsis thaliana. Proc Natl Acad Sci U S A 113(35):E5232–E5241
63. Noctor G, Mhamdi A, Chaouch S, Han Y, Neukermans J, Marquez-Garcia B, Queval G, Foyer CH (2012) Glutathione in plants: an integrated overview. Plant Cell Environ 35:454–484
64. Cairns NG, Pasternak M, Wachter A, Cobbett CS, Meyer AJ (2006) Maturation of Arabidopsis seeds is dependent on glutathione biosynthesis within the embryo. Plant Physiol 141:446–455

65. Oliver MJ, Guo L, Alexander DC, Ryals JA, Wone BWM, Cushman JC (2011) A sister group contrast using untargeted global metabolomics analysis delineates the biochemical regulation underlying desiccation tolerance in *Sporobolus stapfianus*. Plant Cell 23:1231–1248
66. Lepiniec L, Debeaujon I et al (2006) Genetics and biochemistry of seed flavonoids. Annu Rev Plant Biol 57:405–430
67. Peel GJ, Pang YZ et al (2009) The LAP1 MYB transcription factor orchestrates anthocyanidin biosynthesis and glycosylation in Medicago. Plant J 59(1):136–149
68. Verdier J, Zhao J et al (2012) MtPAR MYB transcription factor acts as an on switch for proanthocyanidin biosynthesis in Medicago truncatula. Proc Natl Acad Sci U S A 109(5):1766–1771
69. Liu CG, Jun JH et al (2014) MYB5 and MYB14 play pivotal roles in seed coat polymer biosynthesis in Medicago truncatula. Plant Physiol 165(4):1424–1439
70. Li PH, Chen BB et al (2016) Regulation of anthocyanin and proanthocyanidin biosynthesis by Medicago truncatula bHLH transcription factor MtTT8. New Phytol 210(3):905–921
71. Jun JH, Liu CG et al (2015) The transcriptional repressor MYB2 regulates both spatial and temporal patterns of proanthocyandin and anthocyanin pigmentation in Medicago truncatula. Plant Cell 27(10):2860–2879
72. Pang YZ, Cheng XF et al (2013) Medicago glucosyltransferase UGT72L1: potential roles in proanthocyanidin biosynthesis. Planta 238(1):139–154
73. Zhao J, Dixon RA (2009) MATE transporters facilitate vacuolar uptake of Epicatechin 3′-O-glucoside for proanthocyanidin biosynthesis in Medicago truncatula and Arabidopsis. Plant Cell 21(8):2323–2340
74. Zhao J, Huhman D et al (2011) MATE2 mediates vacuolar sequestration of flavonoid glycosides and glycoside malonates in Medicago truncatula. Plant Cell 23(4):1536–1555
75. Li DF, Zhang YQ et al (2011) Transcriptional profiling of Medicago truncatula under salt stress identified a novel CBF transcription factor MtCBF4 that plays an important role in abiotic stress responses. BMC Plant Biol 11:109
76. Arraouadi S, Badri M et al (2012) QTL mapping of physiological traits associated with salt tolerance in Medicago truncatula recombinant inbred lines. Genomics 99(2):118–125
77. Yoder JB, Stanton-Geddes J et al (2014) Genomic signature of adaptation to climate in Medicago truncatula. Genetics 196(4):1263
78. Badri M, Bouhaouel I et al (2016) Variation in tolerance to drought among Tunisian populations of Medicago truncatula. Plant Genet Resour 14(1):41–49
79. Zhang JY, Cruz DE, Carvalho MH, Torres-Jerez I, Kang Y, Allen SN, Huhman DV et al (2014) Global reprogramming of transcription and metabolism in *Medicago truncatula* during progressive drought and after rewatering. Plant Cell Environ 37:2553–2576. https://doi.org/10.1111/pce.12328
80. Tejedor-Cano J, Prieto-Dapena P et al (2010) Loss of function of the HSFA9 seed longevity program. Plant Cell Environ 33(8):1408–1417
81. Terrasson E, Darrasse A et al (2015) Identification of a molecular dialogue between developing seeds of Medicago truncatula and seedborne xanthomonads. J Exp Bot 66(13): 3737–3752

Chapter 14

Functional Genomics and Genetic Control of Compound Leaf Development in *Medicago truncatula*: An Overview

Rujin Chen

Abstract

Diverse forms of leaves are present in nature. However, the regulatory mechanisms that underpin the development of diverse leaf forms remain enigmatic. The initiation of leaf primordia from the periphery of shoot apical meristem (SAM) requires downregulation of the class 1 knotted-like homeobox KNOXI proteins. In plants with simple leaves, this downregulation is permanent, consistent with leaves being determinant organs. In most of plants with compound leaves, the KNOXI proteins are reactivated in developing leaf primordia, and this reactivation is required for the development of compound leaves in these plants. Surprisingly, in *Medicago truncatula* and pea (*Pisum sativum*) that belong to the so-called inverted repeat-lacking clade (IRLC) of legume plants, the KNOXI proteins are not reactivated in leaf primordia and therefore not likely involved in the development of compound leaves in these plants. Instead, the legume *FLORICAULA/LEAFY* orthologues, *UNIFOLIATA* (*UNI*) and *SINGLE LEAFLET1* (*SGL1*), are required for the initiation and development of lateral leaflet primordia in pea and *M. truncatula* plants, *respectively. On the other hand, PALMATE-LIKE PENTAFOLIATA1 (PALM1*) encoding a novel Cys(2) His(2) zinc finger transcription factor is required to suppress a morphogenetic activity at the leaf margin by negatively regulating *SGL1* gene expression, and *FUSED COMPOUND LEAF1* (*FCL1*) encoding a class M KNOX protein is required for the development of the leaf proximo-distal axis and organ boundary separation in *M. truncatula*. Thus, these recent studies have shown that *SGL1/UNI*, *FCL1*, and *PALM1* provide a genetic framework for our understanding of compound leaf development in the legume plants.

Key words Compound leaf, Genetic control, *M. truncatula*, Functional genomics

1 Introduction

In nature, plant diversity is to a large degree reflected by the diverse forms and arrangements of their leaves on stems. Leaves can be classified into simple and compound groups dependent on their complexities. A simple leaf is composed of a single continuous blade. On the other hand, a compound leaf has multiple units, each resembling a single leaf called a leaflet. How different forms of leaves are regulated remain enigmatic.

Luis A. Cañas and José Pío Beltrán (eds.), *Functional genomics in* Medicago truncatula*: Methods and Protocols*, Methods in Molecular Biology, vol. 1822, https://doi.org/10.1007/978-1-4939-8633-0_14,

Both simple and compound leaf primordia are initiated from the periphery of the shoot apical meristem (SAM), a pluripotent structure capable of self-renewal. Previously, it has been shown that the class 1 knotted-like homeobox transcription factors (KNOXIs) are required for the activity and maintenance of SAM [1, 2]. The expression of KNOXI genes is excluded at the incipient leaf primordia. In simple leaf species, the expression of *KNOXI* gene is also excluded in developing leaf primordia [1, 3, 4]. In compound leaf species, including tomato (*Solanum lycopersicum*) and *Cardamine hirsuta*, however, *KNOXI* gene expression is reactivated in developing leaf primordia [3, 5]. This reactivation is required for the development of compound leaves in these species, because ectopic overexpression of *KNOXI* results in super-compound leaves and loss of function of KNOXI producing simplified leaves [5–9]. However, *KNOXI* gene expression in leaf primordia is also detected in species with simple leaves, suggesting that *KNOXI* gene expression is not sufficient for compound leaf development likely due to secondary morphogenesis [3].

In a large group of legume plants, the so-called inverted repeat-lacking clade (IRLC) of the legume family (Fabaceae), including pea (*Pisum sativum*) and *Medicago* species with compound leaves, the expression of *KNOXI* genes and proteins are not detected in developing leaf primordia, suggesting that they are not likely involved in compound leaf development in these species [10–12]. Instead, the legume *LEAFY* orthologues, *UNIFOLIATA* (*UNI*) and *SINGLE LEAFLET1* (*SGL1*), play a role in compound leaf development, since loss-of-function mutations of *UNI* and *SGL1* result in simplified and simple leaves in pea and *M. truncatula*, respectively [13, 14].

We propose that *M. truncatula*, one of the model legume species for plant-microbe symbiosis studies, is also a model species for compound leaf studies in legumes mainly because of the following attributes: (1) it has the simplest form of compound leaves [13]; (2) a large number of compound leaf development mutants have been isolated, and some of them has been characterized at the molecular level [10, 13, 15–21]; (3) genetic interaction studies can be performed easily by cross-pollination between and among mutants [10, 15, 16, 20, 21]; and (4) plants can be relatively easily transformed so that gain of function and partial and complete loss of function of a gene may be performed to allow mechanistic understanding of gene functions [19].

In this chapter, an overview of the current knowledge of processes and genes that are key for compound leaf development in the IRLC legume, *M. truncatula*, will be presented.

2 *M. truncatula SINGLE LEAFLET1* (*SGL1*) Is Required for the Development of Lateral Leaflets

In loss-of-function mutants of the pea (*Pisum sativum*) *FLORICAULA/LEAFY* orthologue *UNIFOLIATA* (*UNI*), compound leaves are simplified [14]. Similarly, in loss-of-function mutants of the *M. truncatula LEAFY/UNI* orthologue *SINGLE LEAFLET1* (*SGL1*), all leaves are simple, with greatly reduced petiole length (up to ~50% of the wild-type level dependent on developmental stages) [13].

Scanning electron microscopic image analysis shows that the initiation of lateral leaflet primordia is blocked in the *sgl1* mutants, indicating that *SGL1* is required for the initiation of lateral leaflet primordia, in addition to its role in petiole development [13].

M. truncatula sgl1 mutants develop cauliflower-like inflorescence and are sterile. Thus, the role of *SGL1* in reproductive growth is conserved with those in other species including *Arabidopsis thaliana* with simple leaves. Indeed, when *SGL1* is expressed under control of the *LFY* promoter, which is only active during transition to reproductive growth, it completely rescues the reproductive defects of *A. thaliana lfy* mutants [13]. On the other hand, when *LFY* is expressed under control of the *SGL1* promoter, which is active in both vegetative and reproductive organs (young leaves and flowers), it rescues both compound leaf and inflorescence defects of *M. truncatula sgl1* mutants, confirming that SGL1 and LFY are functional orthologues [13].

3 *M. truncatula* PALMATE-LIKE PENTAFOLIATA1 (PALM1), a Novel Cys(2)His(2) Zinc Finger Transcription Factor, Regulates Compound Leaf Development by Repressing *SGL1* Gene Expression

We have isolated an allelic series of *M. truncatula* mutants exhibiting compound leaves with five leaflets, a terminal leaflet, and two pairs of lateral leaflets (LLs), clustered at the tip of petiole, in contrast to trifoliate leaves in wild-type plants [21]. Scanning electron microscopic image analysis indicates that an extra pair of LL primordia is developed at the proximal region of LL primordia developed earlier in *M. truncatula palm1* mutants [21]. In addition to the changes in compound leaf patterning, the proximal-distal axis of leaves is also changed in the *M. truncatula palm1* mutants. Compared with wild-type counterparts, the petiole length of mature leaves in 6-week-old *palm1-1* mutant is increased by ~20%, whereas the central rachis length is reduced by ~19%, although extra rachis structures are developed on the two distally oriented LLs in the *palm1* mutants [21].

PALM1 encodes a Cys(2)His(2) zinc finger transcription factor with a previously unknown function. RNA in situ hybridization analysis shows that *PALM1* transcripts are detected in LL primordia as early as the P2 stage, but remain low or are hardly detected in other tissues such as SAM, terminal leaflets, and stipules [21]. The expression pattern of *PALM1* is partially overlapping with *SGL1*, which is expressed in both SAM and entire leaf primordia [13, 21]. Quantitative RT-PCR analysis shows that *SGL* gene expression is significantly upregulated in the shoot apices in the *palm1* mutant compared with wild type. In addition, *SGL1*pro::*uidA* (*GUS*) reporter gene is expressed in all five leaflets, and its expression remains at a high level in expanding leaflets in the *palm1* mutant, in contrast to its expression being primarily in SAM and young leaflets, but greatly attenuated in expanding leaflets in wild type [13, 21].

All leaves developed in *palm1-3;sgl1-1* double mutants are simple, similar to that of the *sgl1* mutants [21]. Thus, the genetic interaction studies show that *sgl1* is epistatic to *palm1*, indicating that *SGL1* is required for the proliferation of LL primordia in the loss-of-function *palm1* mutants [21].

In contrast to *LFY*pro::*GUS* reporter gene, which is not expressed in vegetative tissues, *SGL1*pro::*GUS* reporter gene is expressed in leaf tissues in transgenic *Arabidopsis thaliana* plants [13]. When *PALM1* is ectopically overexpressed, *SGL1*pro::*GUS* reporter gene expression is greatly repressed in leaves as indicated by quantitative RT-PCR and GUS staining analyses [21]. Furthermore, electrophoretic mobility shift assays show that purified PALM1 proteins bind to a specific sequence between −354 and −747 nucleotide in the *SGL1* promoter [21]. Taken together, these results collectively show that PALM1 is a negative regulator of *SGL1* gene expression.

Interestingly, although ectopic overexpression of *PALM1* does not change the simple leaf and flower development in transgenic *A. thaliana* plants, it completely blocks excessive leaf lobing and lamina outgrowth caused by overexpression of the *KNOXI* gene, *KNAT1*, suggesting that PALM1 may also repress leaf morphogenetic processes that are regulated by *KNOXI* genes in a heterologous system [21].

4 *M. truncatula FUSED COMPOUND LEAF1* (*FCL1*), Encoding a Class M KNOX Protein, Regulates LEAF Proximal-Distal Axis Development

We have isolated two allelic deletion mutants named *fcl1-1* and *fcl1-2* from a fast neutron bombardment (FNB)-induced mutant collection in *M. truncatula* cv. Jemalong A17 background [10]. *fcl1* mutants exhibit greatly reduced petiole length (~50% of wild-type level dependent on developmental stages) and various compound leaf patterning defects including simple leaf, leaflet fusion, and clustering of leaflets at the tip of petiole without rachis. Similar to that of *sgl1* mutants, the reduction in the petiole length in the *fcl1* mutants is likely caused by reduced cell division but not cell expansion [10].

FCL1 encodes a class M KNOX protein that contains both KNOX1 and KNOX2 domains (called the MEINOX domain) but lacks the homeodomain [10]. FCL1 shares a high amino acid sequence similarity with KNATM-B (63% similarities and 43% identities) and PTS/TKD1 (60% similarities and 48% identities), two class M KNOX proteins identified in *A. thaliana* and tomato, respectively [10, 22, 23].

Consistent with its role in compound leaf patterning and petiole development, RNA in situ hybridization analysis shows that *FCL1* transcripts are detected in leaf primordia at P0 and early P1 stages [10]. At later stages, FCL1 transcripts are detected at boundaries between leaf primordia and SAM and at the proximal adaxial domain of developing leaf primordia. On the other hand, *FCL1* transcripts are not detected in SAM. Expression of FCL1-GFP fusion proteins is detected in both the cytoplasm and nucleus in tobacco leaf epidermal cells, suggesting that FCL1 may move freely between nucleus and cytoplasm likely due to its small size and no known subcellular localization signals in its sequence [10].

Genetic studies show that *sgl1;fcl1* double mutants exhibit simple leaves similar to that of *sgl1* mutants and completely lack petioles, indicating that *sgl1* is epistatic to *fcl1* in compound leaf patterning, and *SGL1* and *FCL1* play an additive role in petiole development [10]. At the molecular level, SGL1 and FCl1 do not appear to regulate each other's expression, and they also do not interact with each other in yeast two-hybrid assays [10].

On the other hand, *fcl1;palm1* double mutants exhibit compound leaf phenotypes, including patterning and petiole development similar to that of *fcl1* mutants, indicating an epistatic interaction between *fcl1* and *palm1* [10]. Interestingly, upregulation of *SGL1* gene expression detected in the *palm1* single mutants is not affected in the *fcl1;palm1* double mutants, although *fcl1;palm1* double mutants do not develop extra leaflets [10]. These results suggest that, although *FCL1* is not required for the upregulation of *SGL1* expression, it is required for the development of extra leaflet primordia in the *palm1* mutants.

How FCL1 regulates compound leaf patterning and petiole development in *M. truncatula* remains enigmatic. Studies in *A. thaliana* have shown that the class M KNOX protein KNATM-B interferes the function of the class 1 KNOX proteins (KNOXI) by interacting with their binding partners, the homeodomain BELL proteins, including BEL1, SAW1, SAW2, and PNY, but not the PNY paralogue PNF [22]. Interestingly, FCL1 interacts strongly with a subset of the BELL proteins from *A. thaliana*, including

PNF, SAW1, and BEL1, but not with SAW2 and PNF [10]. Thus, FCL1 interacts with BELL proteins with slightly different specificities from KNATM-B. In addition, in contrast to KNATM-B, which interacts strongly with KNAT1/BP and weakly with KNAT3 and KNAT4 in yeast two-hybrid assays, FCL1 does not interact with KNAT1 and other KNOX proteins from *A. thaliana*. In *M. truncatula* and pea (*P. sativum*) that belong to the IRLC clade of legumes, the class 1 KNOX proteins are not likely involved in regulating compound leaf development since they are not expressed in developing leaf primordia. It is plausible that FCL1 regulates compound leaf patterning and petiole development through interacting with a subset of the homeodomain BELL proteins, and this process may not involve the class 1 KNOX proteins [10].

Acknowledgments

This work was supported in part by National Science Foundation (DBI-0703285 and IOS-1127155), Oklahoma Center for the Advancement of Science and Technology (PS12-036 and PS16-034), Noble Research Institute, LLC, and Lanzhou University.

References

1. Long JA, Moan EI, Medford JI, Barton MK (1996) A member of the KNOTTED class of homeodomain proteins encoded by the *STM* gene of Arabidopsis. Nature 379(6560):66–69
2. Clark SE, Jacobsen SE, Levin JZ, Meyerowitz EM (1996) The *CLAVATA* and *SHOOT MERISTEMLESS* loci competitively regulate meristem activity in *Arabidopsis*. Development 122(5):1567–1575
3. Bharathan G et al (2002) Homologies in leaf form inferred from *KNOXI* gene expression during development. Science 296(5574):1858–1860
4. Lincoln C, Long J, Yamaguchi J, Serikawa K, Hake S (1994) A *knotted1*-like homeobox gene in Arabidopsis is expressed in the vegetative meristem and dramatically alters leaf morphology when overexpressed in transgenic plants. Plant Cell 6(12):1859–1876
5. Hay A, Tsiantis M (2006) The genetic basis for differences in leaf form between *Arabidopsis thaliana* and its wild relative *Cardamine hirsuta*. Nat Genet 38(8):942–947
6. Janssen BJ, Lund L, Sinha N (1998) Overexpression of a homeobox gene, *LeT6*, reveals indeterminate features in the tomato compound leaf. Plant Physiol 117(3):771–786
7. Parnis A et al (1997) The dominant developmental mutants of tomato, *Mouse-ear* and *Curl*, are associated with distinct modes of abnormal transcriptional regulation of a *Knotted* gene. Plant Cell 9(12):2143–2158
8. Chen JJ, Janssen BJ, Williams A, Sinha N (1997) A gene fusion at a homeobox locus: alterations in leaf shape and implications for morphological evolution. Plant Cell 9(8):1289–1304
9. Hareven D, Gutfinger T, Parnis A, Eshed Y, Lifschitz E (1996) The making of a compound leaf: genetic manipulation of leaf architecture in tomato. Cell 84(5):735–744
10. Peng J et al (2011) Regulation of compound leaf development in *Medicago truncatula* by *fused compound leaf1*, a class M *KNOX* gene. Plant Cell 23(11):3929–3943
11. Champagne CE et al (2007) Compound leaf development and evolution in the legumes. Plant Cell 19(11):3369–3378
12. Hofer J, Gourlay C, Michael A, Ellis TH (2001) Expression of a class 1 *knotted1*-like homeobox gene is down-regulated in pea compound leaf primordia. Plant Mol Biol 45(4):387–398
13. Wang H et al (2008) Control of compound leaf development by *FLORICAULA/LEAFY* ortholog *SINGLE LEAFLET1* in *Medicago truncatula*. Plant Physiol 146(4):1759–1772

14. Hofer J et al (1997) *UNIFOLIATA* regulates leaf and flower morphogenesis in pea. Curr Biol 7(8):581–587
15. Peng J, Berbel A, Madueño F, Chen R (2017) AUXIN RESPONSE FACTOR3 Regulates Compound Leaf Patterning by Directly Repressing *PALMATE-LIKE PENTAFOLIATA1* Expression in *Medicago truncatula*. Front Plant Sci 8:1630. https://doi.org/10.3389/fpls.2017.01630
16. Ge L, Peng J, Berbel A, Madueno F, Chen R (2014) Regulation of Compound Leaf Development by *PHANTASTICA* in *Medicago truncatula*. Plant Physiol 164(1):216–228
17. Ge L, Chen R (2014) PHANTASTICA regulates leaf polarity and petiole identity in Medicago truncatula. Plant Signal Behav 9(3):e28121
18. Cheng X et al (2012) NO APICAL MERISTEM (MtNAM) regulates floral organ identity and lateral organ separation in Medicago truncatula. New Phytol 195(1):71–84
19. Chen J et al (2012) Conserved genetic determinant of motor organ identity in *Medicago truncatula* and related legumes. Proc Natl Acad Sci U S A 109(29):11723–11728
20. Peng J, Chen R (2011) Auxin efflux transporter MtPIN10 regulates compound leaf and flower development in *Medicago truncatula*. Plant Signal Behav 6(10):1537–1544
21. Chen J et al (2010) Control of dissected leaf morphology by a Cys(2)His(2) zinc finger transcription factor in the model legume *Medicago truncatula*. Proc Natl Acad Sci U S A 107(23):10754–10759
22. Magnani E, Hake S (2008) *KNOX* lost the *OX*: the *Arabidopsis KNATM* gene defines a novel class of KNOX transcriptional regulators missing the homeodomain. Plant Cell 20(4):875–887
23. Kimura S, Koenig D, Kang J, Yoong FY, Sinha N (2008) Natural variation in leaf morphology results from mutation of a novel *KNOX* gene. Curr Biol 18(9):672–677

Chapter 15

Root Development in *Medicago truncatula*: Lessons from Genetics to Functional Genomics

Hélène Proust, Caroline Hartmann, Martin Crespi, and Christine Lelandais-Brière

Abstract

This decade introduced "omics" approaches, such as genomics, transcriptomics, proteomics, and metabolomics in association with reverse and forward genetic approaches, developed earlier, to try to identify molecular pathways involved in the development or in the response to environmental conditions as well as in animals and plants. This review summarizes studies that utilized "omics" strategies to unravel the root development in the model legume *Medicago truncatula* and how external factors such as soil mineral status or the presence of bacteria and fungi affect root system architecture in this species. We also compare these "omics" data to the knowledges concerning the *Arabidopsis thaliana* root development, nowadays considered as the model of allorhiz root systems. However, unlike legumes, this species is unable to interact with soil nitrogen-fixing rhizobia and arbuscular-mycorrhizal (AM) fungi to develop novel root-derived symbiotic structures. Differences in root organization, development, and regulatory pathways between these two model species have been highlighted.

Key words Root system architecture, Legume, Transcription factor, MicroRNA, Peptides, Hormones, Root apical meristem, Lateral root, Nitrogen

1 Introduction

Roots play major roles in plant biology as they ensure the anchorage in the soil, the mineral nutrition, and the interactions with various soil-living beneficial and pathogenic organisms. Two types of root system architectures (RSA) exist in adult angiosperm plants. The homorhiz RSA of monocotyledon species is typified by postembryonic adventitious roots, while an allorhiz RSA is observed in eudicotyledons [1]. Classically, an allorhizic or taproot system is composed of a primary root (PR) that corresponds to the embryonic root and a network of postembryonic lateral roots (LRs) that emerge first from the PR and then from the primary LRs. The length of the PR as well as the number, the root tip angle, and

Luis A. Cañas and José Pío Beltrán (eds.), *Functional genomics in* Medicago truncatula*: Methods and Protocols*, Methods in Molecular Biology, vol. 1822, https://doi.org/10.1007/978-1-4939-8633-0_15,

the length of the LRs define the RSA, which gives an idea about the volume of soil explored by the plant [2]. RSA is genetically determined but also influenced by environmental factors [3, 4]. For all species examined to date, RSA is highly plastic; indeed, root growth and development are always strongly impacted by environmental cues such as mineral and water availabilities but also by interactions with fungal and/or bacterial species. Moreover, it is now clear that cross talks between biotic and mineral signaling but also between different mineral signaling pathways exist and control RSA [5].

The way by which the plant increases soil exploration in response to environmental conditions varies strongly across species but generally depends on soil mineral composition [5]. The enhanced growth of PR or LRs and/or the organogenesis of new LRs support the increase of the RSA volume. However, even if key players are common for PR and LR growth, we know that they have differential growth dynamics at least in certain environmental conditions [6–8]. In addition, root and shoot growths are coordinated by systemic signals such as hormones, small RNAs, and/or short peptides that convey information about carbon status in the shoot and the availability of mineral resources and water in the root.

Root anatomy is characterized by several layers of concentric tissues. The epidermis is the first outside layer in contact with soil. Inside, according to the species, the cortex tissue can be constituted by one or several cell layers. More depth, the unicellular endodermis layer regulates water and nutrient transfer into the stele, through the Casparian strip. The pericycle cells constitute the outside stele layer; the latter encompasses cells, generally in front of the xylem poles, that could divide again and form the LR primordia. Inside the stele, vascular tissues composed of several cell types are found; xylem and phloem vessels transport water, nutrients, photo-assimilates, and hormones.

Along its apical-basal axis, the root tip is organized into successive regions: the root cap protecting the root apical meristem (RAM) where cells divide, the elongation (EZ), and the differentiation (DZ) zones. The latter may be visualized by the presence of long epidermal cells specialized in water and mineral uptake, the root hairs. The boundaries between RAM, EZ, and DZ are not strict but rather hazy. In dicots species, there are three RAM organizations: closed, basic open, and intermediate open [9]. Closed RAM has been largely studied in the model species *Arabidopsis thaliana*; in this species, specific initials (stem cells) are connected to each root cell type to define its lineage. In contrast, legume species, like pea or *Medicago truncatula*, display a basic open RAM, where no identifiable initials cannot be clearly connected to specific root cell files [10]. RAM organization can evolve during plant life cycle. For example, in *A. thaliana* [11] but also in other plants [12], the RAM organization changes from a closed RAM organization to an open one when the PR reaches its determinate size. The RAM type seems to have also an influence on the type of border cells released from the lateral part of the root cap. Indeed, plants with open RAM release living border cells in soil where they can continue to divide while closed RAM plants deliver dead cells [13].

In *A. thaliana*, LR development comprises several stages: priming of the founder cells, initiation of the LR primordium, which progresses through eight stages until emergency. LR priming occurs in a developmental window behind the root tip, at the level of the EZ, where cells of the lateral root cap are subjected to programmed cell death. The systematic cell death of lateral root cap cells produces a periodic influx of auxin that induces expression oscillations of specific genes able to prime cells which could subsequently become the future branching sites [14]. When the LR development program is launched, some pericycle cells in front of xylem poles can gain meristematic activity. By using endodermal cell ablation, it has recently been shown that endodermal cells mechanically inhibit the division of pericycle cells via a restriction of cell enlargement. During LR initiation, auxin activates a network that acts at least on cell wall remodeling genes at endodermal cells in front of xylem poles. These cell wall modifications remove mechanical constraints and thus permit the reactivation of the cell cycle in pericycle cells [15]. Auxin also acts directly on pericycle cells to generate formative asymmetric divisions at the origin of the LR primordia [16]. The LR morphogenesis seems to be directed by the mechanical properties of overlying tissues rather than by a strict pattern of division plans [17]. Thereafter, LR emergence results from constraints linked to the enlargement of primordium cells, and then the process is followed by the activation of the LR meristem allowing the new lateral organ to grow [18]. Unlike *A. thaliana*, legume root systems can interact with soil nitrogen-fixing rhizobia and arbuscular-mycorrhizal (AM) fungi to develop symbiotic relationships. These mutualistic interactions imply the formation of new specialized structures, the nodules or the arbuscules, which allow the plant to cope with N- or P-limiting conditions, respectively. Early steps of these two symbioses are regulated by lipo-chitooligosaccharide (LCO) signal molecules produced by the bacteria or the fungus, called the Nod and the Myc factors, respectively. Their signaling pathways integrate root developmental regulatory networks to allow infection and formation of the new structures originating from or inside the cortical cells. Interestingly, applications of these LCOs on the root system stimulate LR development [19], confirming the tight interconnections between their respective signaling pathways and root developmental programs.

RSA is controlled by intrinsic (genetic and hormonal) regulators and by external environmental conditions. Several studies mainly in *A. thaliana* have dissected the genetic networks that control RAM functions, root growth, and LR organogenesis even in a cell-specific manner [20–22]. These last years, in addition to transcriptions factors (TF), new regulators of gene expression such as microRNAs (miRNA) emerged as important players of root development in plants including legumes [23, 24]. Most of these small riboregulators are posttranscriptional repressors of key developmental TFs [25]. Moreover, small peptides that could mediate short- and long-distance signals from root to shoot or shoot to root have also been implicated in the RSA [26, 27], in the symbiotic nodulation [28, 29], and in the mineral ion homeostasis through the endodermal Casparian strip [30]. Genomic approaches also allowed to better integrate the different hormone signaling pathways and their cross talks in the different stages of root development [20, 31].

In this review, we will first describe how recent global omics approaches in *M. truncatula* have allowed identifying novel putative regulators of root development and their associated networks. We will then focus on the known genetic determinants of root development in *M. truncatula* (Table 1), highlighting the similarities and specificities in comparison to *A. thaliana*. Finally, we will present the actual knowledges on the hormonal and environmental control of root development in this model legume species.

1.1 Attempts to Define a Root Developmental Regulatory Map by Omics Approaches

Since two decades, global in silico and experimental approaches using functional genomics have given access to large sets of regulatory genes such as TFs and miRNAs involved in root development in *M. truncatula*.

Proteomic analyses have been performed in roots grown under standard conditions [67, 68], infected with beneficial microorganisms or pathogenic/parasitic agents [69, 70] or under different treatments [71, 72]. In addition, more dedicated root proteomes have focused on various subcellular compartments [70, 73, 74] or small proteins/peptides [75]. However, to our knowledge, the specificity of the root proteome in comparison to other organs was rarely investigated [67, 68]. According to these studies, a large part of the root-specific or enriched proteins were associated with defense reactions and stress responses, even in absence of biotic or abiotic stress. In addition, several were regulated by abscisic acid (ABA), a phytohormone associated both with abiotic stress responses and root development in plants [67, 76]. Recently, Marx et al. [68] used high-resolution tandem mass spectrometry to establish a complete atlas of proteins and posttranslational modifications in *M. truncatula* nodule cells and its rhizobial symbiont *Sinorhizobium meliloti* and integrated them into regulatory networks linking transcriptome [77] and proteome complexity. In this study, comparison of vegetative and reproductive organs showed that, among the 12,992 proteins identified in roots, only 155 accumu-

Table 1
Genes functionally linked to root development in *Medicago truncatula*

Subgroup	Gene	Protein family	Role	Identification	Expression analysis	Functional analysis	microRNA targets	References
Transcription factors								
	WOX5	WUSCHEL related homeobox	Meristem maintenance	*At* homology	Prom:GUS			[32]
	PLT1 to PLT4	AP2/EREBP TF	Meristem maintenance	*At* homology	Prom:GUS	RNAi		[33]
	KNOX3	Class II KNOX homeodomain TF		*At* homology	Prom:GUS	OE, RNAi		[34]
	GRF2, 4, 6	Growth-regulating factor	Cell division	*At* homology	Prom:GUS	RNAi		[35]
	ARF2,3,4	Auxin-response-factor	LR emergence	miR390/TAS3 siRNA target	Prom:GUS-GFP			[36]
	RR5, 9, 11	Response regulator	LR number	*At* homology	Prom:GUS			[37]
	MtMIG1	MADS-box	Growth	*Mt* other family members homology	Prom:GUS	OE		[38]
	NAC696	NAC transcripftion factor	Growth	Transcriptomic	HIS	RNAi, OE		[39]
	HB1	HD-zip transcription factor	Growth/LR emergence	Transcriptomic	Prom:GUS	Tilling mutant, OE		[40]

(continued)

Table 1 (continued)

Subgroup	Gene	Protein family	Role	Identification	Expression analysis	Functional analysis	microRNA targets	References
	LOB1	Lateral organ boundary	Growth/LR emergence	DNA binding assay with *HB1*	Prom:GUS	Mutated-promoter:cDNA		[41]
	CBF4	C-repeat binding factor	Growth	Transcriptomic		OE in Arabidopsis		[42]
microRNA								
	miR156		Cell elongation	*At* homology		OE	SPL/SBP TFs, WD40 TF	[43]
	miR166		Vasculature differentiation	*At* homology	HIS	OE, mimicry	HD-ZIP III TFs (MtCNA1, MtCNA2 and MtHB8)	[44]
	miR393		LR growth	*At* homology			TIR1/AFB auxin receptor	[45]
	miR160		Meristem organization, root cap differentiation, growth, gravitropism	*At* homology			ARF10, ARF16, ARF17	[36]
	miR390/ tas3 tasiRNA		LR emergence	*At* homology	Prom:GUS	OE, mimicry	ARF2, ARF3, ARF4	[46]

Hormone/ peptide signaling							
	MtCHK1/ CRE1	Histidine kinase	Growth, LR development	EMS mutant collection	Prom:GUS	EMS mutant, RNAi	[149– 151]
	MtHK2,3,4	Histidine kinase		*At* homology		RNAi	[47–49]
	LOG1	Cytokinin riboside 5′-monophosphate phosphoribohydrolase	Growth, vasculature formation, lateral root development	*At* homology	Prom:GUS	OE, RNAi	[50]
	MtDELLA 1, 2, 3	Interactor of GA receptor	Growth	*At* homology	Prom:GUS; GFP translational fusion	Tnt1 insertion mutant	[51]
	MtBRI1	Leucine-rich repeat receptor protein kinase	Growth, LR development	*At* homology		OE, Tnt1 insertion mutant	[52]
	CEP1	C-terminally encoded peptide	Growth/LR emergence	*At* homology	Prom:GUS	OE, RNAi, application of synthetic peptide	[53, 54]
	CRA2	Leucine-rich repeat receptor protein kinase	Growth/LR number	*At* homology	Prom:GUS	Tnt1 insertion mutant	[55]

(continued)

Table 1
(continued)

Subgroup	Gene	Protein family	Role	Identification	Expression analysis	Functional analysis	microRNA targets	References
	SUNN	Leucine-rich repeat receptor protein kinase	LR number/PR growth	*At* homology	Prom:GUS	Tnt1 insertion mutant		[56–58]
Cell cycle/ mitosis								
	CYC A2	Cyclin	Cell division	*At* homology	Prom:GUS	OE, antisens		[59]
	MtCDC16	E3 ubiquitin ligase	Cell division	*At* homology	Prom:GUS	RNAi		[60]
Other categories								
	TR185	ND	LR development	Gamma-ray mutagenesis				[61]
	CRA1	ND	Growth	T-DNA insertion mutant collection		Mutant		[62]
	MtNPF1.7/ LATD/ NIP	Nitrate transporter	Growth	EMS mutant screen		Mutant		[63]
	MtNPF6.8/ MtNRT1–1	Nitrate/ABA transporter	Growth	QTL	Prom:GUS, GFP translational fusion	RNAi		[64, 65]
	PT4	Phosphate transporter	LR number	*Lj* homology		Tilling mutant, RNAi		[66]

At Arabidopsis thaliana, *Mt Medicago truncatula*, *Lj Lotus japonicas*, *OE* overexpression, *HIS* in situ hybridization, *ND* not determined in *M. truncatula*, *TF* transcription factor, *RNAi* RNA interference, *PR* primary root, *LR* lateral root

lated exclusively in this organ. These proteins mostly corresponded to enzymes involved in secondary metabolism, cell wall structure, defense reaction/stress responses, as well as several kinases. Interestingly, a RALF1 peptide (RAPID ALKALINIZATION FACTOR 1) was identified whose *A. thaliana* homologue is well known to be mostly active in root tips, where it suppresses cell elongation by activating the receptor FERONIA [78]. This peptide-receptor complex then triggers phosphorylation of the plasma membrane H^+–ADENOSINE TRIPHOSPHATASE 2 (AHA2) mediating the inhibition of proton transport and extracellular alkalinization. Unfortunately, in these proteomic studies, very few known regulatory proteins involved in development, such as TFs, have been detected, certainly because of their strict tissue specificity or low expression levels. Despite this limit, a root-specific CCAAT box binding TF homologous to AtNF-YB1/AtHAP3 has been detected using proteomics [67]. In *A. thaliana*, the overexpression of *AtHAP3b* promoted root cell elongation [79]. Moreover, two novel TFs, a WRKY23-like and a basic HELIX LOOP HELIX bHLH35-like protein, were described as root specific by Marx et al. [68].

Genome sequencing of the three model legumes – soybean [80], *Lotus japonicus* [81], and *M. truncatula* [82] – as well as increasing transcriptomic resources – authorized the in silico identification of around 70 TF families characterized by their conserved domains [83]. These families were registered in integrated databases such as the LegumeTFDB [83] and the PlantTFDB [84]. A platform of quantitative RT-PCR (qRT-PCR) was also constructed for *M. truncatula* TFs [85]. Moreover, complete expression profiles of 1427 TFs, present on the Affymetrix *Medicago* Gene Chip, are available in a specific section of the *M. truncatula* Gene Expression Atlas (https://mtgea.noble.org/v1/list_tf.php, 77). The high sensitivity of the transcriptomic analyses thus provided the opportunity to detect larger sets of transcriptional regulators in *M. truncatula* roots. For instance, Beneditto et al. [77] used the Affymetrix GeneChip Medicago Genome Array to compare eight organs of adult 4-week-old plants. Among 32,098 transcripts detected in roots, they found that 908 were enriched in this organ in comparison to others (more than twofold). More recently, Roux et al. [86] developed an RNASeq approach to compare transcriptomic profiles of roots and mature nitrogen-fixing nodules. They identified 30,213 mRNAs expressed in both organs out of which only 30 mRNAs were strictly root-specific. This was expected as large parts of the transcriptome of the nitrogen-fixing nodules are linked to signaling pathways or development processes involved in LR development, even if nodule-specific genes were detected [87]. Thus, several members of TF families involved in root development [such as WUS HOMEOBOX-CONTAINING (WOX), GRAS (for GAI, RGA, SCR) including SCARECROW (SCR),

NAC (for NO APICAL MERISTEM, ATAF, CUP-SHAPED COTYLEDON), NFY (for NUCLEAR TRANSCRIPTION FACTOR Y), or ARF (for AUXIN RESPONSE FACTOR) families] were also expressed during nodule formation. More than 800 transcripts were highly enriched in roots versus nodules, including 41 TFs (with a log2 fold change higher than 4, data from Roux et al., [86]). These root-enriched TFs included orthologues of essential regulators of root development in *A. thaliana*, such as the GRAS SHORTROOT (SHR) TF involved in stem cell renewal and radial patterning [88, 89], ROOT HAIR DEFICIENT 6 (RHD6) acting in root hair differentiation [90, 91], and NAC15/ BEARSKIN 1(BSK1) involved in root cap maturation [92]. The presence of TFs associated with the root cap or the root hairs in this class is likely linked to the fact that these two cell types are absent in indeterminate nodules. However, the establishment of developmental gene regulatory networks by meta-analysis in *M. truncatula* roots remains challenging first because available public data were obtained using different technologies, from dedicated macroarrays to recent RNASeq, but also because root RNAomes are highly variable according to the developmental stage or environmental conditions.

Recent RNASeq approaches have enabled the discovery of large sets of nonprotein-coding RNAs [93]. These RNAs (ncRNA) are crucial regulators of gene expression and key actors in plant development and adaptation to the environment. In plants, they are classified into long ncRNAs (more than 200 nt and generally without an ORF longer than 100 codons) and short ncRNAs (miRNAs and small-interfering RNAs, 20 to 24 nt in length). As for *A. thaliana*, Wang et al. [93] reported that long ncRNA populations are more diverse and have higher expression levels in roots than in leaves (18,275 and 11,501 candidates, respectively). Most are not conserved in *A. thaliana*. Interestingly, in contrast to mRNAs, a high proportion of long ncRNAs seem to be root-specific (8711 lncRNAs). In this study, gene networks associating lncRNAs with their overlapping or neighboring genes were constructed, thanks to co-regulation analyses. However, additional functional data and tissue-specific studies are still lacking to validate these networks. Indeed, the unique lncRNA with a known biological function in *M. truncatula* is *MtENOD40*, a key regulator of nodulation. This gene functions both by its RNA molecule that relocalizes a nuclear RNA-binding protein MtRBP1 into the cytoplasm [94] and by two short encoded peptides regulating sucrose synthase activity [95, 96].

In contrast to long ncRNAs, several high-throughput sequencing studies have characterized the microRNA repertoire in roots of the Jemalong A17 reference plants, grown under standard conditions or under various abiotic or biotic treatments [97–106]. MicroRNAs (miRNAs) generally repress the expression of specific target genes by posttranscriptional degradation or by translation inhibition [107]. Their biogenesis results from the processing of long ncRNA precursors with a hairpin structure. Up to now, a total of 710 *M. truncatula* miRNA genes belonging to 261 families are registered in the miRNA database miRBase v21 (http://www.mirbase.org/). Formey et al. [105] showed that, like for long ncRNAs, miRNA populations in roots are very distinct from those observed in leaves. Indeed, 47 miRNAs (14%), including conserved and novel candidates, were specifically detected in roots. In an attempt to further characterize the *M. truncatula* root miRNAome, we have compared the miRNAs detected in four different biological experiments performed on roots of plants grown under standard conditions [99, 100, 105, 106]. This allowed the identification of 96 miRNA families that were detected in at least two distinct studies. This set included all plant conserved miRNAs previously known to regulate key developmental TFs [107]. Interestingly, members of the miR156 and miR166 families were the most abundant in *M. truncatula* root libraries. In *A. thaliana*, miR156 negatively regulates specific members of the conserved SQUAMOSA PROMOTER BINDING PROTEIN (SBP) family of TFs, notably involved in vegetative-adult transition and flowering. Some miR156-repressed SBPs also regulate adventitious and LR organogenesis and growth [108, 109]. Surprisingly, although its functions in shoot growth and flowering are conserved in legumes, this conserved miRNA seems to play contradictory roles in root development in different legume species. Indeed, its ectopic expression led to an enhanced root growth and a maintenance of nodulation in alfalfa (*Medicago sativa*) [110], while shorter roots and reduced nodulation were described in *Lotus japonicus* [111]. In *Medicago* species, in addition to SBPs, novel targets were identified by 5′RACE-PCR in *M. truncatula* root tips [112] or by transcriptomic analysis of miR156-overexpressing (OE) alfalfa plants [43], suggesting a diversification of its function. In the future, it may be interesting to investigate the putative roles of these species-specific regulations in root development.

MIR166 controls SAM functions and shoot branching in several species [113]. In roots, this miRNA is also an essential regulator of xylem differentiation through the dose-dependent regulation of its class III homeodomain-leucine zipper (HD-Zip III) TF targets [44, 114]. In addition, in the RAM, *AtmiR166*, regulated by the well-known SHORTROOT/SCARECROW module, controls root growth through cross talks of at least six hormones [115]. Nonetheless, Singh et al. have shown that auxin decreases the accumulation of the mature miR165/166 [116]. In *M. truncatula*, in situ hybridization experiments revealed that *MtmiR166* and three HD-ZIP III targets, *MtCNA1*, *MtCNA2*, and *MtHB8*, are expressed in the RAM and in vascular tissues of the root differentiation zone [44]. They were also detected in the apical region of emerging LRs, where the meristem is activated. Although no obvious defect in root growth was noticed, overexpression of a *MtmiR166a* precursor containing two tandem copies of miR166 in transgenic roots led to a reduced expression of its targets concomitant with defects in lateral organogenesis (both LR and nodules). In accordance with the function of miR166 in xylem differentiation, this phenotype was linked to an ectopic formation of xylem poles and an aberrant disposition of vascular cells in the root stele [44].

Taken together, proteomics and transcriptomic approaches associated with genome-wide gene annotations have provided the identification of conserved and novel regulators of root development in *M. truncatula*. Nevertheless, the comprehension of the conserved and specific regulatory mechanisms involved in root growth and development in *M. truncatula* will likely require additional root cell tissue-specific studies and appropriate experiments allowing to integrate data from the different omic approaches.

1.2 Genetic Determinants of M. truncatula Root Growth and Development

1.2.1 Root Apical Meristem and Growth

Divisions of the meristematic cells in the RAM are at the origin of root growth. In *A. thaliana*, the stem cell niche comprises the quiescent center (QC) surrounded by the stem cells or initials which are at the origin of all root file tissues. Stem cell identities are maintained by the QC, whose identity is maintained by auxin [117, 118] and complex genetic networks (reviewed in Drisch and Stahl [119]). Daughter cells of the initials proliferate through the meristematic zone (MZ) until they reach the transition zone (TZ) [120]. At the level of TZ, cells stop to divide and start to elongate and then differentiate. In *M. truncatula* due to the basic-open RAM organization, visualization of the different root tip regions is difficult.

In *A. thaliana*, the CLAVATA3 (CLV3) and CLE40 peptides and their respective receptor kinases (CLV1 and CRINKLY4) are responsible for the maintenance of the SAM and the RAM, through the regulation of the homeodomain WUSCHEL (WUS) and WUSCHEL-RELATED HOMEOBOX 5 (WOX5) TFs, respectively. In root tips of *M. truncatula*, cell files converge to a small region behind the root cap that must correspond to the stem cell niche of the basic-open meristem. Indeed, quite a few cells in this region express both the auxin-sensitive synthetic promoter DR5 fused to the *GUS* reporter gene (DR5:*GUS*) and the *WUSCHEL-RELATED HOMEOBOX 5* (*WOX5*) gene [33], two markers characterizing the QC cells [121, 122]. In *M. truncatula*, the maintenance of RAM stem cell populations could also be controlled by the MtCLE39 peptide, encoding a peptide homologue to the AtCLE40 [30, 123].

The PLETHORA (PLT) subfamily is also essential for root stem cell maintenance [124]. These APETALA2-domain containing TFs are required for specification and maintenance of stem cells within the RAM, where *PLT* expression is controlled by the auxin maximum [124]. In *M. truncatula*, expression profiles of *MtPLT1* to *MtPLT4* in the root tip were always more large and diffuse than those observed in the *A. thaliana* closed meristem [33, 125, 126]. Moreover, in contrast to *A. thaliana*, *MtPLT3* and *MtPLT4* showed a shoot-ward extension of expression [33]. Interestingly, the *PLT* gene *cis*-regulator elements seem to be conserved between the two species. Indeed, the transcriptional *pAtPLT1 to 4:GUS* fusions showed similar patterns of expression in *A. thaliana* and *M. truncatula* RAMs despite the differences of their organization [126].

The use of a *CYCLIN-A2:GUS* construct as a marker of the cell cycle as well as microscopy analyses after propidium iodide staining of root tips have allowed to visualize the zone of active cell proliferation (MZ) behind root stem cell niche in the *M. truncatula* RAM [35, 59]. In *A. thaliana*, the conserved GROWTH-REGULATING FACTORS (GRF) TFs regulate organ growth by controlling cell division and entry into endoreduplication cycles [127–129]. Rodriguez et al. showed that the *AtGRF* genes are expressed in the dividing cells of the MZ where they repress the expression of *AtPLT* genes, thus modulating their spatial expression gradient in the RAM [130]. Conversely, PLTs activate *AtMIR396* expression in the stem cells. This conserved miRNA posttranscriptionally cleaves the *GRF* transcripts in these cells. The miR396/GRF/PLT regulatory module may thus control the boundary between the QC and the surrounding stem cells in the RAM. In *M. truncatula* root tips, *MtmiR396* cleaves the transcripts of at least six *GRF* and two *bHLH79*-like genes [35]. Promoter-GUS analyses showed that *MtMIR396a* and *b* genes display different expression profiles in root tips with *MtMIR396a* expressed in the MZ and *MtMIR396b* more restricted to the root cap, the stem cell niche, and in root tissues located upper of the TZ. In turn, *MtGRF4*, *5*, and *6* expression levels were high in the MZ and gradually decreased upward from the TZ. In transgenic roots with higher *miR396b* levels or with downregulation of three *MtGRF* genes, the RAM length was significantly reduced due to less active meristematic cell proliferation. In contrast, roots with lower miR396 activity, due to its sequestration by an artificial mimicry RNA, showed a higher *GRF* expression and a slightly enhanced RAM size. In addition to this strong effect in root growth and RAM size maintenance, ectopic expression of *MtMIR396b* affected PR growth and altered the formation of arbuscules during AM symbiosis. However, no obvious effect on LR or nodule development was observed, even if *GRFs* were expressed in their respective apical meristems [35].

In an attempt to detect RAM-specific regulators, Holmes et al. compared the transcriptomes of the MZ/EZ (3 mm from the root tip) and the upper differentiated tissues (1 cm from the RAM) using the Affymetrix Medicago GeneChip [131]. They identified 324 RAM-enriched transcripts, including cell cycle genes such as the *CYCLIN A2* and several TFs. Quantitative RT-PCR experiments confirmed the RAM enrichment of certain TFs involved in the control of cell division or elongation, like GRFs [35, 128], the orthologue of the *Arabidopsis* BABYBOOM AP2/ERF protein [132], and a DEVIL/ROTUNDIFOLIA (DVL/ROT) gene [133]. In addition, novel RAM-enriched TFs, including two bHLHs, a plant AT-rich sequence- and zinc-binding (PLATZ) protein and a LATERAL ORGAN BOUNDARIES (LOB)-domain containing protein, were identified, whose functions remain to be investigated.

RAM formation and maintenance also require the action of SHORT-ROOT (SHR) and SCARECROW (SCR) [89, 118, 120]. Transcripts corresponding to *MtSHR* and *MtSCR* genes have been identified in whole root [134]. These latter authors reported that *MtSCR1* was equivalently expressed in all regions of the root tip, whereas *MtSHR1* transcript mainly accumulated in the meristem in comparison to the EZ. Like many RAM markers, *MtSCR1* was also expressed in the symbiotic nodule meristem [135]. This persistent meristematic zone located at the apex of *M. truncatula* nodules supports their indeterminate growth and is a specificity of legume species from temperate regions such as pea or *Medicago* species. However, precise cell-specific expression profiles and functions of *MtSCR1* and *MtSHR1* in root and nodule meristems have still to be clarified.

Interestingly, Limpens et al. also reported that one of the transcripts detected at high level in the nodule meristem encoded an ortholog of *AtIAA14/SLR* (SOLITARY ROOT) [135]. This protein was mainly studied for its repressive action on LR development in *A. thaliana* through its inhibition of two auxin response factors ARF7 and ARF19 [136–138]. These data thus suggested that IAA14 may be involved in the maintenance and identity of the nodule meristem by repressing its switch to a root meristem. The *nodule root* (NOOT) gene, an orthologue of the *Arabidopsis* ANKYRIN-BTB/POZ protein BLADE OF PETIOLE 1 (BOP1) gene, also acts in the control of the nodule identity. Indeed, in the *noot* mutants, root-like structures emerge from the nodules [139, 140]. Interestingly, these roots seem to originate from vascular initials [139], which also correspond to the site of *WOX5* expression [32].

The maintenance of the TZ position is very important to maintain meristem size and ensure root growth [141]. In *Arabidopsis*, this is done by cytokinins (CKs) through the repression of *PIN FORMED (PIN)* auxin efflux transporter genes by the SHORT HYPOCOTYL2 (SHY2/IAA3) TF. Recently, in the TZ, Sabatini's lab has shown that CKs act not only on auxin polar transport but also on auxin degradation. This double action generates a minimum of auxin at the TZ which is essential for its maintenance [142]. Thus, the RAM is located between a maximum of auxin in the QC and a minimum of auxin in the TZ. The PIN transporters are essential for long-distance and local polar auxin transport in plants. In *M. truncatula*, qRT-PCR experiments showed that the 11 *MtPIN* genes are expressed at different levels in root tips [143]. *MtPIN2* was the most highly expressed gene; according to several authors, this *AtPIN2* orthologue could act in root gravitropism by controlling auxin transport from the tip to the EZ [131, 143]. It is interested to note that expression of ten *MtPINs* was lower in nodules when compared to root tips. Only *MtPIN9* was upregulated in nodules [143] suggesting an essential role in their development. However, other *MtPIN* genes are involved in nodule development; indeed, repression of *MtPIN2* to *MtPIN4* expression by RNA interference (RNAi) in *Agrobacterium rhizogenes*-mediated transgenic roots led to a strong reduction of the number of nodules [144]. Additional studies are thus required to dissect the roles of these auxin transporters in *Medicago* roots. Indeed, to our knowledge, the only *mtpin* mutant characterized so far in *M. truncatula* are the *smooth leaf margin1* (*slm1*) mutants, isolated from the Tnt1 retrotransposon-tagged mutant population. *SLM1* corresponds to the *MtPIN10* gene. However, Zhou et al. focused on shoot and reproductive organs, and, like in Huo et al., the root phenotype of the *slm1* mutant was not described [144, 145].

1.2.2 Lateral Roots

The patterning of LR formation in *M. truncatula* has been studied by Herrbach et al. using both DR5:*GUS*, DR5:*VENUS-N7* reporter lines to detect auxin responses and a *Lupinus albus* p*SCR*:*GUS* construct to specifically visualize the endodermal cells at sites of LR formation [146]. These authors observed seven stages until LR emergency in *M. truncatula*. The stage I encompasses an anticlinal division of pericycle cells (Stage Ia) followed by a second round of anticlinal division in pericycle cells but also in endodermal cells (Stage Ib). Endoderm divisions are not detectable in *A. thaliana* but have also been observed in monocots [37]. A further difference with *A. thaliana* exists at stage III where some cortical cells divide. The contribution of cortical cells to LR development is common in legumes [147] but is not specific to this family. It is generally assumed that cortical cell divisions could facilitate the growth of the LR primordium through peripheral root tissues and/or the emergence of LRs in species that present several layers of cortical cells [147]. In *A. thaliana*, auxin promotes LRs in synergy with brassinosteroids (BR) and ethylene and in antagonism with CKs. In addition, ABA represses the emergence of the LR primordium both in auxin-dependent and auxin-independent ways. The genetic networks associated with the different stages of LR development are well decrypted and are mainly under the control of these hormones [148, 149]. In *M. truncatula*, the antagonism between auxin and CK to regulate LR organogenesis is conserved [150]. Furthermore, large sets of AUX/IAA and ARF genes are expressed in roots [151, 152]. However, unlike *Arabidopsis*, various legume and nonlegume plants, including *Medicago*, respond to ABA by increasing their LR number. As most of the known genetic determinants of LR development identified in *M. truncatula* are directly or indirectly linked to hormonal regulations, they will be described in the next chapter.

1.3 Hormonal Regulations of M. truncatula root Development

The hormonal regulation of RSA is very complex and involves the cross talks between several hormones [20, 31]. These regulations in *M. truncatula* have mainly been described by Imin et al. and Bensmihen et al. [134, 150].

1.3.1 Auxin

As for other plants, auxin is a major regulator of RSA in *M. truncatula*. At the root level, this hormone modulates almost all fundamental cellular processes including cell division. The anaphase-promoting complex/cyclosome (APC/C) is a particular E3 ubiquitin ligase specifically involved in the degradation of proteins that control the cell cycle [60]. In *M. truncatula* roots, the silencing of the APC/C subunit *MtCDC16* gene, which is expressed in root tips, reduced auxin sensibility resulting on a decrease of both PR growth and the number of LRs [62]. In the *compact root architecture 1* (*cra1*) mutant, isolated from a *M. truncatula* T-DNA mutant collection [153], overaccumulation of flavonoids and defects in lignin biosynthesis affected auxin polar transport and cell elongation leading to the very compact root phenotype. In spite of that, the number of LRs and nodules in the *cra1* mutant was similar to wild-type plants [154].

All stages of LR formation are controlled by auxin [149] through the action of different AUXIN/INDOLE-3-ACETIC ACID (AUX/IAA)-AUXIN RESPONSE FACTOR (ARF) modules [155]. For instance, in *Arabidopsis*, the IAA28-ARFs module is involved in cell fate specification of the LR founder cells; during LR initiation, the SOLITARY ROOT (SLR)/IAA14-ARF7-ARF19 regulatory node controls the asymmetric cell divisions of the LR founder cells, and the BODENLOS/IAA12-MONOPTEROS/ARF5 module acts during LR organogenesis. In *M. truncatula*, many of the 17 AUX/IAA and 24 ARF genes are highly expressed in roots in comparison to other organs [156, 157], but more cell-specific expression profiles and functional data are lacking. Indeed, auxin action on *M. truncatula* root development was mainly investigated through the study of several miRNAs, miR393, miR160, miR167, as well as miR390 and its related trans-acting siRNAs (tasiRNAs). In all angiosperm species studied so far, these miRNAs mediate the direct or indirect repression of key components of auxin signaling such as the TIR1/AFB auxin receptor (*MIR393*) [158] or specific members of the ARF TFs – *ARF6/8* (for *miR167*)

[159], ARF10/16/17 (for *miR160*) [160], or ARF2/3/4 (for the *miR390*-associated tasiRNAs) [161]. As for *A. thaliana* [36, 160], the expression of *MtmiR160* under different constitutive promoters strongly repressed PR growth and disturbed gravitropism; this phenotype was a consequence of a disorganization of the RAM and severe defects in root cap differentiation [45]. For the other miRNAs, conserved and specific roles in LR development were shown. Indeed, overexpression of *MtmiR393* in hairy roots resulted in a global decrease of lateral organ formation (both LRs and nodules), probably due to the reduced expression of the *TIR1* target gene [46]. In contrast, *MtmiR390*-OE affected *MtARF3/4* RNA levels but promoted LR growth probably by an increase of auxin sensibility at the emergency step [162]. Surprisingly, unlike in *A. thaliana* [163], ectopic expression of *MtmiR160* had no significant effect on lateral or adventitious root development, despite the strong repression of their *MtARF* targets [45]. Similarly, D'haeseleer et al. reported that *MtmiR164*-OE in transgenic roots did not affect LR numbers despite lower nodule numbers [164]. In *A. thaliana*, this auxin-regulated miRNA posttranscriptionally represses the *NAC1* TF gene when the induction of its expression affected LR emergence [165]. However, in *M. truncatula*, neither loss of function nor overexpression of *MtNAC1* affected LR or nodule formation. Thus, these authors suggested that in *M. truncatula MtmiR164* may regulate nodule formation independently of *NAC1*, but it was not involved in LR development. Globally, a diversity of auxin regulations in legumes does not strictly follow the networks identified in *Arabidopsis*.

1.3.2 Cytokinins

Like auxins, CKs are major regulators of root growth and development. In *A. thaliana* roots, these hormones control growth and vasculature differentiation and repress LR development [166]. In *M. truncatula* roots, the expression profile of the synthetic reporter, two-component signaling sensor (TCS)::*GFP*, which reflects CK responses, mainly encompasses the QC, the root cap, and some pericycle cells at LR branching sites, a profile slightly different from *Arabidopsis* [33, 167, 168]. CK signaling consists of a two-component phosphorelay system, which begins with sensing and initiation of signaling by histidine kinase (HK) receptors. Once these receptors are activated, their phosphoryl group is transferred to histidine phosphotransfer (HP) proteins which are translocated to the nucleus, where they activate or inhibit specific TFs, the response regulators (type A and type B RR). In *M. truncatula*, in addition to the conserved roles in roots, it is now clear that CK plays essential roles in promoting nodule development [169]. Indeed, the Nod factor signaling cascade leads to CK accumulation [168, 170]. Then, CKs, perceived by the CRE1 receptor, promote nodulation by enhancing the expression of specific nodulation GRAS and NF-Y TFs required for nodule organogenesis. CKs also exert a negative effect on local auxin transport [171]. The CRE1-dependent CK signaling could thus be required for the local accumulation of auxin in cortical cells in early stages of nodule development [47]. In agreement with this positive role on nodulation, the *cre1* mutant presented a reduction of nodule numbers [48, 49]. However, an increase of LRs was noticed in the mutant [48], suggesting a dual role of the CRE1-dependent signaling. Recently, three additional CK receptors have been identified in *M. truncatula*, and their expression patterns in roots were analyzed using promoter-GUS fusions [49]. The *MtCHK1/MtCRE1* showed the strongest expression and was detected in the stele, the RAM meristematic zone, and LR primordia. *MtCHK3* was only expressed in the vascular cylinder, while *MtCHK2* and *MtCHK4* were expressed in the dividing RAM cells and LR primordia. Redundant functions between CK receptors may thus occur both in the RAM and during LR development.

Several other components of CK homeostasis or signaling have been studied in *M. truncatula* roots and nodules. For instance, the *LONELY GUY* (*LOG1*) gene encodes a cytokinin riboside 5′-monophosphate phosphoribohydrolase, which converts some inactive CKs into their active forms. In *A. thaliana*, *log* mutants showed altered root and shoot morphologies and a lower sensitivity to CK during lateral root formation [50]. In *M. truncatula*, *pMtLOG1:GUS* staining was detected in nodule primordia, in LR primordia, and in root tips. Interestingly, in the *cre1–1* mutant, *MtLOG1* expression was abolished in the nodules but not in the LR primordia, suggesting that its expression during LR organogenesis is independent of CRE1 signaling. Repression of *MtLOG* genes by RNAi led to a slight decrease in PR length, a twofold increase in LR density, but a strong reduction of nodule numbers. These are in agreement with an antagonist role of CKs in the development of these two lateral organs. Finally, when *MtLOG1* was overexpressed in roots, a decrease of LR number and the formation of longer and thicker roots were observed, correlating with an expansion of the vascular cylinder [172].

The KNOTTED1-like homeobox (KNOX) protein SHOOTMERISTEMLESS (STM) is known to activate the *ISOPENTENYL TRANSFERASE* (*IPT*) genes involved in CK biosynthesis in the shoot apical meristem [34, 173]. KNOX TFs may thus activate CK biosynthesis required for nodulation and/or root development. Among the ten KNOX genes identified in *M. truncatula*, expression of *MtKNOX3* was strongly induced in roots inoculated with the symbiotic bacteria *Sinorhizobium meliloti* (Sm) and was high in the nodule primordia. Moreover, Di Giacomo et al. found that *MtKNOX3* transcript levels rapidly increased in response to exogenous CKs, whereas expression of its putative ortholog *AtKNAT3* was repressed [174]. This indicates a difference between *MtKNOX3* and *AtKNAT3* signaling in roots. Like in *A. thaliana*, *MtKNOX3* promoter activity was detected in root tips. However, it was also detected in the LR primordia, in contrast to *AtKNAT3* [175]. Transgenic roots with ectopic *MtKNOX3* expression formed spontaneous nodule-like bumps in absence of bacterial inoculation, similar to those observed in legume plants with constitutive activity of CK receptor. However, although *MtKNOX3* was expressed in LR primordia, no difference in LR formation was reported in *MtKNOX3*-OE or *MtKNOX3*-RNAi roots [174].

Regarding CK signaling, Op den Camp et al. studied the expression and the roles of type A RR genes in Mt. roots inoculated or not with *Sm*-purified Nod factors [147]. Six MtRR-A genes were expressed in roots, including the two paralogues *MtRR9* and *MtRR11*. The p*MtRR9*:*GUS* activity was restricted to the root meristematic zone, while p*MtRR11*:*GUS* was detected in the epidermis, cortex, and endodermis at the differentiation zone. RNAi roots with a concomitant repression of *MtRR9*, *MtRR11*, and *MtRR5* gene expression showed approximately 33% fewer LRs when compared to the control plants. Moreover, nodulation efficiency was impaired, indicating that these type A RR genes are required both for nodule and root formation. In *A. thaliana*, the auxin-responsive *AtHP6* gene encodes a CK signaling repressor during vasculature differentiation and LR development in *A. thaliana* [176, 177]. It was proposed that HP6-dependent regulated genes may modulate the orientation of pericycle cell divisions during LR initiation by affecting the localization of the auxin efflux carrier PIN1. The *pAtHP6*:*GUS* construct expressed in *M. truncatula* transgenic roots mediated GUS staining not only in the stele and in LR primordia but also in nodule primordia [178]. In addition, *AtHP6* and *MtHP6* promoters have conserved cis-regulatory motifs. Taken together, these data reveal that several CK signaling pathways, dependent or independent of the CRE1 receptor, are involved in the regulation of root growth and LR development in *M. truncatula* and appear as common regulators of nodule and LR development even though they may have either similar or opposite downstream effects.

1.3.3 Abscisic Acid

Abscisic acid (ABA) mediates responses to different abiotic factors, such as the nitrate availability in the soil, drought, or salt, which shape the RSA by regulating the production of LRs and controlling root growth. Interestingly, this hormone may play different roles in plant species. In particular, ABA negatively controls LR development in *A. thaliana*, whereas exogenous applications of low concentrations of ABA on *M. truncatula* roots stimulate LR initiation [40], emergence [179], and activation of the LR meristem [180]. These low ABA concentrations probably act by cross talk with auxin signaling and could decrease auxin sensibility [63].

In opposite, high ABA concentrations inhibit initiation and emergency of LRs [179]. The link between ABA and root development was mainly investigated in *M. truncatula* plants in the context of a salt stress (see next chapter). However, in 2005, a screen for nodulation defective mutants allowed to isolate the *ltad/nip* mutant [181]. In addition to strong defects in nodulation, this mutant showed a severe root phenotype characterized by shorter PR and LRs which stopped to grow just after emergence. Interestingly, exogenous ABA was able to restore a proper PR growth in the mutant *ltad/nip* suggesting deficient ABA synthesis or responses in this mutant [180]. More recent data proposed that this rescue may occur through a decreased accumulation of reactive oxygen species, known to inhibit root cell elongation [182].

1.3.4 Gibberellins

In plants, gibberellins (GAs) regulate root growth by controlling cell division and elongation through the degradation of DELLA proteins [38]. The roles of GAs in nodule formation and arbuscular-mycorrhizal (AM) symbiosis [51] have recently been investigated. By using exogenous active GA_3 treatments and mutants of the three *MtDELLA* genes, Fonouni-Farde et al. showed that GA and DELLA act negatively and positively on nodule formation, respectively [183]. During AM symbiosis, fungi induced the specific expression of the *MtMIG1* gene (for *M*ADS-box protein required for *i*nfectious *g*rowth *1*) in root cortical cells, whereas this TF was only expressed in the stele of non-infected roots. Ectopic expression of *MtMIG1* in roots caused an increase of both the number and the size (width and length) of the cortical cells. In addition, *MtMIG1*-OE plants were insensitive exogenous GA_3 [51]. Thus, the MIG1 pathway may interact with GA signaling triggered by the action of the MtDELLA1 protein [184, 185]. This study clearly showed that MIG1 and MtDELLA1 are positive regulators of cortical radial cell expansion [51].

1.3.5 Brassinosteroids

Plant brassinosteroids (BR) mutants display a short root phenotype [186]. Indeed, BRs are known to control cell division and elongation. In roots, the effects of BR signaling depend on the type of root cells [187]. In an interaction with auxin, BRs promote LR initiation in *A. thaliana* [188]; this is done by modifications of auxin transport (at low concentrations) or by the induction of *AUX/IAA* genes (at high concentrations) [52]. In *M. truncatula*, Cheng et al. have recently shown that low concentrations of exogenous BRs promote PR elongation and LR development, whereas opposite phenotypes are observed under high concentrations [189]. Moreover, according to Ferguson et al., the *mtbri1* mutant affected in BR perception presented a reduction in nodule number and size when, in pea, the modification of BR signaling only led to a lower number of nodules [190]. Three additional *mtbri1* mutants with a Tnt1 insertion, *mtbri1-1*, *mtbri1-2*, and *mtbri1-3*, were characterized [189]; these dwarf mutants displayed severe pleiotropic phenotypes including a decreased efficiency of symbiotic nitrogen fixation but also reduced PR growth and LR number. A potential link between root growth, BR signaling, and the *MtMIR396/GRF* regulatory module, described before, has been proposed by Bazin et al. [35]. Indeed, qRT-PCR and promoter-GUS analyses performed in roots treated with growth-repressive concentrations of BRs showed that the activation of *MtmiR396b* expression was concomitant with the posttranscriptional degradation of the *MtGRF5* transcripts, suggesting that part of the action of BRs on root growth may be mediated by this GRF and its role in cell proliferation.

1.4 Environmental Regulation of RSA

The root system is an integrator of environmental signals such as nutrient availability, drought, or presence of metals in soil, which strongly affect the root architecture. These environmental conditions impinge on hormonal regulations to modulate the final RSA in the soil.

1.4.1 Nitrogen Status

Nitrates regulate the RSA in a majority of plants. Generally, at high concentration, nitrate affects root development through the inhibition of PR growth and regulation of LR emergence [191, 192]. However, the effects depend on how and where the nitrate is sensed; systemic high concentrations of nitrate generally repress, whereas high nitrate concentrations locally stimulate LR development [61, 191].

In *M. truncatula*, the TR185 mutant, obtained by gamma-ray mutagenesis, exhibits a highly branched root phenotype and is insensitive to exogenous nitrate supply [193]. This mutant shows a constitutive N-starvation status which can explain this phenotype. Grafting experiments showed that the inducing signal is transmissible from shoots. The comparison of shoot amino acid concentrations in the mutant and WT lines suggested that the glutamine and asparagine could be signals in this systemic regulation. Finally, a transcriptomic analysis performed on the TR185 mutant revealed changes in the expression of genes involved in different pathways such as cell wall modification, auxin signaling, or the phenylpropanoid pathway which is important for the biosynthesis of numerous secondary metabolites [194].

Different studies have shown that two nitrate transporters, LATD/NIP (MtNPF1.7) [61, 64, 181, 195] and MtNRT1-1 (MtNPF6.8) [65, 196, 197], affect root growth and development in *M. truncatula*. Bagchi et al. (2012) showed that the *LATD/NIP* gene, already described in this review due to its link with ABA, encodes a nitrate transporter mainly expressed in the root tip (meristem and elongation zone). Moreover, the *ltad/nip* roots are insensitive to addition of nitrate in the medium suggesting that LATD/NIP could be a nitrate sensor [61]. This study highlighted the direct link between nitrate and ABA signaling in roots. A quantitative genetic approach was also used to determine QTLs correlated with the nitrate-induced inhibition of PR growth [65, 196]. The authors thus detected a QTL corresponding to a nitrate transporter gene, *MtNPF6.8*. Roots with reduced expression of *MtNPF6.8* were insensitive to the growth inhibitory effect of exogenous nitrate. As, in WT plants, cortical cells were smaller in presence of nitrogen than in control conditions, these results suggested that nitrate may mainly inhibit PR growth through the control of root cell elongation, via MtNPF6.8 action. Interestingly, MtNPF6.8 appears to be able to transport ABA (according to experiments in Xenopus oocytes), again highlighting the tight links between nitrate and ABA signaling pathways to regulate root growth [196].

The RSA is additionally controlled through other signaling molecules such as small RNAs or peptides. In plants, several miRNAs, including those involved in auxin responses, are differentially expressed in function of the nitrate content and are known to behave as mobile signals of the nitrogen status in the plant [23, 158, 198, 199]. However, to our knowledge, these miRNAs were not studied in legume roots in the specific context of N starvation. Small peptides such as CLE (CLAVATA3/EMBRYON SURROUNDING REGION) or CEP (C-TERMINALLY ENCODED PEPTIDE) peptides are essential regulators of RSA in response to N status in *A. thaliana* [27, 200, 201]. For instance, the AtCEP5 peptide represses LR formation. However, unexpectedly, the *atxip1* line mutated in the putative CEP5 receptor *AtXIP1* was affected in LR initiation [53]. The authors thus suggested that CEP5 would act as an antagonist and thus inhibit its receptor XIP1.

In *M. truncatula*, MtCEP1, the closest homologue of AtCEP5, was also shown to inhibit LR emergence [54, 202]. *MtCEP1* expression was upregulated in response to nitrogen starvation [54], suggesting that it could be a signal in the inhibition of LR development. Transcriptomic analysis performed on plants treated or not with the CEP1 peptide revealed a modification in expression of different key genes involved in LR development such as the *MtLOB29* MYB TF [203] and *SUPERROOT2* cytochrome P450 [204], confirming a putative role of this peptide in the control of LR organogenesis [54]. Recently, it was observed that the regulation of LR development by MtCEP1 depends on signaling pathways different from the Nod factor and the ETHYLENE INSENSITIVE2 (EIN2)/SICKLE (SKL) nodulation pathways [55]. In 2014, Huault et al. described a *cra2* mutant (*compact root architecture 2*) exhibiting more LRs but shorter roots [205]. This mutant was isolated from a forward genetic screen of the *M. truncatula* Tnt1 insertional mutant collection, together with six other mutant lines showing a normal shoot morphology and a "compact root system architecture." In the *cra2* mutant, the root phenotype occurred independently of nutrient availability, suggesting an involvement of this gene in nutrient sensing. Like *AtXIP1*, the

MtCRA2 gene encodes a LEUCINE-RICH REPEAT/ RECEPTOR-LIKE KINASE (LRR/RLK) and could thus be a good candidate to be the CEP1 peptide receptor. However, unlike in *Arabidopsis*, the increased root branching in the *cra2* mutant suggested that this peptide receptor may be activated by the CEP1 peptide leading to the nitrate-dependent repression of LR formation [206].

In legumes, CLE peptides were essentially associated with SAM and RAM functions and with the autoregulation of nodulation mechanism which allows the control of nodule numbers by the plant [29, 56]. The *SUNN* gene (*SUPER NUMERIC NODULES*) encodes a leucine-rich repeat/receptor-like kinase receptor of CLE peptides with homology to *AtCLAVATA1*, which is known to regulate the root architecture in response to nitrogen through CLE 3/4/7 peptides [200]. During nodulation, SUNN binds to the MtCLE12 and MtCLE13 peptides. The *sunn* mutant, in addition to its super-nodulating phenotype [57], is also affected in RSA: this mutant exhibits a shorter PR and more abundant LRs [58, 207]. This phenotype was linked to a defect in sensing local nitrogen accumulation in soil. Indeed, while N supply reduced the LR density and increased the LR length in the WT, no difference was observed in the *sunn* mutant. The LR density was dependent on the concentration of nitrate in WT shoots when no correlation was detected for the *sunn* mutant. Thus, SUNN modulates the amount of shoot-root auxin transport in function of N concentration in shoots. Under nitrate starvation, low concentration of nitrate in shoots will induce a repression of shoot-root auxin transport through SUNN activity, allowing the formation of LRs.

1.4.2 Phosphorus Availability

Phosphorus (P), one of the lowest available elements in soils, belongs to essential macroelements for plant development. In most land plants, excluding *A. thaliana*, arbuscular-mycorrhizal (AM) symbioses help to uptake inorganic P (Pi) in soil. Pi availability deeply affects the RSA, and phosphorus starvation leads to a reduction of PR growth associated with an increase of LR number. In *M. truncatula*, this response is abolished in the *Mtpt4* mutant. *MtPT4* encodes an AM-inducible phosphate transporter. Interestingly, this gene is expressed in root tips of non-mycorrhizal roots. Thus, *MtPT4* may modulate the root developmental response to phosphate independently of AM symbiosis [208]. Gene expression analysis indicated that the expression of the *MtTIR1* auxin receptor was not induced in response to phosphate starvation in the *Mtpt4* mutant, suggesting that this gene regulates root branching through an auxin-related pathway. Phosphorus availability also impacts plant responses to strigolactones (SL). The effect of GR24 (an active SL) on LRs in *M. truncatula* was observed on 19-day-old plants in phosphate-limiting conditions. GR24 application at a concentration of 10–7 M (phosphate-limiting condition) caused a transient activation of LR development in seedlings, while an opposite effect was observed in adult plants [209]. According to Bensmihen, this repressive role of SL in adult plants was confirmed by the characterization of the *Mtccd7* and *Mtccd8* mutants, impaired in SL accumulation [150, 210]. Indeed, mutant plants grown under P starvation developed more LR than the WT. Presumably SLs modulate LR organogenesis through interaction with auxin signaling [150].

1.4.3 Drought and Salt Stress

Hydric stresses, including salt stress and drought, strongly affect crop yields and plant development. Several studies have been performed in legumes and more especially *M. truncatula* to determine how roots can adapt to these constraints. *M. truncatula* is mainly cultivated in the Mediterranean area, an arid or semiarid region with saline coast soils [42], and it has been reported that even low concentrations of NaCl inhibit root growth in some genotypes [211]. Transcriptomic analyses of root tips treated by salt [39, 211–213] revealed the differential expression of members of many TF families such as MYB, WRKY, HD-Zip, NAC, bHLH, or DREB/CBF (DEHYDRATION-RESPONSIVE ELEMENT-BINDING PROTEIN/C-REPEAT BINDING FACTOR), which is consistent with that observed in other plant species. Different functional approaches were used to confirm the role of some salt-regulated TFs. For instance, the expression of a salt-responsive CBF TF gene, *MtCBF4*, was also induced by drought and cold in an ABA-dependent manner. *A. thaliana* plants overexpressing *MtCBF4* exhibited a better root growth than the wild type in presence of low concentrations of salt, suggesting that this TF may have a conserved role in root tolerance to salt stress [211]. In 2012, RNAi against another salt-induced NAC TF gene, *MtNAC696*, showed that this TF modulated RSA in response to salt stress. Indeed, growth of the knockdown roots was less inhibited by salt than WT roots [41]. Another TF particularly important for root growth regulation in response to salt is *MtHB1* [179, 214]. This HD-Zip TF seems to especially affect the LR emergence in response to salt. Indeed, the inhibition of LR emergence by salt was abolished in a *Tnt1* insertion loss of function mutant. Using several approaches, like electrophoretic mobility shift assay and chromatin immunoprecipitation-PCR experiments, Ariel et al. (2010b) showed that MtHB1 directly represses the expression of the auxin-induced *MtLBD1*, a LOB-like (lateral organ boundaries) TF, by binding to its promoter. In *A. thaliana*, also, certain *LBD* genes are involved in LR formation through auxin signaling [215].

Salt stress response has also been investigated through the analyses of different *M. truncatula* or *M. sativa* ecotypes presenting contrasted responses to this constraint [66, 216]. Long et al. (2016) compared the proteomes of the salt-tolerant *M. sativa* cv. Zhongmu-1 with that of the *M. truncatula* A17 roots and identified several differentially regulated DNA-/RNA-binding proteins but no known TF. Transcriptomes were also obtained from two genotypes of *M. truncatula*, the TN1.11 salt-adapted genotype and the cv. Jemalong A17 considered as salt-sensitive [216]. Several TFs differentially regulated in the two genotypes were identified that may be candidate genes for local ecotype adaptation to salt stress. Interestingly, the *MtbHLH658* transcripts were overaccumulated in the salt-adapted genotype compared to the sensitive one. Moreover, *MtbHLH658* overexpression in the sensitive genotype was sufficient to improve its tolerance to salt stress, suggesting a key role of this gene in salt stress adaptation.

2 Conclusions

These last years, "omics" approaches facilitated significant progress in understanding how soil environmental conditions affect *M. truncatula* root development. However, both the scarcity of mutant lines and the difficulty to generate stable transgenic lines had delayed our knowledges in this species. In the future, the CRISPR technology, adapted to *M. truncatula*, must reduce these limits. Moreover, very few single tissue cells analyses using fluorescent-activated cell sorting (FACS) or INTACT have been done. These latter techniques in animals and *A. thaliana* allow the understanding of complex differentiation and/or development processes at the level of cell types; indeed, it is now clear that specific cells encompass distinct transcriptome, proteome, and metabolome. Recently, in *A. thaliana*, concomitant analyses of transcriptome and epigenome of RAM and root cap columella cells revealed the epigenomic diversities in root tip cells [217].

The interactions between plant and soil involve complex physiological and developmental processes that contribute to define a specific RSA. The future challenge must be the genome-wide multi-omics data integration at the level of interactions of both protein-proteins and nucleic acid-proteins. This will certainly play an important part in the identification of molecular mechanisms that realize cross talk between hormones and development processes. However, the construction of these networks needs a standardization of culture conditions in the different laboratories and a clear definition of *M. truncatula* root phenomics.

References

1. Esau K (1965) Plant Anatomy, 2nd edn. John Wiley, New York
2. Osmont KS, Sibout R, Hardtke CS (2007) Hidden branches: developments in root system architecture. Annu Rev Plant Biol 58:93–113
3. Pacheco-Villalobos D, Hardtke CS (2012) Natural genetic variation of root system architecture from Arabidopsis to Brachypodium: towards adaptive value. Philos Trans R Soc Lond Ser B Biol Sci 367(1595):1552–1558. https://doi.org/10.1098/rstb.2011.0237

4. Rellán-Álvarez R, Lobet G, Dinneny JR (2016) Environmental control of root system biology. Annu Rev Plant Biol 67:619–642
5. Gruber BD, Giehl RF, Friedel S et al (2013) Plasticity of the Arabidopsis root system under nutrient deficiencies. Plant Physiol 163(1):161–179. https://doi.org/10.1104/pp.113.218453
6. Duan L, Dietrich D, Ng CH et al (2013) Endodermal ABA signaling promotes lateral root quiescence during salt stress in Arabidopsis seedlings. Plant Cell 25(1):324–341. https://doi.org/10.1105/tpc.112.107227
7. Roychoudhry S, Del Bianco M, Kieffer M et al (2013) Auxin controls gravitropic setpoint angle in higher plant lateral branches. Curr Biol 23(15):1497–1504. https://doi.org/10.1016/j.cub.2013.06.034
8. Tian H, De Smet I, Ding Z (2014) Shaping a root system: regulating lateral versus primary root growth. Trends Plant Sci 19(7):426–431. https://doi.org/10.1016/j.tplants.2014.01.007
9. Groot EP, Doyle JA, Nichol SA et al (2004) Phylogenetic distribution and evolution of root apical meristem organization in dicotyledonous angiosperms. Int J Plant Sci 165:97–105
10. Rost TL (2011) The organization of roots of dicotyledonous plants and the positions of control points. Ann Bot 107(7):1213–1222. https://doi.org/10.1093/aob/mcq229
11. Baum SF, Dubrovsky JG, Rost TL (2002) Apical organization and maturation of the cortex and vascular cylinder in Arabidopsis thaliana (Brassicaceae) roots. Am J Bot 89(6):908–920. https://doi.org/10.3732/ajb.89.6.908
12. Chapman K, Groot EP, Nichol SA et al (2002) Primary root growth and the pattern of root apical meristem organization are coupled. J Plant Growth Regul 21(4):287–295
13. Hamamoto L, Hawes MC, Rost TL (2006) The production and release of living root cap border cells is a function of root apical meristem type in dicotyledonous angiosperm plants. Ann Bot 97(5):917–923
14. Moreno-Risueno MA, Van Norman JM, Moreno A et al (2010) Oscillating gene expression determines competence for periodic Arabidopsis root branching. Science 329(5997):1306–1311. https://doi.org/10.1126/science.1191937
15. Marhavý P, Montesinos JC, Abuzeineh A et al (2016) Targeted cell elimination reveals an auxin-guided biphasic mode of lateral root initiation. Genes Dev 30(4):471–483. https://doi.org/10.1101/gad.276964.115
16. Casimiro I, Marchant A, Bhalerao RP et al (2001) Auxin transport promotes Arabidopsis lateral root initiation. Plant Cell 13(4):843–852
17. Lucas M, Kenobi K, von Wangenheim D et al (2013) Lateral root morphogenesis is dependent on the mechanical properties of the overlaying tissues. Proc Natl Acad Sci U S A 110(13):5229–5234. https://doi.org/10.1073/pnas.1210807110
18. Malamy JE, Benfey PN (1997) Organization and cell differentiation in lateral roots of Arabidopsis thaliana. Development 124(1):33–44
19. Oláh B, Brière C, Bécard G (2005) Nod factors and a diffusible factor from arbuscular mycorrhizal fungi stimulate lateral root formation in Medicago truncatula via the DMI1/DMI2 signalling pathway. Plant J 44(2):195–207
20. Jung JK, McCouch S (2013) Getting to the roots of it: genetic and hormonal control of root architecture. Front Plant Sci 4:186. https://doi.org/10.3389/fpls.2013.00186 eCollection 2013
21. Wachsman G, Sparks EE, Benfey PN (2015) Genes and networks regulating root anatomy and architecture. New Phytol 208(1):26–38
22. Slovak R, Ogura T, Satbhai SB et al (2016) Genetic control of root growth: from genes to networks. Ann Bot 117(1):9–24
23. Khan GA, Declerck M, Sorin C et al (2011) MicroRNAs as regulators of root development and architecture. Plant Mol Biol 77(1–2):47–58. https://doi.org/10.1007/s11103-011-9793-x Epub 2011 May 24
24. Couzigou JM, Combier JP (2016) Plant microRNAs: key regulators of root architecture and biotic interactions. New Phytol 212(1):22–35
25. Samad AFA, Sajad M, Nazaruddin N et al (2017) MicroRNA and transcription factor: key players in plant regulatory network. Front Plant Sci 8:565
26. Delay C, Imin N, Djordjevic MA (2013) Regulation of Arabidopsis root development by small signaling peptides. Front Plant Sci 4:352
27. Araya T, von Wirén N, Takahashi H (2016) CLE peptide signaling and nitrogen interactions in plant root development. Plant Mol Biol 91(6):607–615. https://doi.org/10.1007/s11103-016-0472-9 Epub 2016 Mar 19
28. Djordjevic MA, Mohd-Radzman NA, Imin N (2015) Small-peptide signals that control root nodule number, development, and symbiosis. J Exp Bot 66(17):5171–5181
29. Hastwell AH, de Bang TC, Gresshoff PM et al (2017) CLE peptide-encoding gene

families in Medicago truncatula and Lotus japonicus, compared with those of soybean, common bean and Arabidopsis. Sci Rep 7(1):9384. https://doi.org/10.1038/s41598-017-09296-w
30. Nakayama T, Shinohara H, Tanaka M et al (2017) A peptide hormone required for Casparian strip diffusion barrier formation in Arabidopsis roots. Science 355(6322):284–286. https://doi.org/10.1126/science.aai9057
31. Pacifici E, Polverari L, Sabatini S (2015) Plant hormone cross-talk: the pivot of root growth. J Exp Bot 66(4):1113–1121
32. Osipova M, Dolgikh E, Lutova L (2011) Peculiarities of meristem-specific WOX5 gene expression during nodule organogenesis in legumes. Russ J Dev Biol 42:226–237
33. Franssen HJ, Xiao TT, Kulikova O et al (2015) Root developmental programs shape the Medicago truncatula nodule meristem. Development 142(17):2941–2950. https://doi.org/10.1242/dev.120774
34. Yanai O, Shani E, Dolezal K (2005) Arabidopsis KNOXI proteins activate cytokinin biosynthesis. Curr Biol 15(17):1566–1571
35. Bazin J, Khan GA, Combier JP et al (2013) miR396 affects mycorrhization and root meristem activity in the legume Medicago truncatula. Plant J 74(6):920–934. https://doi.org/10.1111/tpj.12178
36. Wang JW, Wang LJ, Mao YB et al (2005) Control of root cap formation by MicroRNA-targeted auxin response factors in Arabidopsis. Plant Cell 17(8):2204–2216
37. Orman-Ligeza B, Parizot B, Gantet PP et al (2013) Post-embryonic root organogenesis in cereals: branching out from model plants. Trends Plant Sci 18(8):459–467. https://doi.org/10.1016/j.tplants.2013.04.010
38. Ubeda-Tomás S, Federici F, Casimiro I et al (2009) Gibberellin signaling in the endodermis controls Arabidopsis root meristem size. Curr Biol 19(14):1194–1199. https://doi.org/10.1016/j.cub.2009.06.023 Epub 2009 Jul 2
39. Li D, Su Z, Dong J et al (2009) An expression database for roots of the model legume Medicago truncatula under salt stress. BMC Genomics 10:517. https://doi.org/10.1186/1471-2164-10-517
40. Liang Y, Harris JM (2005) Response of root branching to abscisic acid is correlated with nodule formation both in legumes and nonlegumes. Am J Bot 92(10):1675–1683. https://doi.org/10.3732/ajb.92.10.1675
41. de Zélicourt A, Diet A, Marion J et al (2012) Dual involvement of a Medicago truncatula NAC transcription factor in root abiotic stress response and symbiotic nodule senescence. Plant J 70(2):220–230. https://doi.org/10.1111/j.1365-313X.2011.04859.x Epub 2012 Jan 10
42. Duan L, Sebastian J, Dinneny JR (2015) Salt-stress regulation of root system growth and architecture in Arabidopsis seedlings. Methods Mol Biol 1242:105–122. https://doi.org/10.1007/978-1-4939-1902-4_10
43. Gao R, Austin RS, Amyot L et al (2016) Comparative transcriptome investigation of global gene expression changes caused by miR156 overexpression in Medicago sativa. BMC Genomics 17:658. https://doi.org/10.1186/s12864-016-3014-6
44. Boualem A, Laporte P, Jovanovic M et al (2008) MicroRNA166 controls root and nodule development in Medicago truncatula. Plant J 54(5):876–887
45. Bustos-Sanmamed P, Mao G, Deng Y et al (2013) Overexpression of miR160 affects root growth and nitrogen-fixing nodule number in *Medicago truncatula*. Funct Plant Biol 40(12):1208–1220
46. Mao G, Turner M, Yu O et al (2013) miR393 and miR164 influence indeterminate but not determinate nodule development. Plant Signal Behav 8(10):e26753. https://doi.org/10.4161/psb.26753
47. Ng JL, Hassan S, Truong TT et al (2015) Flavonoids and Auxin transport inhibitors rescue symbiotic nodulation in the Medicago truncatula Cytokinin perception mutant cre1. Plant Cell 27(8):2210–2226. https://doi.org/10.1105/tpc.15.00231
48. Gonzalez-Rizzo S, Crespi M, Frugier F (2006) The Medicago truncatula CRE1 cytokinin receptor regulates lateral root development and early symbiotic interaction with Sinorhizobium meliloti. Plant Cell 18(10):2680–2693
49. Murray JD, Karas BJ, Sato S et al (2007) A cytokinin perception mutant colonized by rhizobium in the absence of nodule organogenesis. Science 315(5808):101–104
50. Kuroha T, Tokunaga H, Kojima M et al (2009) Functional analyses of LONELY GUY cytokinin-activating enzymes reveal the importance of the direct activation pathway in Arabidopsis. Plant Cell 21(10):3152–3169. https://doi.org/10.1105/tpc.109.068676
51. Heck C, Kuhn H, Heidt S et al (2016) Symbiotic fungi control plant root cortex development through the novel GRAS transcription factor MIG1. Curr Biol 26(20):2770–2778. https://doi.org/10.1016/j.cub.2016.07.059 Epub 2016 Sep 15
52. Kim H, Park PJ, Hwang HJ et al (2006) Brassinosteroid signals control expression of the AXR3/IAA17 gene in the cross-talk

point with auxin in root development. Biosci Biotechnol Biochem 70(4):768–773
53. Roberts I, Smith S, Stes E et al (2016) CEP5 and XIP1/CEPR1 regulate lateral root initiation in Arabidopsis. J Exp Bot 67(16):4889–4899. https://doi.org/10.1093/jxb/erw231
54. Imin N, Mohd-Radzman NA, Ogilvie HA et al (2013) The peptide-encoding CEP1 gene modulates lateral root and nodule numbers in Medicago truncatula. J Exp Bot 64(17):5395–5409. https://doi.org/10.1093/jxb/ert369
55. Mohd-Radzman NA, Laffont C, Ivanovici A et al (2016) Different pathways act downstream of the CEP peptide receptor CRA2 to regulate lateral root and nodule development. Plant Physiol 171(4):2536–2548. https://doi.org/10.1104/pp.16.00113 Epub 2016 Jun 24
56. Saur IM, Oakes M, Djordjevic MA et al (2011) Crosstalk between the nodulation signaling pathway and the autoregulation of nodulation in Medicago truncatula. New Phytol 190(4):865–874. https://doi.org/10.1111/j.1469-8137.2011.03738.x Epub 2011 Apr 20
57. Penmetsa RV, Frugoli JA, Smith LS et al (2003) Dual genetic pathways controlling nodule number in Medicago truncatula. Plant Physiol 131(3):998–1008
58. Schnabel E, Journet EP, de Carvalho-Niebel F et al (2005) The Medicago truncatula SUNN gene encodes a CLV1-like leucine-rich repeat receptor kinase that regulates nodule number and root length. Plant Mol Biol 58(6):809–822
59. Roudier F, Fedorova E, Lebris M et al (2003) The Medicago species A2-type cyclin is auxin regulated and involved in meristem formation but dispensable for endoreduplication-associated developmental programs. Plant Physiol 131(3):1091–1103
60. Heyman J, De Veylder L (2012) The anaphase-promoting complex/cyclosome in control of plant development. Mol Plant 5(6):1182–1194. https://doi.org/10.1093/mp/sss094
61. Yendrek CR, Lee YC, Morris V et al (2010) A putative transporter is essential for integrating nutrient and hormone signaling with lateral root growth and nodule development in Medicago truncatula. Plant J 62(1):100–112. https://doi.org/10.1111/j.1365-313X.2010.04134.x Epub 2010 Jan 20
62. Kuppusamy KT, Ivashuta S, Bucciarelli B et al (2009) Knockdown of CELL DIVISION CYCLE16 reveals an inverse relationship between lateral root and nodule numbers and a link to auxin in Medicago truncatula. Plant Physiol 151(3):1155–1166. https://doi.org/10.1104/pp.109.143024
63. Gonzalez AA, Agbévénou K, Herrbach V et al (2015) Abscisic acid promotes pre-emergence stages of lateral root development in Medicago truncatula. Plant Signal Behav 10(1):e977741. https://doi.org/10.4161/15592324.2014.977741
64. Bagchi R, Salehin M, Adeyemo OS et al (2012) Functional assessment of the Medicago truncatula NIP/LATD protein demonstrates that it is a high-affinity nitrate transporter. Plant Physiol 160(2):906–916. https://doi.org/10.1104/pp.112.196444 Epub 2012 Aug 2
65. Morère-Le Paven MC, Viau L et al (2011) Characterization of a dual-affinity nitrate transporter MtNRT1.3 in the model legume Medicago truncatula. J Exp Bot 62(15):5595–5605. https://doi.org/10.1093/jxb/err243 Epub 2011 Aug 23
66. Long R, Li M, Zhang T et al (2016) Comparative proteomic analysis reveals differential root proteins in Medicago sativa and Medicago truncatula in response to salt stress. Front Plant Sci 7:424. https://doi.org/10.3389/fpls.2016.00424 eCollection 2016
67. Mathesius U, Keijzers G, Natera SH et al (2001) Establishment of a root proteome reference map for the model legume Medicago truncatula using the expressed sequence tag database for peptide mass fingerprinting. Proteomics 1(11):1424–1440
68. Marx H, Minogue CE, Jayaraman D et al (2016) A proteomic atlas of the legume Medicago truncatula and its nitrogen-fixing endosymbiont Sinorhizobium meliloti. Nat Biotechnol 34:1198–1205. https://doi.org/10.1038/nbt.3681
69. Bestel-Corre G, Dumas-Gaudot E, Poinsot V et al (2002) Proteome analysis and identification of symbiosis-related proteins from Medicago truncatula Gaertn. by two-dimensional electrophoresis and mass spectrometry. Electrophoresis 23(1):122–137
70. Abdallah C, Valot B, Guillier C et al (2014) The membrane proteome of Medicago truncatula roots displays qualitative and quantitative changes in response to arbuscular mycorrhizal symbiosis. J Proteome 108:354–368. https://doi.org/10.1016/j.jprot.2014.05.028
71. Van Noorden GE, Kerim T, Goffard N et al (2007) Overlap of proteome changes in Medicago truncatula in response to auxin and Sinorhizobium meliloti. Plant Physiol 144(2):1115–1131 Epub 2007 Apr 27
72. Holmes P, Farquharson R, Hall PJ et al (2006) Proteomic analysis of root meristems

and the effects of acetohydroxyacid synthase-inhibiting herbicides in the root of Medicago truncatula. J Proteome Res 5(9):2309–2316
73. Valot B, Gianinazzi S, Eliane DG (2004) Sub-cellular proteomic analysis of a Medicago truncatula root microsomal fraction. Phytochemistry 65(12):1721–1732
74. Lee J, Lei Z, Watson BS (2013) Sub-cellular proteomics of Medicago truncatula. Front Plant Sci 4:112. https://doi.org/10.3389/fpls.2013.00112 eCollection 2013
75. Zhang K, McKinlay C, Hocart CH et al (2006) The Medicago truncatula small protein proteome and peptidome. J Proteome Res 5:3355–3367
76. Watson BS, Asirvatham VS, Wang L (2003) Mapping the proteome of barrel medic (Medicago truncatula). Plant Physiol 131(3):1104–1123
77. Benedito VA, Torres-Jerez I, Murray JD et al (2008) A gene expression atlas of the model legume Medicago truncatula. Plant J 55(3):504–513. https://doi.org/10.1111/j.1365-313X.2008.03519.x
78. Du C, Li X, Chen J et al (2016) Receptor kinase complex transmits RALF peptide signal to inhibit root growth in Arabidopsis. Proc Natl Acad Sci U S A 113(51):E8326–E8334
79. Ballif J, Endo S, Kotani M et al (2011) Over-expression of HAP3b enhances primary root elongation in Arabidopsis. Plant Physiol Biochem 49(6):579–583. https://doi.org/10.1016/j.plaphy.2011.01.013
80. Schmutz J, Cannon SB, Schlueter J (2010) Genome sequence of the palaeopolyploid soybean. Nature 463(7278):178–183
81. Sato S, Nakamura Y, Kaneko T (2008) Genome structure of the legume, Lotus japonicus. DNA Res 15(4):227–239
82. Young ND, Debellé F, Oldroyd GE et al (2011) The Medicago genome provides insight into the evolution of rhizobial symbioses. Nature 480:520–524
83. Mochida K, Yoshida T, Sakurai T et al (2010) LegumeTFDB: an integrative database of Glycine max, Lotus japonicus and Medicago truncatula transcription factors. Bioinformatics 26(2):290–291. https://doi.org/10.1093/bioinformatics/btp645
84. Jin J, Zhang H, Kong L et al (2013) PlantTFDB 3.0: a portal for the functional and evolutionary study of plant transcription factors. Nucleic Acids Res 42(Database issue):D1182–D1187. https://doi.org/10.1093/nar/gkt1016
85. Kakar K, Wandrey M, Czechowski T (2008) A community resource for high-throughput quantitative RT-PCR analysis of transcription factor gene expression in Medicago truncatula. Plant Methods 4:18. https://doi.org/10.1186/1746-4811-4-18
86. Roux B, Rodde N, Jardinaud MF (2014) An integrated analysis of plant and bacterial gene expression in symbiotic root nodules using laser-capture microdissection coupled to RNA sequencing. Plant J 77(6):817–837. https://doi.org/10.1111/tpj.12442
87. Desbrosses GJ, Stougaard J (2011) Root nodulation: a paradigm for how plant-microbe symbiosis influences host developmental pathways. Cell Host Microbe 10(4):348–358. https://doi.org/10.1016/j.chom.2011.09.005
88. Cui H, Levesque MP, Vernoux T (2007) An evolutionarily conserved mechanism delimiting SHR movement defines a single layer of endodermis in plants. Science 316(5823):421–425
89. Sozzani R, Cui H, Moreno-Risueno MA (2010) Spatiotemporal regulation of cell-cycle genes by SHORTROOT links patterning and growth. Nature 466(7302):128–132
90. Masucci JD, Schiefelbein JW (1994) The rhd6 mutation of Arabidopsis thaliana alters root-hair initiation through an auxin- and ethylene-associated process. Plant Physiol 106(4):1335–1346
91. Menand B, Yi K, Jouannic S et al (2007) An ancient mechanism controls the development of cells with a rooting function in land plants. Science 316(5830):1477–1480
92. Kamiya M, Higashio SY, Isomoto A et al (2016) Control of root cap maturation and cell detachment by BEARSKIN transcription factors in Arabidopsis. Development 143(21):4063–4072
93. Wang TZ, Liu M, Zhao MG (2015) Identification and characterization of long non-coding RNAs involved in osmotic and salt stress in Medicago truncatula using genome-wide high-throughput sequencing. BMC Plant Biol 15:131. https://doi.org/10.1186/s12870-015-0530-5
94. Campalans A, Kondorosi A, Crespi M (2004) Enod40, a short open reading frame-containing mRNA, induces cytoplasmic localization of a nuclear RNA binding protein in Medicago truncatula. Plant Cell 16(4):1047–1059
95. Röhrig H, Schmidt J, Miklashevichs E et al (2002) Soybean ENOD40 encodes two peptides that bind to sucrose synthase. Proc Natl Acad Sci U S A 99(4):1915–1920
96. Röhrig H, John M, Schmidt J (2004) Modification of soybean sucrose synthase by S-thiolation with ENOD40 peptide A. Biochem Biophys Res Commun 325(3):864–870

97. Lelandais-Brière C, Naya L, Sallet E et al (2009) Genome-wide Medicago truncatula small RNA analysis revealed novel microRNAs and isoforms differentially regulated in roots and nodules. Plant Cell 21(9):2780–2796. https://doi.org/10.1105/tpc.109.068130
98. Branscheid A, Devers EA, May P et al (2011) Distribution pattern of small RNA and degradome reads provides information on miRNA gene structure and regulation. Plant Signal Behav 6(10):1609–1611. https://doi.org/10.4161/psb.6.10.17305
99. Devers EA, Branscheid A, May P et al (2011) Stars and symbiosis: microRNA- and microRNA*-mediated transcript cleavage involved in arbuscular mycorrhizal symbiosis. Plant Physiol 156(4):1990–2010. https://doi.org/10.1104/pp.111.172627
100. Wang T, Chen L, Zhao M et al (2011) Identification of drought-responsive microRNAs in Medicago truncatula by genome-wide high-throughput sequencing. BMC Genomics 12:367
101. Chen L, Wang T, Zhao M et al (2012) Ethylene-responsive miRNAs in roots of Medicago truncatula identified by high-throughput sequencing at whole genome level. Plant Sci 184:14–19. https://doi.org/10.1016/j.plantsci.2011.11.007 Epub 2011 Nov 17
102. Chen L, Wang T, Zhao M et al (2012) Identification of aluminum-responsive microRNAs in Medicago truncatula by genome-wide high-throughput sequencing. Planta 235:375–386
103. Zhou ZS, Zeng SQ, Liu ZP et al (2012) Genome-wide identification of Medicago truncatula microRNAs and their targets reveals their differential regulation by heavy metal. Plant Cell Environ 35:86–99
104. Eyles RP, Williams PH, Ohms SJ (2013) microRNA profiling of root tissues and root forming explant cultures in Medicago truncatula. Planta 238(1):91–105. https://doi.org/10.1007/s00425-013-1871-7
105. Formey D, Sallet E, Lelandais-Brière C et al (2014) The small RNA diversity from Medicago truncatula roots under biotic interactions evidences the environmental plasticity of the miRNAome. Genome Biol 15(9):457. https://doi.org/10.1186/s13059-014-0457-4
106. Long RC, Li MN, Kang JM et al (2015) Small RNA deep sequencing identifies novel and salt-stress-regulated microRNAs from roots of Medicago sativa and Medicago truncatula. Physiol Plant 154(1):13–27. https://doi.org/10.1111/ppl.12266 Epub 2014 Oct 7
107. Bartel DP (2009) MicroRNAs: target recognition and regulatory functions. Cell 136(2):215–213
108. Yu N, Niu QW, Ng KH et al (2015) The role of miR156/SPLs modules in Arabidopsis lateral root development. Plant J 83(4):673–685. https://doi.org/10.1111/tpj.12919
109. Xu M, Hu T, Zhao J (2016) Developmental functions of miR156-regulated SQUAMOSA PROMOTER BINDING PROTEIN-LIKE (SPL) genes in Arabidopsis thaliana. PLoS Genet 12(8):e1006263. https://doi.org/10.1371/journal.pgen.1006263
110. Aung B, Gruber MY, Amyot L et al (2015) MicroRNA156 as a promising tool for alfalfa improvement. Plant Biotechnol J 13(6):779–790. https://doi.org/10.1111/pbi.12308 Epub 2014 Dec 23
111. Wang Y, Wang Z, Amyot L et al (2015) Ectopic expression of miR156 represses nodulation and causes morphological and developmental changes in Lotus japonicas. Mol Gen Genomics 290(2):471–484
112. Naya L, Khan GA, Sorin C (2010) Cleavage of a non-conserved target by a specific miR156 isoform in root apexes of Medicago truncatula. Plant Signal Behav 5(3):328–331
113. Williams L, Grigg SP, Xie M et al (2005) Regulation of Arabidopsis shoot apical meristem and lateral organ formation by microRNA miR166g and its AtHD-ZIP target genes. Development 132(16):3657–3668
114. Miyashima S, Koi S, Hashimoto T et al (2011) Non-cell-autonomous microRNA165 acts in a dose-dependent manner to regulate multiple differentiation status in the Arabidopsis root. Development 138(11):2303–2313
115. Singh A, Singh S, Panigrahi KC (2014) Balanced activity of microRNA166/165 and its target transcripts from the class III homeodomain-leucine zipper family regulates root growth in Arabidopsis thaliana. Plant Cell Rep 33(6):945–953. https://doi.org/10.1007/s00299-014-1573-z
116. Singh A, Roy S, Singh S (2017) Phytohormonal crosstalk modulates the expression of miR166/165s, target Class III HD-ZIPs, and KANADI genes during root growth in Arabidopsis thaliana. Sci Rep 7(1):3408. https://doi.org/10.1038/s41598-017-03632-w
117. Van den Berg C, Willemsen V, Hendriks G et al (1997) Short-range control of cell differentiation in the Arabidopsis root meristem. Nature 390(6657):287–289
118. Sabatini S, Heidstra R, Wildwater M et al (2003) SCARECROW is involved in positioning the stem cell niche in the Arabidopsis root meristem. Genes Dev 17(3):354–358

119. Drisch RC, Stahl Y (2015) Function and regulation of transcription factors involved in root apical meristem and stem cell maintenance. Front Plant Sci 6:505
120. Dello Ioio R, Linhares FS, Scacchi E et al (2007) Cytokinins determine Arabidopsis root-meristem size by controlling cell differentiation. Curr Biol 17(8):678–682
121. Sabatini S, Beis D, Wolkenfelt H et al (1999) An auxin-dependent distal organizer of pattern and polarity in the Arabidopsis root. Cell 99(5):463–472
122. Sarkar AK, Luijten M, Miyashima S et al (2007) Conserved factors regulate signalling in Arabidopsis thaliana shoot and root stem cell organizers. Nature 446(7137):811–814
123. Berckmans B, Simon R (2016) A feed-forward regulation sets cell fates in roots. Trends Plant Sci 21:373–375. https://doi.org/10.1016/j.tplants.2016.03.002
124. Aida M, Beis D, Heidstra R et al (2004) The PLETHORA genes mediate patterning of the Arabidopsis root stem cell niche. Cell 119(1):109–120
125. Osipova MA, Mortier V, Demchenko KN (2012) Wuschel-related homeobox5 gene expression and interaction of CLE peptides with components of the systemic control add two pieces to the puzzle of autoregulation of nodulation. Plant Physiol 158(3):1329–1341. https://doi.org/10.1104/pp.111.188078
126. Franssen HJ, Kulikova O, Willemsen V et al (2017) Cis-regulatory PLETHORA promoter elements directing root and nodule expression are conserved between Arabidopsis thaliana and Medicago truncatula. Plant Signal Behav 12(2):e1278102. https://doi.org/10.1080/15592324.2016.1278102
127. Liu D, Song Y, Chen Z et al (2009) Ectopic expression of miR396 suppresses GRF target gene expression and alters leaf growth in Arabidopsis. Physiol Plant 136(2):223–236. https://doi.org/10.1111/j.1399-3054.2009.01229.x
128. Rodriguez RE, Mecchia MA, Debernardi JM et al (2010) Control of cell proliferation in Arabidopsis thaliana by microRNA miR396. Development 137(1):103–112. https://doi.org/10.1242/dev.043067
129. Bao M, Bian H, Zha Y et al (2014) miR396a-Mediated basic helix-loop-helix transcription factor bHLH74 repression acts as a regulator for root growth in Arabidopsis seedlings. Plant Cell Physiol 55(7):1343–1353
130. Rodriguez RE, Ercoli MF, Debernardi JM et al (2015) MicroRNA miR396 regulates the switch between stem cells and transit-amplifying cells in Arabidopsis roots. Plant Cell 27(12):3354–3366. https://doi.org/10.1105/tpc.15.00452 Epub 2015 Dec 8
131. Holmes P, Goffard N, Weiller GF et al (2008) Transcriptional profiling of Medicago truncatula meristematic root cells. BMC Plant Biol 8:21. https://doi.org/10.1186/1471-2229-8-21
132. Passarinho P, Ketelaar T, Xing M et al (2008) BABYBOOM target genes provide diverse entry points into cell proliferation and cell growth pathways. Plant Mol Biol 68(3):225–237. https://doi.org/10.1007/s11103-008-9364-y Epub 2008 Jul 29
133. Tsuge T, Tsukaya H, Uchimiya H (1996) Two independent and polarized processes of cell elongation regulate leaf blade expansion in *Arabidopsis thaliana* (L.) Heynh. Development 122:1589–1600
134. Imin N, Nizamidin M, Wu T et al (2007) Factors involved in root formation in Medicago truncatula. J Exp Bot 58(3):439–451
135. Limpens E, Moling S, Hooiveld G et al (2013) Cell- and tissue-specific transcriptome analyses of Medicago truncatula root nodules. PLoS One 8(5):e64377. https://doi.org/10.1371/journal.pone.0064377 Erratum in: PLoS One. 2014;9(1). 10.1371/annotation/d8bce646-4127-410a-b4dd-d2d2779a4745
136. Fukaki H, Tameda S, Masuda H et al (2002) Lateral root formation is blocked by a gain-of-function mutation in the SOLITARY-ROOT/IAA14 gene of Arabidopsis. Plant J 29:153–168
137. Okushima Y, Overvoorde PJ, Arima K et al (2005) Functional genomic analysis of the AUXIN RESPONSE FACTOR gene family members in Arabidopsis thaliana: unique and overlapping functions of ARF7 and ARF19. Plant Cell 17:444–463
138. Wilmoth JC, Wang S, Tiwari SB et al (2005) NPH4/ARF7 and ARF19 promote leaf expansion and auxin-induced lateral root formation. Plant J 43:118–130
139. Couzigou JM, Zhukov V, Mondy S et al (2012) NODULE ROOT and COCHLEATA maintain nodule development and are legume orthologs of Arabidopsis BLADE-ON-PETIOLE genes. Plant Cell 24(11):4498–4510. https://doi.org/10.1105/tpc.112.103747
140. Couzigou JM, Magne K, Mondy S (2016) The legume NOOT-BOP-COCH-LIKE genes are conserved regulators of abscission, a major agronomical trait in cultivated crops. New Phytol 209(1):228–240. https://doi.org/10.1111/nph.13634
141. Dello Ioio R, Nakamura K, Moubayidin L (2008) A genetic framework for the control

of cell division and differentiation in the root meristem. Science 322(5906):1380–1384. https://doi.org/10.1126/science.1164147
142. Di Mambro R, De Ruvo M, Pacifici E (2017) Auxin minimum triggers the developmental switch from cell division to cell differentiation in the Arabidopsis root. Proc Natl Acad Sci U S A 114(36):E7641–E7649. https://doi.org/10.1073/pnas.1705833114
143. Sańko-Sawczenko I, Łotocka B, Czarnocka W (2016) Expression analysis of PIN genes in root tips and nodules of Medicago truncatula. Int J Mol Sci 17(8):E1197. https://doi.org/10.3390/ijms17081197
144. Huo X, Schnabel E, Hughes K et al (2006) RNAi phenotypes and the localization of a protein::GUS fusion imply a role for *Medicago truncatula* PIN genes in nodulation. J Plant Growth Regul 25(2):156–165
145. Zhou C, Han L, Wang ZY (2011) Potential but limited redundant roles of MtPIN4, MtPIN5 and MtPIN10/SLM1 in the development of *Medicago truncatula*. Plant Signal Behav 6(11):1834–1836. https://doi.org/10.4161/psb.6.11.17508 Epub 2011 Nov 1
146. Herrbach V, Remblière C, Gough C et al (2014) Lateral root formation and patterning in *Medicago truncatula*. J Plant Physiol 171(3–4):301–310. https://doi.org/10.1016/j.jplph.2013.09.006
147. Op den Camp RH, De Mita S, Lillo A et al (2011) A phylogenetic strategy based on a legume-specific whole genome duplication yields symbiotic cytokinin type-A response regulators. Plant Physiol 157(4):2013–2022. https://doi.org/10.1104/pp.111.187526
148. Fukaki H, Tasaka M (2009) Hormone interactions during lateral root formation. Plant Mol Biol 69(4):437–449. https://doi.org/10.1007/s11103-008-9417-2
149. Du Y, Scheres B (2017) Lateral root formation and the multiple roles of auxin. J Exp Bot 69(2):155–167. https://doi.org/10.1093/jxb/erx223
150. Bensmihen S (2015) Hormonal control of lateral root and nodule development in legumes. Plants (Basel) 4(3):523–547
151. Chen Q, Dai X, De-Paoli H et al (2014) Auxin overproduction in shoots cannot rescue auxin deficiencies in Arabidopsis roots. Plant Cell Physiol 55:1072–1079
152. Chen Q, Liu Y, Maere S et al (2015) A coherent transcriptional feed-forward motif model for mediating auxin-sensitive PIN3 expression during lateral root development. Nat Commun 6:8821. https://doi.org/10.1038/ncomms9821
153. Scholte M, d'Erfurth I, Rippa S et al (2002) T-DNA tagging in the model legume Medicago truncatula allows efficient gene discovery. Mol Breed 10:203–215
154. Laffont C, Blanchet S, Lapierre C et al (2010) The compact root architecture1 gene regulates lignification, flavonoid production, and polar auxin transport in *Medicago truncatula*. Plant Physiol 153(4):1597–1607. https://doi.org/10.1104/pp.110.156620
155. Goh T, Kasahara H, Mimura T et al (2012) Multiple AUX/IAA-ARF modules regulate lateral root formation: the role of Arabidopsis SHY2/IAA3-mediated auxin signalling. Philos Trans R Soc Lond Ser B Biol Sci 367(1595):1461–1468
156. Shen C, Yue R, Yang Y et al (2014) Genome-wide identification and expression profiling analysis of the Aux/IAA gene family in *Medicago truncatula* during the early phase of Sinorhizobium meliloti infection. PLoS One 9(9):e107495. https://doi.org/10.1371/journal.pone.0107495 eCollection 2014
157. Shen C, Yue R, Sun T (2015) Genome-wide identification and expression analysis of auxin response factor gene family in *Medicago truncatula*. Front Plant Sci 6:73. https://doi.org/10.3389/fpls.2015.00073 eCollection 2015
158. Vidal EA, Araus V, Lu C et al (2010) Nitrate-responsive miR393/AFB3 regulatory module controls root system architecture in Arabidopsis thaliana. Proc Natl Acad Sci U S A 107:4477–4482
159. Ru P, Xu L, Ma H et al (2006) Plant fertility defects induced by the enhanced expression of microRNA167. Cell Res 16(5):457–465
160. Mallory AC, Bartel DP, Bartel B (2005) MicroRNA-directed regulation of Arabidopsis AUXIN RESPONSE FACTOR17 is essential for proper development and modulates expression of early auxin response genes. Plant Cell 17(5):1360–1375
161. Fahlgren N, Montgomery TA, Howell MD et al (2006) Regulation of AUXIN RESPONSE FACTOR3 by TAS3 ta-siRNA affects developmental timing and patterning in Arabidopsis. Curr Biol 16(9):939–944
162. Hobecker KV, Reynoso MA, Bustos-Sanmamed P et al (2017) The MicroRNA390/TAS3 pathway mediates symbiotic nodulation and lateral root growth. Plant Physiol 174(4):2469–2486. https://doi.org/10.1104/pp.17.00464
163. Gutierrez L, Mongelard G, Floková K et al (2012) Auxin controls Arabidopsis adventitious root initiation by regulating jasmonic acid homeostasis. Plant Cell 24(6):2515–2527. https://doi.org/10.1105/tpc.112.099119 Epub 2012 Jun 22
164. D'Haeseleer K, Den Herder G, Laffont C et al (2011) Transcriptional and post-

transcriptional regulation of a NAC1 transcription factor in Medicago truncatula roots. New Phytol 191:647–661
165. Guo HS, Xie Q, Fei JF et al (2005) MicroRNA directs mRNA cleavage of the transcription factor NAC1 to downregulate auxin signals for arabidopsis lateral root development. Plant Cell 17(5):1376–1386
166. Laplaze L, Benkova E, Casimiro I et al (2007) Cytokinins act directly on lateral root founder cells to inhibit root initiation. Plant Cell 19(12):3889–3900
167. Zürcher E, Tavor-Deslex D, Lituiev D et al (2013) A robust and sensitive synthetic sensor to monitor the transcriptional output of the Cytokinin signaling network in Planta. Plant Physiol 161(3):1066–1075
168. van Zeijl A, Liu W, Xiao TT et al (2015) The strigolactone biosynthesis gene DWARF27 is co-opted in rhizobium symbiosis. BMC Plant Biol 15:260. https://doi.org/10.1186/s12870-015-0651-x
169. Gamas P, Brault M, Jardinaud MF et al (2017) Cytokinins in symbiotic nodulation: when, where, what for? Trends Plant Sci 22(9):792–802. https://doi.org/10.1016/j.tplants.2017.06.012
170. Crespi M, Frugier F (2008) De novo organ formation from differentiated cells: root nodule organogenesis. Sci Signal 1(49):re11. https://doi.org/10.1126/scisignal.149re11 Review. Erratum in: Sci Signal. 2009 Jan 13;2(53):er1
171. Plet J, Wasson A, Ariel F et al (2011) MtCRE1-dependent cytokinin signaling integrates bacterial and plant cues to coordinate symbiotic nodule organogenesis in Medicago truncatula. Plant J 65(4):622–633. https://doi.org/10.1111/j.1365-313X.2010.04447.x
172. Mortier V, Wasson A, Jaworek P et al (2014) Role of LONELY GUY genes in indeterminate nodulation on Medicago truncatula. New Phytol 202(2):582–593. https://doi.org/10.1111/nph.12681
173. Jasinski S, Piazza P, Craft J et al (2005) KNOX action in Arabidopsis is mediated by coordinate regulation of cytokinin and gibberellin activities. Curr Biol 15(17):1560–1565
174. Di Giacomo E, Sestili F, Iannelli MA et al (2008) Characterization of KNOX genes in Medicago truncatula. Plant Mol Biol 67(1–2):135–150. https://doi.org/10.1007/s11103-008-9307-7
175. Truernit E, Siemering KR, Hodge S et al (2006) A map of KNAT gene expression in the Arabidopsis root. Plant Mol Biol 60(1):1–20
176. Mähönen AP, Bishopp A, Higuchi M et al (2006) Cytokinin signaling and its inhibitor AHP6 regulate cell fate during vascular development. Science 311(5757):94–98
177. Moreira S, Bishopp A, Carvalho H et al (2013) AHP6 inhibits cytokinin signaling to regulate the orientation of pericycle cell division during lateral root initiation. PLoS One 8(2):e56370. https://doi.org/10.1371/journal.pone.0056370
178. Moreira S, Braga T, Carvalho H et al (2013) The Arabidopsis HP6 gene is expressed in Medicago truncatula lateral roots and root nodule primordia. Plant Signal Behav 8(8):e25262. https://doi.org/10.4161/psb.25262 Faut-il la mettre aussi?
179. Ariel F, Diet A, Verdenaud M et al (2010) Environmental regulation of lateral root emergence in Medicago truncatula requires the HD-Zip I transcription factor HB1. Plant Cell 22(7):2171–2183. https://doi.org/10.1105/tpc.110.074823 Epub 2010 Jul 30
180. Liang Y, Mitchell DM, Harris JM (2007) Abscisic acid rescues the root meristem defects of the Medicago truncatula latd mutant. Dev Biol 304(1):297–307 Epub 2006 Dec 21
181. Bright LJ, Liang Y, Mitchell DM et al (2005) The LATD gene of Medicago truncatula is required for both nodule and root development. Mol Plant-Microbe Interact 18(6):521–532
182. Zhang C, Bousquet A, Harris JM et al (2014) Abscisic acid and lateral root organ defective/NUMEROUS INFECTIONS AND POLYPHENOLICS modulate root elongation via reactive oxygen species in Medicago truncatula. Plant Physiol 166(2):644–658. https://doi.org/10.1104/pp.114.248542 Epub 2014 Sep 5
183. Fonouni-Farde C, Tan S, Baudin M et al (2016) DELLA-mediated gibberellin signalling regulates nod factor signalling and rhizobial infection. Nat Commun 7:12636. https://doi.org/10.1038/ncomms12636
184. Floss DS, Levy JG, Lévesque-Tremblay V et al (2013) DELLA proteins regulate arbuscule formation in arbuscular mycorrhizal symbiosis. Proc Natl Acad Sci U S A 110(51):E5025–E5034. https://doi.org/10.1073/pnas.1308973110 Epub 2013 Dec 2
185. Pimprikar P, Carbonnel S, Paries M et al (2016) A CCaMK-CYCLOPS-DELLA complex activates transcription of RAM1 to regulate Arbuscule branching. Curr Biol 26(8):987–998. https://doi.org/10.1016/j.cub.2016.01.069 Epub 2016 Mar 24
186. Müssig C, Shin GH, Altmann T (2003) Brassinosteroids promote root growth in Arabidopsis. Plant Physiol 133(3):1261–1271 Epub 2003 Oct 2
187. Wei Z, Li J (2016) Brassinosteroids regulate root growth, development, and sym-

biosis. Mol Plant 9(1):86–100. https://doi.org/10.1016/j.molp.2015.12.003 Epub 2015 Dec 15
188. Bao F, Shen J, Brady SR et al (2004) Brassinosteroids interact with auxin to promote lateral root development in Arabidopsis. Plant Physiol 134(4):1624–1631 Epub 2004 Mar 26
189. Cheng X, Gou X, Yin H et al (2017) Functional characterisation of brassinosteroid receptor MtBRI1 in Medicago truncatula. Sci Rep 7(1):9327. https://doi.org/10.1038/s41598-017-09297-9
190. Ferguson BJ, Ross JJ, Reid JB (2005) Nodulation phenotypes of gibberellin and brassinosteroid mutants of pea. Plant Physiol 138(4):2396–2405 Epub 2005 Jul 29
191. Zhang H, Jennings A, Barlow PW et al (1999) Dual pathways for regulation of root branching by nitrate. Proc Natl Acad Sci U S A 96(11):6529–6534
192. Walch-Liu P, Ivanov I, Filleur S (2006) Nitrogen regulation of root branching. Ann Bot 97(5):875–881 Epub 2005 Dec 9
193. Bourion V, Martin C, de Larambergue H et al (2014) Unexpectedly low nitrogen acquisition and absence of root architecture adaptation to nitrate supply in a Medicago truncatula highly branched root mutant. J Exp Bot 65(9):2365–2380. https://doi.org/10.1093/jxb/eru124 Epub 2014 Apr 4
194. Fraser CM, Chapple C (2011) The phenylpropanoid pathway in Arabidopsis. Arabidopsis Book 9:e0152. https://doi.org/10.1199/tab.0152
195. Harris JM, Dickstein R (2010) Control of root architecture and nodulation by the LATD/NIP transporter. Plant Signal Behav 5(11):1365–1369 Epub 2010 Nov 1
196. Pellizzaro A, Clochard T, Cukier C et al (2014) The nitrate transporter MtNPF6.8 (MtNRT1.3) transports abscisic acid and mediates nitrate regulation of primary root growth in Medicago truncatula. Plant Physiol 166(4):2152–2165. https://doi.org/10.1104/pp.114.250811 Epub 2014 Nov 3
197. Pellizzaro A, Alibert B, Planchet E et al (2017) Nitrate transporters: an overview in legumes. Planta 246(4):585–595. https://doi.org/10.1007/s00425-017-2724-6 [Epub ahead of print] Review
198. Hofferek V, Mendrinna A, Gaude N et al (2014) MiR171h restricts root symbioses and shows like its target NSP2 a complex transcriptional regulation in Medicago truncatula. BMC Plant Biol 14:199. https://doi.org/10.1186/s12870-014-0199-1
199. Gifford ML, Dean A, Gutierrez RA et al (2008) Cell-specific nitrogen responses mediate developmental plasticity. Proc Natl Acad Sci U S A 105:803–808
200. Araya T, Miyamoto M, Wibowo J (2014) CLE-CLAVATA1 peptide-receptor signaling module regulates the expansion of plant root systems in a nitrogen-dependent manner. Proc Natl Acad Sci U S A 111(5):2029–2034. https://doi.org/10.1073/pnas.1319953111 Epub 2014 Jan 21
201. Czyzewicz N, Shi C-L, Vu LD et al (2015) Modulation of Arabidopsis and monocot root architecture by CLAVATA3/EMBRYO SURROUNDING REGION 26 peptide. J Exp Bot 66:5229–5243. https://doi.org/10.1093/jxb/erv360
202. Mohd-Radzman NA, Binos S, Truong TT et al (2015) Novel MtCEP1 peptides produced in vivo differentially regulate root development in Medicago truncatula. J Exp Bot 66(17):5289–5300. https://doi.org/10.1093/jxb/erv008 Epub 2015 Feb 22
203. Okushima Y, Fukaki H, Onoda M et al (2007) ARF7 and ARF19 regulate lateral root formation via direct activation of LBD/ASL genes in Arabidopsis. Plant Cell 19(1):118–130
204. Boerjan W, Cervera MT, Delarue M et al (1995) Superroot, a recessive mutation in Arabidopsis, confers auxin overproduction. Plant Cell 7:1405–1419
205. Huault E, Laffont C, Wen J et al (2014) Local and systemic regulation of plant root system architecture and symbiotic nodulation by a receptor-like kinase. PLoS Genet 10(12):e1004891. https://doi.org/10.1371/journal.pgen.1004891 eCollection 2014 Dec
206. De Bang TC, Lay KS, Scheible W-R et al (2017) Small peptide signaling pathways modulating macronutrient utilization in plants. Curr Opin Plant Biol 39:31–39
207. Jin J, Watt M, Mathesius U (2012) The autoregulation gene SUNN mediates changes in root organ formation in response to nitrogen through alteration of shoot-to-root auxin transport. Plant Physiol 159(1):489–500. https://doi.org/10.1104/pp.112.194993 Epub 2012 Mar 7
208. Volpe V, Giovannetti M, Sun XG et al (2016) The phosphate transporters LjPT4 and MtPT4 mediate early root responses to phosphate status in non mycorrhizal roots. Plant Cell Environ 39(3):660–671. https://doi.org/10.1111/pce.12659 Epub 2016 Jan 12
209. De Cuyper C, Fromentin J, Yocgo RE et al (2015) From lateral root density to nodule number, the strigolactone analogue GR24 shapes the root architecture of Medicago truncatula. J Exp Bot 66(1):137–146. https://doi.org/10.1093/jxb/eru404 Epub

2014 Nov 4. Erratum in: J Exp Bot. 2015 Jul;66(13):4091

210. Lauressergues D, André O, Peng J et al (2015) Strigolactones contribute to shoot elongation and to the formation of leaf margin serrations in Medicago truncatula R108. J Exp Bot 66(5):1237–1244. https://doi.org/10.1093/jxb/eru471 Epub 2014 Dec 3
211. Li D, Zhang Y, Hu X et al (2011) Transcriptional profiling of Medicago truncatula under salt stress identified a novel CBF transcription factor MtCBF4 that plays an important role in abiotic stress responses. BMC Plant Biol 11:109. https://doi.org/10.1186/1471-2229-11-109
212. Merchan F, de Lorenzo L, Rizzo SG et al (2007) Identification of regulatory pathways involved in the reacquisition of root growth after salt stress in Medicago truncatula. Plant J 51(1):1–17 Epub 2007 May 3
213. Gruber V, Blanchet S, Diet A et al (2009) Identification of transcription factors involved in root apex responses to salt stress in Medicago truncatula. Mol Gen Genomics 281(1):55–66. https://doi.org/10.1007/s00438-008-0392-8 Epub 2008 Nov 6
214. Ariel FD, Diet A, Crespi M et al (2010) The LOB-like transcription factor Mt LBD1 controls Medicago truncatula root architecture under salt stress. Plant Signal Behav 5(12):1666–1668 Epub 2010 Dec 1
215. Lee HW, Cho C, Kim J (2015) Lateral Organ Boundaries Domain16 and 18 Act Downstream of the AUXIN1 and LIKE-AUXIN3 Auxin Influx Carriers to Control Lateral Root Development in Arabidopsis. Plant Physiol 168(4):1792–1806. https://doi.org/10.1104/pp.15.00578 Epub 2015 Jun 9
216. Zahaf O, Blanchet S, de Zélicourt A et al (2012) Comparative transcriptomic analysis of salt adaptation in roots of contrasting Medicago truncatula genotypes. Mol Plant 5(5):1068–1081. https://doi.org/10.1093/mp/sss009 Epub 2012 Mar 14
217. Kawakatsu T, Stuart T, Valdes M et al (2016) Unique cell-type-specific patterns of DNA methylation in the root meristem. Nat Plants 2(5):16058. https://doi.org/10.1038/nplants.2016.58

Chapter 16

The Multiple Faces of the *Medicago-Sinorhizobium* Symbiosis

Fathi Berrabah, El Hosseyn Ait Salem, Marie Garmier, and Pascal Ratet

Abstract

Medicago truncatula is able to perform a symbiotic association with *Sinorhizobium* spp. This interaction leads to the formation of a new root organ, the nodule, in which bacteria infect the host cells and fix atmospheric nitrogen for the plant benefit. Multiple and complex processes are essential for the success of this interaction from the recognition phase to nodule formation and functioning, and a wide range of plant host genes is required to orchestrate this phenomenon. Thanks to direct and reverse genetic as well as transcriptomic approaches, numerous genes involved in this symbiosis have been described and improve our understanding of this fantastic association. Herein we propose to update the recent molecular knowledge of how *M. truncatula* associates to its symbiotic partner *Sinorhizobium* spp.

Key words *Medicago-Sinorhizobium*, Symbiosis, Symbiotic genes, Nitrogen fixation

1 Introduction to Legume-Rhizobia Symbiosis

Leguminous plants growing in poor nitrogen soils are able to establish symbiosis with soil nitrogen-fixing bacteria referred to as rhizobia. This interaction leads to the formation of new root organs called nodules, within which rhizobia fix atmospheric nitrogen and deliver the organic nitrogen to their host. In return the plant provides a beneficial environment for the symbiont [1]. Two processes are required for nodule formation: (1) infection of the root hair cells by the bacteria via the development of infection threads (ITs) and (2) concomitantly an organogenetic process which is activated and corresponds to reactivation of cortical cell division. Organogenesis is observed at the site of infection and gives the future nodule primordium [2] (Fig. 1). In the nodules, bacteria infect massively and chronically the host cells without induction of defense reactions and differentiate in an intracellular form called the bacteroid. This high level of infection is exceptional in nature, and a growing number of studies shows the complexity of the molecular processes requiring repression of the plant immunity

Luis A. Cañas and José Pío Beltrán (eds.), *Functional genomics in* Medicago truncatula*: Methods and Protocols*, Methods in Molecular Biology, vol. 1822, https://doi.org/10.1007/978-1-4939-8633-0_16,

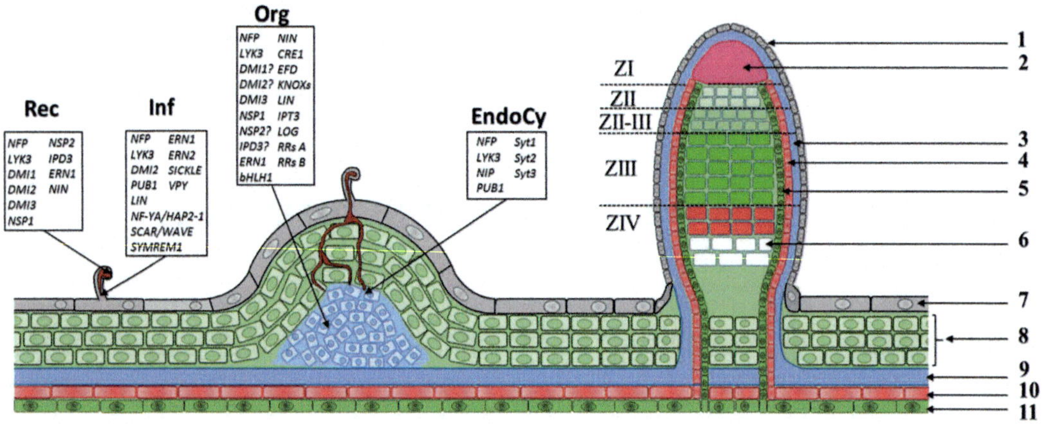

Fig. 1 Nodule formation and structure in the *Medicago-Sinorhizobium* symbiosis. Under nitrogen deficiency, *M. truncatula* interacts with *Sinorhizobium* spp. The recognition phase (Rec) is followed by root infection (Inf). During this Inf phase, the rhizobia penetrate the root hairs by formation of infection threads. In parallel to the infection, the organogenesis (Org) takes place. For Org, cortical cell divisions are reactivated and produce the nodule primordium, and the bacteria are released into the primordium cells (EndoCy). The box indicates the genes associated with the different steps of the early symbiotic association. This interaction leads to the formation of the nodule, composed of a meristem or ZI, an infection zone or ZII, and a fixation zone or ZIII separated from ZII by an interzone or ZII–III. Finally a senescent zone or ZIV is observed in old nodules or stressed plants or when organic nitrogen is added to the growth media after nodulation. Two peripheral vascular bundles (VBs) are present in the root organ and connect the nodule to the root vasculature. These VB tissues are responsible for host-symbiont metabolic exchange. 1, nodule epidermis; 2, meristem; 3, nodule cortex and endodermis; 4, xylem; 5, phloem; 6, empty cells; 7, root epidermis; 8, root cortex; 9, root endodermis; 10, xylem; 11, phloem

during symbiosis [3]. Inside the host cell, the bacteroid nitrogenase fixes atmospheric nitrogen by converting N_2 into $NH_3{}^+$ [4]. As this nitrogenase enzyme is an extremely O_2-sensitive enzyme, the plant host regulates nodule O_2 diffusion via the expression of specific leghemoglobin (Lbs) proteins maintaining low level of free oxygen in the cells [5]. These Lbs confer the pinkish coloration to the nodules that is considered as a marker for healthy and functional symbiotic organs.

Depending on the legume species, two principal types of nodules are found: determinate nodules characterized by spherical shape due to the presence of a diffuse meristem and indeterminate nodules with elongated shape [6]. Different zones are present in these indeterminate nodules (Fig. 1); the apical meristem or zone I (ZI) is responsible for a constant growth of the organ conferring the elongated shape; ZI is followed by infection zone or zone II (ZII) in which bacteria infect the host cells. Next to the ZII is the fixation zone, or zone III (ZIII), in which the differentiated bacteroids start to express the nitrogenase and to fix N_2. This zone is separated from the ZII by the interzone or ZII–III in which undifferentiated bacteroids undergo a terminal differentiation. Finally, when nodules become old, or when the plant is stressed, or when organic nitrogen like nitrate is added to the growth media, a senescent zone, or zone IV (ZIV), is observed. In this zone bacteroids as well as host cells are degraded [7] (Fig. 1). Interestingly, this ZIV can be recolonized by saprophytic rhizobia [8]. Senescence is also observed in determinate nodules, but in this nodule type, it starts from the nodule center and spreads to the periphery of the organ [9].

The interaction between the model plant *Medicago truncatula* and its symbiotic partner *Sinorhizobium* spp. is one of the best-characterized legume-rhizobia associations. This interaction results in the formation of indeterminate nodules. Thanks to mutagenesis approaches such as transposon insertions and fast neutron deletions and reverse genetic approaches using mutant collections and transcriptomic analysis, many genes involved in the *M. truncatula*-*Sinorhizobium* spp. interaction were described. This review summarizes in an integrative model the recent knowledge on the molecular mechanisms underlying this association. In the text, we focus our attention on the best-characterized genes, and Table 1 is a more exhaustive list of the genes associated with the symbiosis (*see* also Fig. 1 for genes involved in early interaction and nodule organogenesis).

2 Rhizobial Recognition Key Events to Successful Interaction

When nitrogen becomes limiting for growth, legumes secrete in the rhizosphere specific secondary metabolites of the flavonoid type that are specifically perceived by rhizobia [10]. Flavonoids bind to and activate the bacterial transcriptional regulators NodD. These activated NodD induce the expression of the *nod* regulon, including the *nodABC* genes responsible for nod factor (NF) backbone biosynthesis [11]. NFs that are key signaling molecules of the symbiosis are lipo-chitooligosaccharides (LCOs) composed of a N-acetylglucosamine (GlcNAc) backbone with fatty acid attached in the nonreducing terminal glucosamine [12]. Different modifications of the NF structure can occur depending on the rhizobial strain and determine the host-rhizobium compatibility and specificity of recognition [13]. The modifications on the NF structure can include modifications on the chitin backbone or changes of the length and degree of unsaturation of the fatty acid chain. However, the NFs are not essential for all leguminous-rhizobia interactions as, for example, *Bradyrhizobium* spp. can nodulate *Aeschynomene evenia* without production of NFs [14].

NFs are perceived in the root epidermis by lysine motif (LysM) receptor-like kinases (RLKs), but the mechanism of recognition is not yet fully understood. Nod factor perception (NFP) and LYsine motif Kinase 3 (LYK3/HCL) are proposed as *M. truncatula* key receptors for the NFs [15–17]. Furthermore, a new LysM-RLK,

Table 1
Mutants and corresponding genes involved in the control of *Medicago-Sinorhizobium* symbiosis

Symbiotic steps	Mutant	Gene product	Biological function	Publication reference
Recognition and symbiotic signaling	*nfp*	NFP	NF perception	[16]
	lyk3	LYK3	NF perception	[17]
	dmi2	DMI2	NF signaling	[19]
	lyr3	LYR3	NF perception	[18]
	lyk4	LYK4	NF perception	[17]
	dmi1	DMI1	Calcium speaking	[21]
	dmi3	DMI3	Calcium speaking	[23]
	ipd3	IPD3	Symbiotic signaling	[25]
	nsp1	NSP1	Symbiotic TF	[27]
	nsp2	NSP2	Symbiotic TF	[26]
	nin	NIN	Symbiotic TF	[29]
	ern1	ERN1	Symbiotic TF	[38]
	ern2	ERN2	Symbiotic TF	[109]
Infection	*RNAi*	PUB1	E3 ligase	[31]
	rit	SCAR/WAV	SCAR/WAV	[32]
	lin	LIN	E3 ligase	[33]
	nf-ya/RNAi	NF-YA/ HAP2-1	CCAAT box-binding FT	[36, 110]
	nip	NIP/LATD	Nitrate transporter	[43, 44, 111]
	sickle	SICKLE	Ethylene receptor	[40]
	RNAi	Syt1/2/3	Synaptotagmin, membrane fusion	[41]
	RNAi/*Tnt1*	SYMREM1	SYMBIOTIC REMORIN 1; lipid microdomain formation	[112]
	RNAi	FLOT2	Membrane shaping	[39]
	RNAi	FLOT4	Membrane shaping	[39]
	RNAi/*Tnt1*	VPY	VAPYRIN; IT formation	[113]
	Cbs	CBS1	Cystathionine-β-synthase: IF formation	[114]
	Rpg	RPG	Coil-coil protein, unknown function	[115]
Organogenesis	*cre1*	CRE1	Cytokinin receptor	[50]
	RNAi	RR4	RR type A	[54]
	RNAi	RR9	RR type B	[116]
	RNAi	LOG	Lonely guy: cytokinin biosynthesis	[117]
	RNAi	CKX1	Cytokinin oxidase; cytokinin degradation	[118]
	RNAi/*bhlh476*	bHLH476	TF	[52]
	RNAi	bHLH1	TF	[58]
	RNAi	KNAT3/4/5	TF	[55]
	RNAi	KNOX3	TF	[56]
	RNAi	ZR1	TF	[119]

(continued)

Table 1
(continued)

Symbiotic steps	Mutant	Gene product	Biological function	Publication reference
Meristem regulation	*noot*	NOOT	TF	[65]
Terminal differentiation	*dnf1*	DNF1	Endopeptidase	[76]
		NCR247	Antimicrobial peptide	[78]
	dnf4	NCR211	Antimicrobial peptide	[79]
	dnf7	NCR169	Antimicrobial peptide	[80]
	rsd	RSD	C_2H_2 TF	[87]
Host selection	Genotype	NFS1	Antimicrobial peptide	[82]
	Genotype	NFS2	Antimicrobial peptide	[81]
Defense repression	*dnf2*	DNF2	PI-PLC like protein	[84]
	nad1	NAD1	Unknown function	[86]
	symCRK	SYMCRK	DUF26-RLK	[85]
Autoregulation of the nodulation (AON)	*sunn*	SUNN	LRR-RLK	[96]
	RNAi	CLE12	CLV3/ embryo-surrounding region	[62]
	RNAi	CLE13	CLV3/ embryo-surrounding region	[62]
		CLV2	LRR-RLK	[99]
	crn	CRN	LRR-RLK	[99]
	rdn1	RDN1	Small uncharacterized peptide	[100]
	RNAi	CEP1	Small signaling peptide	[102]
	cra2	CRA2	LRR-RLK	[101, 103]
Senescence	RNAi	CP6	Cysteine protease	[105]
	RNAi	VPE	Vacuolar processing enzyme	[105]

Certain genes can act at different symbiotic steps. Please refer to the text and Fig. 1 for a more detailed information

LYR3, with high affinity to the NFs has been identified, but the corresponding mutant is not affected in its interaction with rhizobium [18]. Does not Make Infection 2 (DMI2) is a leucine-rich repeat (LRR) RLK necessary during the early steps of the interaction and essential for NF signal transduction [15, 19]. After NF-RLK interaction, a signaling symbiotic pathway is activated and leads to different signaling events characterized by (1) host cell membrane depolarization, (2) reactive oxygen spices (ROS) production, and (3) intracellular calcium (Ca^{+2}) increase referred to as calcium oscillation or calcium spiking [20]. The calcium oscillation requires DMI1, a ligand-gated channel protein localized in the nuclear membrane and responsible for calcium exporting from nucleus to the cytosol [15, 21]. Ca^{2+} is then detected by the calcium-dependent calmodulin kinase DMI3 [15, 22, 23] that phosphorylates interacting protein of DMI3 (IPD3), a coiled-coil domain-containing protein [15, 24] required for nodulation [25]. Nodule signal pathways 1 and 2 (NSP1 and NSP2) are two GRAS transcription factors (TFs) acting downstream of DMI3 [15, 26, 27]. NSP1 and NSP2 form a complex and activate the expression of two key transcriptional regulators: Nodule INception (NIN) [28, 29] and ERF Required for Nodulation 1 (ERN1) [30]. NIN and ERN1 through the early noduline (also known as ENOD) gene induction activate the nodule formation [29, 30].

3 Rhizobia on Road: Infection of the Host Tissues by the Bacteria

After recognition, the symbiont infects the roots through root hairs by formation of infection threads (ITs). Bacterial penetration of the roots follows the deformation of the root hairs, and ITs grow toward the root cortex [15]. Specific RNAi targeting NF receptor (LYK3, LYK4, and NFP) block IT penetration [7, 17], suggesting a role for NF signaling during root infection. Furthermore, the E3 ubiquitin ligase Plant U-box protein 1 (PUB1) regulates negatively this infection process probably via LYK3 degradation [1, 31]. Several *M. truncatula* mutants with anomalies in the infection process have been identified (*see* Table 1). *rit* shows aborted ITs in the root epidermis and the gene affected in this mutant is a member of suppressor of cAMP receptor defect/WASP family verprolin-homologous protein (SCAR/WAVE) family [32]. Similarly, the *lumpy infection* (*lin*) mutant shows IT growth arrest. LIN corresponds to an E3 ubiquitin ligase containing U-box/WD40 domain [33]. The mechanisms of SCAR/WAVE and LIN action are not yet understood. The mutant of the *MtNF-YA* gene also called *HAP2–1* shows disruption of infection thread progression [34]. This gene encodes a CCAAT box-binding transcription factor subunit, and its expression is controlled by NIN [35] and the microRNA169 [36]. *NF-YA/HAP2-1* regulates the *ERN1* expression [37], and as the *ern1* mutant is also affected in IT progression [38], it is proposed that *NF-YA/HAP2-1* controls the infection process at least through *ERN1*. Moreover, suppressing the expression of two *FLOtline* genes, *FLOT2* and *FLOT4*, also arrests IT progression [39]. These proteins are lipid raft markers and may be implicated in recruiting multiprotein complexes [1]. *MtSICKLE* encodes an orthologue of *AtEIN2*, the *Arabidopsis thaliana* ethylene co-receptor. The *sickle* mutant shows a hyperinfection phenotype [40] and is compromised in immunity, with, for example, enhanced sensitivity to the *Phytophthora medicaginis* pathogen [40]. As ethylene is a defense hormone, it is tempting to speculate that via *MtSICKLE*, ethylene regulates negatively the infection via locale defense activation (Fig. 1).

When ITs reach the root cortex, bacteria are released into the targeted future symbiotic cell. Silencing of genes encoding the synaptotagmins MtSyt1/3 or MtSyt2/3, implicated in the membrane trafficking, inhibits releasing of *S. meliloti* into the host cell [41]. NFP, LYK3, and DMI2 are also required for intracellular infection [17, 42] suggesting a role of the NF signaling during the bacterial release. *Numerous Infections and Polyphenolics* (*NIP*) encodes a protein belonging to the nitrate transporter family and shows high affinity to nitrate [43, 44]. Interestingly, the corresponding mutant shows absence of rhizobia release and activation of defense reactions [43].

4 Nodule Organogenesis: Building the Sweet Home

Near to the infection site, the organogenesis is activated and consists of cortical cell dedifferentiation and proliferation to build the nodule primordium, the first hosting tissues for the rhizobia [45]. The NF signaling is required for activation of the organogenetic program, as treatments of roots with NFs induce formation of nodule-like structure [1, 46]. Moreover, constitutive DMI3 activity causes the formation of spontaneous nodules in absence of rhizobia [47]. It is thus possible to postulate that the organogenetic process is strictly under host control [48]. Application of rhizobial LCO on non-inoculated roots triggers accumulation of active forms of cytokinins (CKs) [49]. This class of hormone is involved in the plant development as well as in nodule organogenesis. It has been suggested that CKs coordinate infection and primordium formation [2, 48]. The cytokinin signaling during nodulation depends on the *cytokinin response 1* (*MtCRE1*) receptor [50, 51] and mediates induction of *NIN*, *ERN1*, and *NSP2* genes [51, 52]. CKs also induce nodule formation through the polar auxin transport disruption over flavonoid accumulation [1, 53]. The CK signaling is required for successful interaction, for the expression of the ethylene response factor for nodule differentiation (EFD) [54], and for the expression of some members of the KNOX TF family [55] such as KNOX3 [56] that regulates negatively the CK pathways. This KNOX3 regulation occurs through the repression of the cytokinin biosynthesis genes like *Lonely Guy* 2 (*LOG2*) [56] or through the activation of negative response regulator (RR of type A) like *MtRR4* [54, 55].

The mature nodule is composed of two peripheral vascular bundles (VBs) in which the metabolic exchanges occur. The development of correct vasculature tissues is probably crucial for the functioning of the future nodule. The *lin-4* mutant cannot form infection threads and develops a central vasculature rather than peripheral, indicating that the infection is essential for correct vascular development [57]. Moreover, *MtbHLH1* encodes basic helix-loop-helix TF expressed in nodule primordia and VBs, and the loss of the *bHLH1* function leads to abnormal vascular patterning in the nodules [58].

5 The Fix$^-$ Mutants: Powerful Tools to Understand Nodule Physiology

Most of the mutants described above, altered in the early steps of the interaction, are not able to produce nodules and are called nod$^-$. In contrast, the fix$^-$ mutants are able to form nodules that are not functional. These fix$^-$ mutants are altered during the late steps of the interaction, and the corresponding genes control biological process in the mature nodule. Fix$^-$ mutants represent great opportunity to understand the mechanisms that keep the nodule functional and give access to the genes involved in these steps of the symbiosis. For example, Starker and colleagues found seven fix$^-$ complementation groups referred to as *does not fix nitrogen* (*dnf*) *1–7* during the screen of an A17 ecotype fast neutron mutant collection [59]. Similarly, Pislariu et al. isolated a set of fix$^-$ mutants during the screening of an insertion mutant collection in the R108 background [60]. The next sections describe genes that act in late steps of the symbiotic interaction, some corresponding to the genes affected in the fix$^-$ mutants, and that are involved in the functioning of the symbiotic organ.

6 Meristem Regulation: Expand the Activity Without Forgetting the Identity

M. truncatula forms indeterminate nodules characterized by the presence of a persistent apical meristem. This tissue has an intense mitotic activity and is at the origin of most of the symbiotic tissues of the differentiated nodule. The maintenance of this meristem activity is vital for nodule development. In mature nodules, *MtNF-YA/MtHAP2-1* is expressed in the meristem, and in addition to the infection phenotype, the shutdown of *MtNF-YA/MtHAP2-1* expression leads to the formation of abnormal nodules. J-P Combier and colleagues proposed that *MtNF-YA/MtHAP2-1* controls the nodulation in the mature nodule through the regulation of the meristem [61]. The CLAVATA3/Embryo-surrounding region (CLE) peptides are possibly involved in meristem regulation. The role of the legume *CLE* genes is mostly described during symbiosis for the autoregulation of the nodulation (AON; see below), and the presence of *CLE12* and *CLE13* transcripts in the nodule meristem suggests a potential role in meristem regulation [62]. In *A. thaliana* the shoot meristem maintenance requires stem cell precursors expressing the *WUSCHEL* (*WUS*) transcription factors and its repression by the CLV3 peptide through its

LRR-RLK receptors CLV1/2 in order to regulate the meristem size and activate cell differentiation [63, 64]. The maintenance of the meristem identity is another important part of the meristem regulation. The NOdule-ROot (NOOT1) NBCL gene is an orthologue of *A. thaliana* Blade-On-Petiole (BOP). The *noot1* mutant forms nodules developing roots on their apical region [65]. The description of the mutant phenotype suggests that the VBs are the origin of the emerged roots. Moreover, the *NOOT-1* gene is expressed on apical extremities of VBs suggesting that NOOT represses their root identity to maintain the nodule identity [65].

7 Terminal Differentiation of the Bacteroid: Keep the Symbiont Under Control

After releasing in the symbiotic cells, the bacteroids are surrounded by the host cell membrane called peri-bacteroid membrane (PBM). A peri-bacteroid space (PBS) separates the bacteroids from the PMB. This new organelle is called the symbiosome [4]. In *M. truncatula*, bacteroids undergo a terminal differentiation characterized by endoreduplication of the bacterial genome and an increase of the cell size. This phenomenon is known as the terminal differentiation process [66]. Terminally differentiated bacteroids cannot survive outside their host, and this process is considered by certain scientist as a "slavery process" [67, 68]. This differentiation is not generalizable to all leguminous plants but is restricted to the legume inverted repeat-lacking clade (IRLC) [69]. One hypothesis to explain these differences is that the leguminous plants took different evolution ways to form efficient association. The differentiation process is mediated by small antimicrobial peptides called nodule cysteine rich (NCR) [67, 70, 71]. *M. truncatula* possesses a very large number of these peptides (around 600 genes) [72, 73] with the majority of them being specifically expressed at different stages of the symbiosis [74, 75]. NCRs are targeted to the symbiosomes and reach the PBS [66, 76]. *DNF1*, an endopeptidase subunit targeted to the symbiosome, is required for NCR penetration because the *dnf1* mutant shows undifferentiated bacteroids and defaults in NCR folding [77] (Fig. 2). The bacterial ABC transporter BacA is also essential for the bacteroid intracellular surviving, and the *bacA* mutant shows higher in vitro sensitivity to the NCRs than the WT [71]. The authors of this work speculated that NCRs are toxic when reaching the PBS and BacA is essential for their internalization into the bacteroid where they are detoxified [66, 71]. In the endosymbiont, NCRs can interfere with various cellular functions to mediate differentiation. For example, NCR247 can interact with FtsZ, a protein required for septum formation and bacterial division suggesting that NCR247 blocks the bacterial division by FtsZ inhibition [78]. In contrast, the two fix$^-$ mutants *dnf4* and *dnf7* are mutated in the *NCR211*

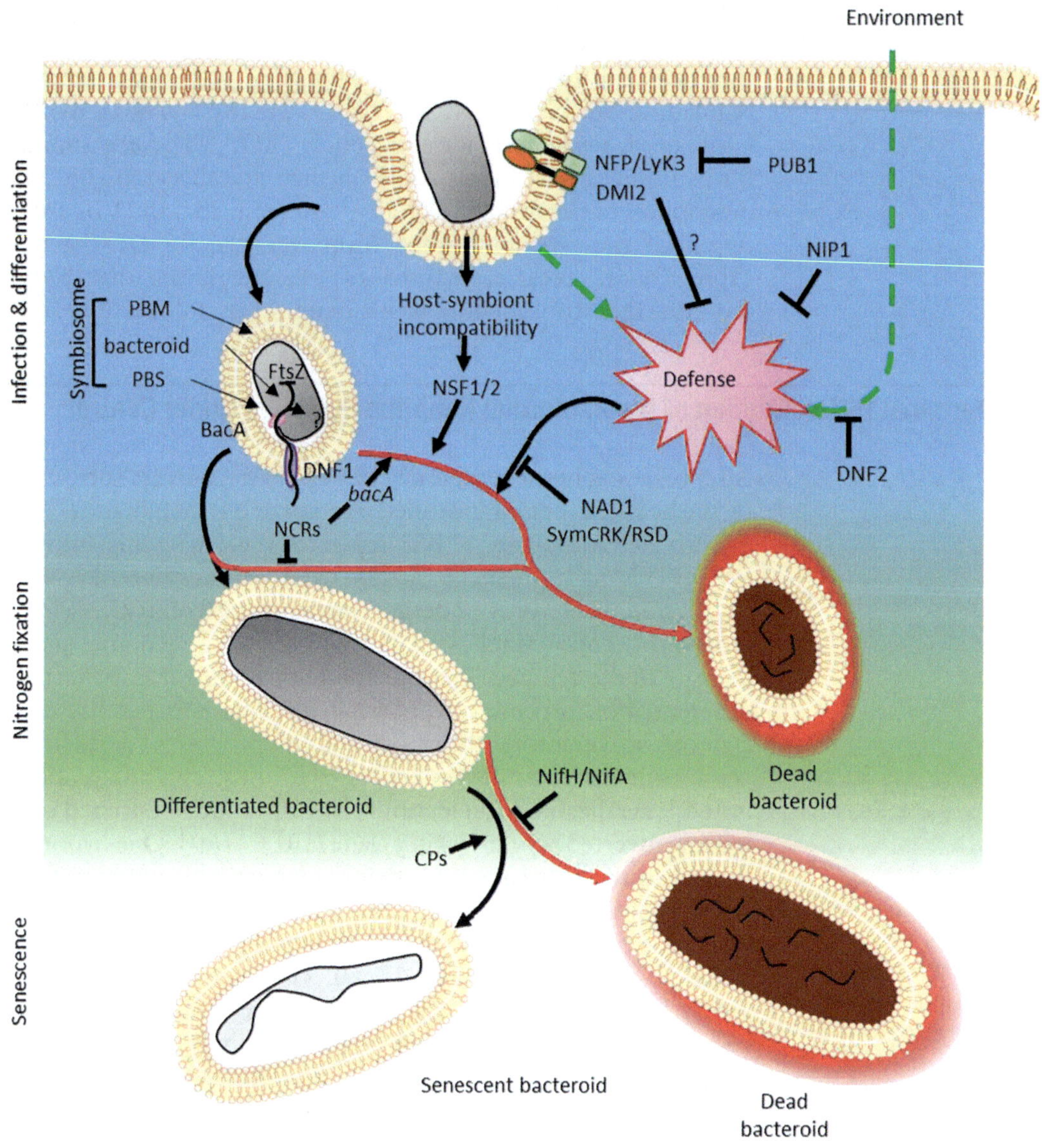

Fig. 2 Endosymbiotic process from infection to senescence. In the ZII, the bacteria infect the cell; NFP/LYK3 and DMI2 are required for the infection via symbiotic signaling and potentially through defense repression. PUB1 negatively regulates the infection by LYK3 degradation. Moreover NIP is also required for bacteria endocytosis and prevents defense reactions. In the host cells, the symbiosome is composed of the bacteroid (intracellular form of the rhizobia) surrounded by the peri-bacteroid membrane (PBM) and separated from it by the peri-bacteroid space (EPS). In the ZII–III, the bacteroid undergoes a terminal differentiation mediated by NCR peptides. DNF1, situated in the peri-bacteroid membrane, is required for NCR folding into the symbiosome. NCRs located in the peri-bacteroid space are internalized into bacteroid via BacA. The *bacA* mutant is unable to resist to NCR toxicity and in this mutant intracellular death of bacteroid is observed. In the bacteria, NCRs can act in different pathways to mediate differentiation. NCR247 can, for example, block the division by FtsZ inhibition; moreover NCRs can protect the bacteria from intracellular lysis. The NCR NFS1 and 2 lead to bacterial lysis of the incompatible rhizobia under host-dependent genotype. During the infection, the repression of the defense reactions is essential and requires symbiotic genes: *SymCRK*/*RSD* and *DNF2* prevent defense

and *NCR169*, respectively [79, 80], and they show increased intracellular death of the undifferentiated (*dnf7*) and differentiated (*dnf4*) bacteroids. Thus in this example, NCR211 and NCR169 protect the bacteria from intracellular death [79, 80]. These observations are in contradiction with the speculated function of NCRs but point to their essential role in symbiosis (Fig. 2). *Rhizobium meliloti* 41 is incompatible with the *M. truncatula* Jemalong A17 background. A comparative analysis between the incompatible A17 and compatible DZA315 *M. truncatula* backgrounds reveals the role of the NCR nitrogen fixation specificity 2 (NFS2) in the bacterial strain specificity and that this NCR has a bacterial lysis property against *R. meliloti 41* after bacteroid differentiation [81]. Furthermore, NCR *NFS1* gene also controls the A17 specificity to *R. meliloti 41* [82] (Fig. 2). These results indicate that NCRs can also play a role in the specificity of the symbiotic host selection.

8 Repression of Immunity Is Necessary for Successful Nodulation

During the symbiotic recognition phase, the repression of the defense reactions is required to reduce the root immune status and to initiate effective rhizobial infection [3]. For example, massive repression of defense genes is observed during early steps of *Medicago-Sinorhizobium* interaction [70]. The role of ethylene in this process is well described [40, 97], but the role of other defense hormones in symbiosis is less known. Various mechanisms participate in this repression of the plant defenses, and NFP-LYK3 can contribute to the regulation of the immunity during the symbiotic interaction [3]. Moreover, recent studies show the co-localization of NFP and LYK3 in the ZII of the mature nodule [42], suggesting that they may participate in the regulation of the defense reactions in this zone (Fig. 2). A recent study shows that two LysM RLKs, LYK9 and LYR4, participate in the recognition of the chitin oligomeric Microbe-Associated Molecular Pattern (MAMP) and activate defense in response to pathogens. The corresponding mutants do not show alteration in the symbiotic ability suggesting that *M. truncatula*, via certain LysM receptors, is able to distinguish the "good" from the "bad" microbes [83]. Most of the studies concerning the symbiotic regulation of the immunity focused

Fig. 2 (continued) reactions elicited by bacterial infection or differentiation and by the environment, respectively. Furthermore *NAD1* also represses defense reactions. In ZIII, bacteroid fixes N_2 for the plant benefit. The absence of N_2 fixation observed in fix$^-$ mutant such as *nifA* or *nifH* leads to the killing of the bacteria by the host. The absence of N_2 fixation probably induces nitrogen/carbon (N/C) imbalance and activates early senescence or other unknown processes to kill the bacteroids. Finally, in the zone IV, a destruction of the bacteroids is observed following a developmental program

on the early steps of the interaction, and few things are known about the repression of defense reactions in the mature nodules.

Recently, several genes were described that regulate negatively the nodule immunity. We will refer to the corresponding genes as *Nodule Deficient in Immune Repression* (*NDIR*). The corresponding class of *ndir* mutants shows typical feature of defense induction in their nodules like accumulation of phenolic compounds and induction of defense genes expression [43, 84–86]. The *DNF2*, *SymCRK*, and *RSD* genes encode a putative phospholipase C phosphatidylinositol-dependent protein, an RLK containing a Domain Unknown Function 26 (DUF26) and a C_2H_2 transcription factor, respectively. The corresponding mutants belong to the *ndir* class and form fix$^-$ nodules with induction of defense leading to intracellular death of the bacteroid [84, 85, 87, 88] (Fig. 2). When the *dnf2* mutant is grown in media without potential priming agent, the formation of functional nodules is observed [89] indicating that *DNF2* represses defense reactions in an environment-dependent manner. By contrast, the environmental conditions do not affect the *symCRK* phenotype [85], but *symCRK* and *rsd* inoculation with the bacterial *bacA* mutant produces nodules devoid of defense reactions. This suggests that the correct infection or differentiation of the bacteroids is required to induce defenses in *symCRK* and *rsd* [88]. By analyzing the *dnf2-symCRK* double mutant phenotype and comparing *DNF2*, *SymCRK*, and *RSD* expressions in different bacterial and plant mutant backgrounds, we have speculated for successive steps required for defense suppression after bacterial internalization [88]. *Nodule activation defense 1* (*NAD1*) is also involved in prevention of the defense reactions in the nodules. *NAD1* encodes a small transmembrane protein with unknown function [86], and the corresponding mutant displays typical features of the *ndir* class. Furthermore, *DNF2*, *SymCRK*, and *RSD* are downregulated in this mutant [86] suggesting that *NAD1* may act before these genes.

9 Host Regulation of the Symbiosis: Control the Nodulation to Optimize the Benefit

Symbiotic nitrogen fixation by the rhizobia requires large amount of energy provided by the plant [4] suggesting that a tight regulation of this process maximizes the profits for the host. Leguminous plants have developed different mechanisms to prevent inefficient interactions or to establish interactions with rational utilization of energy. Bacterial mutants impaired in *nifA* or *nifH* genes corresponding, respectively, to a transcriptional activator [90, 91] and a subunit of the nitrogenase [92] are not able to produce functional nitrogenase and do not fix nitrogen [70, 93]. These mutants show intracellular death of the differentiated bacteroid [88] suggesting that the host is able to control bacteroid survival when they are not efficient. One hypothesis is that the absence of nitrogen fixation generates a nitrogen/carbon (N/C) imbalance, that the plant senses this N/C balance perturbation and kills the inefficient bacteroids. In certain cases, the plant can suppress the formation and functioning of the nodules, for example, when nitrate is supplied to nodulated plants. In this case, the plant host stops the nodule functioning and activates a senescence process leading to tissue recycling (see below). Actually, the control of the C/N balance is not understood and needs more investigation.

The host adjusts the number of the nodules depending on its ability to provide carbon to the bacteroid and on its nitrogen requirement [94]. This regulation occurs during the infection process and is called the autoregulation of the nodulation (AON). It acts at a systemic level and requires shoot-root communication [45, 95]. *SUperNumerary nodules* (*SUNN*) gene encodes a LRR-RLK CLV1-like protein [96] and is expressed in root and shoot tissues. The *sunn* mutant shows high density of small poorly efficient nodules [96, 97], and this phenotype is shoot controlled [97]. The current hypothesis proposes the presence of a long-distance messenger, produced in the roots during the symbiotic interaction that is exported to the shoot, where it is recognized by *MtSUNN*. A second signal is then produced in the shoots and comes back to the roots to regulate negatively the infection process through a local action [98]. In comparison with *A. thaliana* where CLV1 forms heterodimer with CLV2 and recognizes the CLV3 peptides [64], it is proposed that CLV3/Embryo-surrounding region peptides (CLE12 and CLE13) produced in dividing nodule cells play the role of the root signal [62]. Moreover, SUNN can interact with the homologues of the CLV1-interacting proteins MtCLV2 and MtCORYN (MtCRN). It is thus speculated that MtCLV2 and MtCRN can form a protein complex with SUNN to regulate the AON [99]. The root determined nodulation 1 (RDN1) belongs to a family of uncharacterized peptide specific to plants and algae, expressed in roots, and that negatively controls the nodule number. RDN1 is proposed to interact with SUNN in the AON process and to regulate the nodule number in roots [100]. Furthermore, the C-terminally Encoded Peptides (CEPs) were recently shown to play a role in AON regulation and to interact with the Compact Root Architecture 2 (CRA2) LRR-RLK. They regulate negatively the lateral root emergence and positively the nodule formation at a local level [101–103].

10 Senescence Process at the End of the Road

The senescence of the nodule occurs when the plant suffers from environmental stresses, when nitrogen is added to the medium, or when the nodule is old [9]. The senescence results in the apparition of a senescent zone (or ZIV) where the bacteroids start dying before the host cell is degraded. This process is progressive and characterized by cellular constituent recycling [9]. Transcriptional analyses revealed differences in gene expression between developmental senescence and dark-induced senescence. These results show the presence of common and different patterns of gene expression suggesting the presence of common and specialized host responses between developmental and induced senescence [104]. In both types of senescence, induction of certain Cysteine Proteases (CPs1–6) is observed, and they were proposed as nodule senescence markers [9, 104]. Interestingly, downregulation of *MtCP6* and of the *Vacuolar Processing Enzyme* (*MtVPE*) delay the senescence program. The corresponding proteins accumulate in the vacuole of uninfected cells and in the symbiosome, suggesting that CPs participate to the cell recycling during senescence [105]. Nitric oxide (NO) is produced in the nodules by the Nitrite Reductase (NR) activity, also responsible for nitrite to NO conversion [106]. Nitric oxide is proposed to stimulate the senescence in the nodule because symbiosis with a *S. meliloti* strain overexpressing a flavohemoglobin encoded by the *hmp* gene (*hmp*$^+$) responsible for NO detoxification shows a delay of nodule senescence [107]. The mechanism of NO action is actually unknown but we can speculate that it acts via cell signaling activation [108]. It should be noted that the *ndir* mutants show also typical early senescent phenotypes. As other fix$^-$ mutants show early senescence without defense induction as *dnf7*, *bacA*, or *nifH* and *nifA* [80, 88, 89], this suggests that the immune deficiency in these *ndir* mutants is the origin of the senescence and not the reverse.

11 Concluding Remarks

The *Medicago-Sinorhizobium* symbiosis is a complex interaction with multiple aspects that involves various processes and numerous genes. Since the discovery of the NFs, our understanding of the symbiosis has been largely improved using transcriptomic, genetic, and reverse genetic approaches. However, despite these recent advances, certain aspects of the interaction remain unclear. These include the role of immunity and signaling defense pathways. The role of the specificity of recognition toward specific rhizobial strains is also an emerging field that will allow understanding how plants specifically select their symbiont. This specific selection may allow developing cultivars that would select the most efficient N_2-fixing partners. Thus further investigations are still required to better describe this symbiosis and better use it in sustainable agriculture.

References

1. Oldroyd GED, Murray JD, Poole PS, Downie JA (2011) The rules of engagement in the Legume-Rhizobial symbiosis. Annu Rev Genet 45:119–144
2. Oldroyd GED, Downie JA (2008) Coordinating nodule morphogenesis with Rhizobial infection in Legumes. Annu Rev Plant Biol 59:519–546
3. Gourion B, Berrabah F, Ratet P, Stacey G (2015) Rhizobium – legume symbioses: the crucial role of plant immunity. Trends Plant Sci 20:186–194
4. Udvardi M, Poole PS (2013) Transport and metabolism in Legume-Rhizobia symbioses. Annu Rev Plant Biol 64:781–805
5. Ott T, van Dongen JT, Günther C, Krusell L, Desbrosses G, Vigeolas H et al (2005) Symbiotic leghemoglobins are crucial for nitrogen fixation in legume root nodules but not for general plant growth and development. Curr Biol 15:531–535
6. Vasse J, De Billy F, Camut S, Truchet G (1990) Correlation between ultrastructural differentiation of bacteroids and nitrogen fixation in alfalfa nodules. J Bacteriol 172:4295–4306
7. Popp C, Ott T (2011) Regulation of signal transduction and bacterial infection during root nodule symbiosis. Curr Opin Plant Biol 14:458–467
8. Timmers AC, Soupène E, Auriac MC, de Billy F, Vasse J, Boistard P, Truchet G (2000) Saprophytic intracellular rhizobia in alfalfa nodules. Mol Plant-Microbe Interact 13:1204–1213
9. Van de Velde W, Guerra JC, De Keyser A, De Rycke R, Rombauts S, Maunoury N et al (2006) Aging in legume symbiosis. A molecular view on nodule senescence in *Medicago truncatula*. Plant Physiol 141:711–720
10. Liu C-W, Murray J (2016) The role of flavonoids in nodulation host-range specificity: an update. Plants (Basel) 5:33
11. Jones KM, Kobayashi H, Davies BW, Taga ME, Walker GC (2007) How rhizobial symbionts invade plants: the *Sinorhizobium-Medicago* model. Nat Rev Microbiol 5:619–633
12. Fliegmann J, Bono JJ (2015) Lipo-chitooligosaccharidic nodulation factors and their perception by plant receptors. Glycoconj J 32:455–464
13. Dénarié J, Debellé F, Promé J-CC (1996) Rhizobium lipo-chitooligosaccharide nodulation factors: signaling molecules mediating recognition and morphogenesis. Annu Rev Biochem 65:503–535
14. Arrighi JF, Cartieaux F, Brown SC, Rodier-Goud M, Boursot M, Fardoux J et al (2012) *Aeschynomene evenia*, a model plant for studying the molecular genetics of the nod-independent rhizobium-legume symbiosis. Mol Plant-Microbe Interact 25:851–861
15. Oldroyd GE (2013) Speak , friend , and enter : signalling systems that promote beneficial symbiotic associations in plants. Nat Rev Microbiol 11:252–263
16. Madsen EB, Madsen LH, Radutoiu S, Olbryt M, Rakwalska M, Szczyglowski K et al (2003) A receptor kinase gene of the LysM type is involved in legume perception of rhizobial signals. Nature 425:637–640
17. Limpens E, Franken C, Smit P, Willemse J, Bisseling T, Geurts R (2003) LysM domain receptor kinases regulating rhizobial nod factor-induced infection. Science 302:630–633
18. Fliegmann J, Canova S, Lachaud C, Uhlenbroich S, Gasciolli V, Pichereaux C (2013) Lipo-chitooligosaccharidic symbiotic signals are recognized by LysM receptor-like kinase LYR3 in the legume *Medicago truncatula*. ACS Chem Biol 8:1900–1906
19. Endre G, Kereszt A, Kevei Z, Mihacea S, Kalo P, Kiss GB (2002) A receptor kinase gene regulating symbiotic nodule development. Nature 417:962–966
20. Oldroyd GED, Downie JA (2006) Nuclear calcium changes at the core of symbiosis signaling. Curr Opin Plant Biol 9(4):351–357
21. Ané JM, Kiss GB, Riely BK, Penmetsa RV, Oldroyd GE, Ayax C et al (2004) *Medicago truncatula* DMI1 required for bacterial and fungal symbioses in legumes. Science 303:1364–1367
22. Mitra RM, Gleason CA, Edwards A, Hadfield J, Downie JA, Oldroyd GE, Long SR (2004) A Ca^{2+}/calmodulin-dependent protein kinase required for symbiotic nodule development: gene identification by transcript-based cloning. Proc Natl Acad Sci U S A 101:4701–4705
23. Lévy J, Bres C, Geurts R, Chalhoub B, Kulikova O, Duc G et al (2004) A putative Ca^{2+} and calmodulin-dependent protein kinase required for bacterial and fungal symbioses. Science 303:1361–1364
24. Messinese E, Mun JH, Yeun LH, Jayaraman D, Rougé P, Barre A et al (2007) A novel nuclear protein interacts with the symbiotic DMI3 calcium- and calmodulin-dependent protein kinase of *Medicago truncatula*. Mol Plant-Microbe Interact 20:912–921
25. Horváth B, Yeun LH, Domonkos A, Halász G, Gobbato E, Ayaydin F et al (2011) *Medicago truncatula* IPD3 is a member of the common symbiotic signaling pathway required for rhi-

zobial and mycorrhizal symbioses. Mol Plant-Microbe Interact 24:1345–1358
26. Kaló P, Gleason C, Edwards A, Marsh J, Mitra RM, Hirsch S et al (2005) Nodulation signaling in legumes requires NSP2, a member of the GRAS family of transcriptional regulators. Science 308:1786–1789
27. Smit P, Raedts J, Portyanko V, Debellé F, Gough C, Bisseling T, Geurts R (2005) NSP1 of the GRAS protein family is essential for rhizobial nod factor-induced transcription. Science 308:1789–1791
28. Hirsch S, Kim J, Muñoz A, Heckmann AB, Downie JA, Oldroyd GED (2009) GRAS proteins form a DNA binding complex to induce gene expression during nodulation signaling in *Medicago truncatula*. Plant Cell 21:545–557
29. Marsh JF, Rakocevic A, Mitra RM, Brocard L, Sun J, Eschstruth A et al (2007) *Medicago truncatula* NIN is essential for rhizobial-independent nodule organogenesis induced by autoactive calcium/calmodulin-dependent protein kinase. Plant Physiol 144:324–335
30. Cerri MR, Frances L, Laloum T, Auriac MC, Niebel A, Oldroyd GE et al (2012) *Medicago truncatula* ERN transcription factors: regulatory interplay with NSP1/NSP2 GRAS factors and expression dynamics throughout rhizobial infection. Plant Physiol 160:2155–2172
31. Mbengue M, Camut S, de Carvalho-Niebel F, Deslandes L, Froidure S, Klaus-Heisen D et al (2010) The *Medicago truncatula* E3 ubiquitin ligase PUB1 interacts with the LYK3 symbiotic receptor and negatively regulates infection and nodulation. Plant Cell 22:3474–3488
32. Miyahara A, Richens J, Starker C, Morieri G, Smith L, Long S et al (2010) Conservation in function of a SCAR/WAVE component during infection thread and root hair growth in *Medicago truncatula*. Mol Plant-Microbe Interact 23:1553–1562
33. Kiss E, Oláh B, Kaló P, Morales M, Heckmann AB, Borbola A et al (2009) LIN, a novel type of U-box/WD40 protein, controls early infection by rhizobia in legumes. Plant Physiol 151:1239–1249
34. Laporte P, Lepage A, Fournier J, Catrice O, Moreau S, Jardinaud MF et al (2014) The CCAAT box-binding transcription factor NF-YA1 controls rhizobial infection. J Exp Bot 65:481–494
35. Soyano T, Kouchi H, Hirota A, Hayashi M (2013) NODULE INCEPTION directly targets NF-Y subunit genes to regulate essential processes of root nodule development in *Lotus japonicus*. PLoS Genet 9:e1003352
36. Combier JP, De Billy F, Gamas P, Niebel A, Rivas S (2008) Trans-regulation of the expression of the transcription factor MtHAP2-1 by a uORF controls root nodule development. Genes Dev 22:1549–1559
37. Laloum T, Baudin M, Frances L, Lepage A, Billault-Penneteau B, Cerri MR et al (2014) Two CCAAT-box-binding transcription factors redundantly regulate early steps of the legume-rhizobia endosymbiosis. Plant J 79:757–768
38. Middleton PH, Jakab J, Penmetsa RV, Starker CG, Doll J, Kaló P et al (2007) An ERF transcription factor in *Medicago truncatula* that is essential for nod factor signal transduction. Plant Cell 19:1221–1234
39. Haney CH, Long SR (2010) Plant flotillins are required for infection by nitrogen-fixing bacteria. Proc Natl Acad Sci U S A 107:478–483
40. Penmetsa RV, Uribe P, Anderson J, Lichtenzveig J, Gish JC, Nam YW et al (2008) The *Medicago truncatula* ortholog of Arabidopsis EIN2, sickle, is a negative regulator of symbiotic and pathogenic microbial associations. Plant J 55:580–595
41. Gavrin A, Kulikova O, Bisseling T, Fedorova EE (2017) Interface symbiotic membrane formation in root nodules of *Medicago truncatula*: the role of synaptotagmins MtSyt1, MtSyt2 and MtSyt3. Front Plant Sci 8:1–10
42. Moling S, Pietraszewska-Bogiel A, Postma M, Fedorova E, Hink MA, Limpens E et al (2014) Nod factor receptors form heteromeric complexes and are essential for intracellular infection in medicago nodules. Plant Cell 26:4188–4199
43. Veereshlingam H, Haynes JG, Penmetsa RV, Cook DR, Sherrier DJ, Dickstein R (2004) Nip, a symbiotic *Medicago truncatula* mutant that forms root nodules with aberrant infection threads and plant defense-like response. Plant Physiol 136:3692–3702
44. Bagchi R, Salehin M, Adeyemo OS, Salazar C, Shulaev V, Sherrier DJ, Dickstein R (2012) Functional assessment of the *Medicago truncatula* NIP/LATD protein demonstrates that it is a high-affinity nitrate transporter. Plant Physiol 160:906–916
45. Suzaki T, Yoro E, Kawaguchi M (2015) Leguminous plants: inventors of root nodules to accommodate symbiotic bacteria. Int Rev Cell Mol Biol 316:111–158
46. Stacey G, Libault M, Brechenmacher L, Wan J, May GD (2006) Genetics and functional genomics of legume nodulation. Curr Opin Plant Biol 9:110–121
47. Tirichine L, Imaizumi-Anraku H, Yoshida S, Murakami Y, Madsen LH, Miwa H et al (2006) Deregulation of a Ca^{2+}/calmodulin-dependent kinase leads to spontaneous nodule development. Nature 441:1153–1156

48. Crespi M, Frugier F (2009) *De novo* organ formation from differentiated cells: root nodule organogenesis. Sci Signal 2:er1
49. van Zeijl A, Op den Camp RH, Deinum EE, Charnikhova T, Franssen H, Op den Camp HJ et al (2015) Rhizobium lipochitooligosaccharide signaling triggers accumulation of cytokinins in *Medicago truncatula* roots. Mol Plant 8:1213–1226
50. Gonzalez-rizzo S, Crespi M, Frugier F (2006) The *Medicago truncatula* CRE1 cytokinin receptor regulates lateral root development and early symbiotic interaction with Sinorhizobium meliloti. Plant Cell 18:2680–2693
51. Gamas P, Brault M, Jardinaud M, Frugier F (2017) Cytokinins in symbiotic nodulation: when, where, what for? Trends Plant Sci 22:792–802
52. Ariel F, Brault-Hernandez M, Laffont C, Huault E, Brault M, Plet J et al (2012) Two direct targets of cytokinin signaling regulate symbiotic nodulation in *Medicago truncatula*. Plant Cell 24:3838–3852
53. Ng JL, Hassan S, Truong TT, Hocart CH, Laffont C, Frugier F, Mathesius U (2015) Flavonoids and auxin transport inhibitors rescue symbiotic nodulation in the *Medicago truncatula* cytokinin perception mutant cre1. Plant Cell 27:2210–2226
54. Vernie T, Billy D, Plet J, Combier J, Rogers C (2008) EFD is an ERF transcription factor involved in the control of nodule number and differentiation in *Medicago truncatula*. Plant Cell 20:2696–2713
55. Di Giacomo E, Laffont C, Sciarra F, Iannelli MA, Frugier F, Frugis G (2017) KNAT3/4/5-like class 2 KNOX transcription factors are involved in *Medicago truncatula* symbiotic nodule organ development. New Phytol 213:822–837
56. Azarakhsh M, Kirienko AN, Zhukov VA, Lebedeva MA, Dolgikh EA, Lutova LA (2015) KNOTTED1-LIKE HOMEOBOX 3: a new regulator of symbiotic nodule development. J Exp Bot 66:7181–7195
57. Guan D, Stacey N, Liu C, Wen J, Mysore KS, Torres-Jerez I et al (2013) Rhizobial infection is associated with the development of peripheral vasculature in nodules of *Medicago truncatula*. Plant Physiol 162:107–115
58. Godiard L, Lepage A, Moreau S, Laporte D, Verdenaud M, Timmers T, Gamas P (2011) MtbHLH1, a bHLH transcription factor involved in *Medicago truncatula* nodule vascular patterning and nodule to plant metabolic exchanges. New Phytol 191:391–404
59. Starker CG, Parra-Colmenares AL, Smith L, Mitra RM, Long SR (2006) Nitrogen fixation mutants of *Medicago truncatula* fail to support plant and bacterial symbiotic gene expression. Plant Physiol 140:671–680
60. Pislariu CI, Murray JD, Wen J, Cosson V, Muni RR, Wang M et al (2012) A *Medicago truncatula* tobacco retrotransposon insertion mutant collection with defects in nodule development and symbiotic nitrogen fixation. Plant Physiol 159:1686–1699
61. Combier JP, Frugier F, de Billy F, Boualem A, El-Yahyaoui F, Moreau S et al (2006) MtHAP2-1 is a key transcriptional regulator of symbiotic nodule development regulated by microRNA169 in *Medicago truncatula*. Genes Dev 20:3084–3088
62. Mortier V, Den Herder G, Whitford R, Van de Velde W, Rombauts S, D'Haeseleer K et al (2010) CLE peptides control *Medicago truncatula* nodulation locally and systemically. Plant Physiol 153:222–237
63. Mayer KFX, Schoof H, Haecker A, Lenhard M, Jürgens G, Laux T (1998) Role of WUSCHEL in regulating stem cell fate in the Arabidopsis shoot meristem. Cell 95:805–815
64. Williams L, Fletcher JC (2005) Stem cell regulation in the Arabidopsis shoot apical meristem. Curr Opin Plant Biol 8:582–586
65. Couzigou JM, Zhukov V, Mondy S, Abu el Heba G, Cosson V, Ellis TH et al (2012) NODULE ROOT and COCHLEATA maintain nodule development and are legume orthologs of Arabidopsis BLADE-ON-PETIOLE genes. Plant Cell 24:4498–4510
66. Kereszt A, Mergaert P, Maróti G, Kondorosi É (2011) Innate immunity effectors and virulence factors in symbiosis. Curr Opin Microbiol 14:76–81
67. Mergaert P, Uchiumi T, Alunni B, Evanno G, Cheron A, Catrice O et al (2006) Eukaryotic control on bacterial cell cycle and differentiation in the Rhizobium-legume symbiosis. Proc Natl Acad Sci U S A 103:5230–5235
68. Alunni B, Gourion B (2016) Terminal bacteroid differentiation in the legume-rhizobium symbiosis: nodule-specific cysteine-rich peptides and beyond. New Phytol 211:411–417
69. Czernic P, Gully D, Cartieaux F, Moulin L, Guefrachi I, Patrel D et al (2015) Convergent evolution of endosymbiont differentiation in dalbergioid and inverted repeat-lacking clade legumes mediated by nodule-specific cysteine-rich peptides. Plant Physiol 169:1254–1265
70. Maunoury N, Redondo-Nieto M, Bourcy M, Van de Velde W, Alunni B, Laporte P et al (2010) Differentiation of symbiotic cells and endosymbionts in *Medicago truncatula* nodulation are coupled to two transcriptome-switches. PLoS One 5:e9519
71. Haag AF, Baloban M, Sani M, Kerscher B, Pierre O, Farkas A et al (2011) Protection of *Sinorhizobium* against host cysteine-rich

antimicrobial peptides is critical for symbiosis. PLoS Biol 9:e1001169
72. Mergaert P, Nikovics K, Kelemen Z, Maunoury N, Kondorosi A, Kondorosi E (2003) A novel family in *Medicago truncatula* consisting of more than 300 nodule-specific genes coding for small , secreted polypeptides with conserved cysteine motifs. Plant Physiol 132:161–173
73. Alunni B, Kevei Z, Redondo-Nieto M, Kondorosi A, Mergaert P, Kondorosi E (2007) Genomic organization and evolutionary insights on GRP and NCR genes, two large nodule-specific gene families in *Medicago truncatula*. Mol Plant-Microbe Interact 20:1138–1148
74. Guefrachi I, Nagymihaly M, Pislariu CI, Van de Velde W, Ratet P, Mars M et al (2014) Extreme specificity of NCR gene expression in *Medicago truncatula*. BMC Genomics 15:712
75. Nallu S, Silverstein KAT, Samac DA, Bucciarelli B, Vance CP, Vandenbosch KA (2013) Regulatory patterns of a large family of defensin-like genes expressed in nodules of *Medicago truncatula*. PLoS One 8:e60355
76. Van de Velde W, Zehirov G, Szatmari A, Debreczeny M, Ishihara H, Kevei Z et al (2010) Plant peptides govern terminal differentiation of bacteria in symbiosis. Science 327:1122–1126
77. Wang D, Griffitts J, Starker C, Fedorova E, Limpens E, Ivanov S et al (2010) A nodule-specific protein secretory pathway required for nitrogen-fixing symbiosis. Science 327:1126–1129
78. Farkas A, Maróti G, Durgő H, Györgypál Z, Lima RM, Medzihradszky KF et al (2014) *Medicago truncatula* symbiotic peptide NCR247 contributes to bacteroid differentiation through multiple mechanisms. Proc Natl Acad Sci U S A 111:5183–5188
79. Kim M, Chen Y, Xi J, Waters C, Chen R, Wang D (2015) An antimicrobial peptide essential for bacterial survival in the nitrogen-fixing symbiosis. Proc Natl Acad Sci U S A 112:15238–15243
80. Horváth B, Domonkos Á, Kereszt A, Szűcs A, Ábrahám E, Ayaydin F et al (2015) Loss of the nodule-specific cysteine rich peptide, NCR169, abolishes symbiotic nitrogen fixation in the *Medicago truncatula dnf7* mutant. Proc Natl Acad Sci U S A 112:15232–15237
81. Wang Q, Yang S, Liu J, Terecskei K, Ábrahám E, Gombár A et al (2017) Host-secreted antimicrobial peptide enforces symbiotic selectivity in *Medicago truncatula*. Proc Natl Acad Sci U S A 114:6854–6859
82. Yang S, Wang Q, Fedorova E, Liu J, Qin Q, Zheng Q et al (2017) Microsymbiont discrimination mediated by a host-secreted peptide in *Medicago truncatula*. Proc Natl Acad Sci U S A 114:6848–6853
83. Bozsoki Z, Cheng J, Feng F, Gysel K, Vinther M, Andersen KR et al (2017) Receptor-mediated chitin perception in legume roots is functionally separable from Nod factor perception. Proc Natl Acad Sci U S A 114:E8118–E8127
84. Bourcy M, Brocard L, Pislariu CI, Cosson V, Mergaert P, Tadege M et al (2013) *Medicago truncatula* DNF2 is a PI-PLC-XD-containing protein required for bacteroid persistence and prevention of nodule early senescence and defense-like reactions. New Phytol 197:1250–1261
85. Berrabah F, Bourcy M, Eschstruth A, Cayrel A, Guefrachi I, Mergaert P et al (2014) A nonRD receptor-like kinase prevents nodule early senescence and defense-like reactions during symbiosis. New Phytol 203:1305–1314
86. Wang C, Yu H, Luo L, Duan L, Cai L, He X et al (2016) NODULES WITH ACTIVATED DEFENSE 1 is required for maintenance of rhizobial endosymbiosis in *Medicago truncatula*. New Phytol 212:176–191
87. Sinharoy S, Torres-Jerez I, Bandyopadhyay K, Kereszt A, Pislariu CI, Nakashima J et al (2013) The C2H2 transcription factor regulator of symbiosome differentiation represses transcription of the secretory pathway gene VAMP721a and promotes symbiosome development in *Medicago truncatula*. Plant Cell 25:3584–3601
88. Berrabah F, Ratet P, Gourion B (2015) Multiple steps control immunity during the intracellular accommodation of rhizobia. J Exp Bot 66:1977–1985
89. Berrabah F, Bourcy M, Cayrel A, Eschstruth A, Mondy S, Ratet P, Gourion B (2014) Growth conditions determine the DNF2 requirement for symbiosis. PLoS One 9:e91866
90. Fischer HM (1994) Genetic regulation of nitrogen fixation in rhizobia. Microbiol Rev 58:352–386
91. Dixon R, Kahn D (2004) Genetic regulation of biological nitrogen fixation. Nat Rev Microbiol 2:621–631
92. Rubio LM, Ludden PW (2008) Biosynthesis of the cofactor of nitrogenase. Annu Rev Microbiol 62:93–111
93. Hirsch M, Bang M, Ausubel FM (1983) Ultrastructural analysis of ineffective alfalfa nodules formed by Nif-Tn5 mutants of *Rhizobium meliloti*. J Bacteriol 155:367–380
94. Kosslak RM, Bohlool BB (1984) Suppression of nodule development of one side of a split-root system of soybeans caused by prior inoculation of the other side. Plant Physiol 75:125–130

95. Mortier V, Holsters M, Goormachtig S (2012) Never too many? How legumes control nodule numbers. Plant Cell Environ 35:245–258
96. Schnabel E, Journet EP, De Carvalho-Niebel F, Duc G, Frugoli J (2005) The *Medicago truncatula* SUNN gene encodes a CLV1-like leucine-rich repeat receptor kinase that regulates nodule number and root length. Plant Mol Biol 58:809–822
97. Penmetsa RV, Frugoli JA, Smith LS, Long SR, Cook DR (2003) Dual genetic pathways controlling nodule number in *Medicago truncatula*. Plant Physiol 131:998–1008
98. Oka-Kira E, Kawaguchi M (2006) Long-distance signaling to control root nodule number. Curr Opin Plant Biol 9:496–502
99. Crook AD, Schnabel EL, Frugoli JA (2016) The systemic nodule number regulation kinase SUNN in *Medicago truncatula* interacts with MtCLV2 and MtCRN. Plant J 88:108–119
100. Schnabel EL, Kassaw TK, Smith LS, Marsh JF, Oldroyd GE, Long SR, Frugoli JA (2011) The ROOT DETERMINED NODULATION1 gene regulates nodule number in roots of *Medicago truncatula* and defines a highly conserved, uncharacterized plant gene family. Plant Physiol 157:328–340
101. Huault E, Laffont C, Wen J, Mysore KS, Ratet P, Frugier F (2014) Local and systemic regulation of plant root system architecture and symbiotic nodulation by a receptor- like kinase. PLoS Genet 10:e1004891
102. Imin N, Mohd-Radzman NA, Ogilvie HA, Djordjevic MA (2013) The peptide-encoding CEP1 gene modulates lateral root and nodule numbers in *Medicago truncatula*. J Exp Bot 64:5395–5409
103. Mohd-Radzman NA, Laffont C, Ivanovici A, Patel N, Reid D, Stougaard J et al (2016) Different pathways act downstream of the CEP peptide receptor CRA2 to regulate lateral root and nodule development. Plant Physiol 171:2536–2548
104. Pérez Guerra JC, Coussens G, De Keyser A, De Rycke R, De Bodt S, Van De Velde W et al (2010) Comparison of developmental and stress-induced nodule senescence in *Medicago truncatula*. Plant Physiol 152:1574–1584
105. Pierre O, Hopkins J, Combier M, Baldacci F, Engler G, Brouquisse R et al (2014) Involvement of papain and legumain proteinase in the senescence process of *Medicago truncatula* nodules. New Phytol 202:849–863
106. Horchani F, Prévot M, Boscari A, Evangelisti E, Meilhoc E, Bruand C et al (2011) Both plant and bacterial nitrate reductases contribute to nitric oxide production in *Medicago truncatula* nitrogen-fixing nodules. Plant Physiol 155:1023–1036
107. Cam Y, Pierre O, Boncompagni E, Hérouart D, Meilhoc E, Bruand C (2012) Nitric oxide (NO): a key player in the senescence of *Medicago truncatula* root nodules. New Phytol 196:548–560
108. Moreau M, Lindermayr C, Durner J, Klessig DF (2010) NO synthesis and signaling in plants – where do we stand? Physiol Plant 138:372–383
109. Cerri MR, Frances L, Kelner A, Fournier J, Middleton PH, Auriac MC et al (2016) The symbiosis-related ERN transcription factors act in concert to coordinate rhizobial host root infection. Plant Physiol 171:1037–1054
110. Rey T, Laporte P, Bonhomme M, Jardinaud MF, Huguet S, Balzergue S et al (2016) MtNF-YA1, a central transcriptional regulator of symbiotic nodule development, is also a determinant of *Medicago truncatula* susceptibility toward a root pathogen. Front Plant Sci 7:1837
111. Harris JM, Dickstein R (2010) Control of root architecture and nodulation by the LATD / NIP transporter. Plant Signal Behav 5:1365–1369
112. Lefebvre B, Timmers T, Mbengue M, Moreau S, Hervé C, Tóth K et al (2010) A remorin protein interacts with symbiotic receptors and regulates bacterial infection. Proc Natl Acad Sci U S A 107:2343–2348
113. Murray JD, Muni RR, Torres-Jerez I, Tang Y, Allen S, Andriankaja M et al (2011) Vapyrin, a gene essential for intracellular progression of arbuscular mycorrhizal symbiosis, is also essential for infection by rhizobia in the nodule symbiosis of *Medicago truncatula*. Plant J 65:244–252
114. Sinharoy S, Liu C, Breakspear A, Guan D, Shailes S, Nakashima J et al (2016) A *Medicago truncatula* cystathionine-β-synthase-like domain-containing protein is required for rhizobial infection and symbiotic nitrogen fixation. Plant Physiol 170:2204–2217
115. Arrighi J-F, Godfroy O, de Billy F, Saurat O, Jauneau A, Gough C (2008) The *RPG* gene of *Medicago truncatula* controls Rhizobium-directed polar growth during infection. Proc Natl Acad Sci U S A 105:9817–9822
116. Op den Camp RH, De Mita S, Lillo A, Cao Q, Limpens E, Bisseling T, Geurts R (2011) A phylogenetic strategy based on a legume-specific whole genome duplication yields symbiotic cytokinin type-a response regulators. Plant Physiol 157:2013–2022
117. Mortier V, Wasson A, Jaworek P, De Keyser A, Decroos M, Holsters M et al (2014) Role of LONELY GUY genes in indeterminate nodulation on *Medicago truncatula*. New Phytol 202:582–593

118. Lohar DP, Schaff JE, Laskey JG, Kieber JJ, Bilyeu KD, Bird DM (2004) Cytokinins play opposite roles in lateral root formation, and nematode and Rhizobial symbioses. Plant J 38:203–214

119. Hopkins J, Pierre O, Kazmierczak T, Gruber V, Frugier F, Clement M (2014) Mt ZR1, a PRAF protein, is involved in the development of roots and symbiotic root nodules in *Medicago truncatula*. Plant Cell Environ 37:658–669

Chapter 17

Functional Genomics and Flowering Time in *Medicago truncatula*: An Overview

James L. Weller and Richard C. Macknight

Abstract

Flowering time is an important trait that influences adaptation and yield in many crop legumes. Both the inherent earliness of flowering and the degree to which it is responsive to environmental factors determine both the eco-geographic range across which crops can be successfully grown and the seasonal cycles most suitable for production. This chapter will provide a brief review of studies investigating the genetic control of flowering time in *Medicago truncatula*.

Key words Flowering time, Photoperiod, Vernalization, Florigen, *Medicago*, Legume

1 Introduction

Taxonomically, the major legume species fall into two main clades, which show a characteristic difference in how they respond to environmental variables for initiation of flowering. In phaseoloid or warm season legumes such as soybean, common bean, cowpea, and pigeon pea, flowering is accelerated by short days, whereas in the galegoid or temperate legumes, flowering is promoted by exposure to long days and many species also flower earlier in response to extended cold (vernalization).

Historically, the understanding of flowering time control in these groups has been based mainly on studies in soybean and pea as representative short-day and long-day species [1]. These studies began with characterization of natural variation, and in the case of pea, this also subsequently extended to the use of mutagenesis for forward genetic analysis. However, both systems have limitations. The palaeotetraploid nature of the soybean genome has limited its use for gene discovery through forward genetics, and pea has a very large genome that has not yet been sequenced and has presented significant challenges for identification of causal genes by positional cloning.

Luis A. Cañas and José Pío Beltrán (eds.), *Functional genomics in* Medicago truncatula*: Methods and Protocols*, Methods in Molecular Biology, vol. 1822, https://doi.org/10.1007/978-1-4939-8633-0_17,

The development of *Medicago truncatula* and *Lotus japonicus* as model legumes and generation of genomic resources has opened up new options for functional genomics of developmental processes, including control of flowering. In *Medicago*, the release of early genome and transcriptome versions [2] facilitated early surveys of flowering-related gene content and comparative phylogenetic analyses of gene families [3]. Mutant platforms for reverse genetics [4–6] have also provided new options for systematic reverse genetic analysis of flowering time. As a result, *M. truncatula* has received increasing attention both as a system in its own right and as a valuable complement to other legume systems where flowering time has been of long-standing fundamental and/or applied interest. In addition, extensive collections of natural diversity and investment in whole-genome resequencing have opened the possibility for use in genome-wide association studies and population and ecological genetic approaches (e.g., [7–11]).

2 Flowering Physiology in Medicago

Standard *Medicago* cultivars respond to both long days and vernalization in a facultative manner. A three-week vernalization treatment at 4 °C was reported to be sufficient for maximal promotion of flowering in cv Jemalong [12]. In three different cultivars (Jemalong, Jester, and R108), either vernalization or growth in LD conditions resulted in significant promotion of flowering relative to unvernalized plants grown in short days, but maximal promotion required both vernalization and LD [12, 13]. These observations suggest that responses to photoperiod and vernalization have both independent and interacting components. However, across the species as a whole there is wide natural variation for flowering time, and it is likely that there is substantial diversity in the relative effectiveness of these factors and in the nature of their interactions, which has not been systematically explored to date.

3 Genomic Survey of Flowering-Related Genes

Hecht et al. [3] conducted the first systematic analysis of flowering-related genes in an early version of the *Medicago* genome (v3.0) complemented by EST databases. Most *Arabidopsis* genes with a known role in flowering time control were found to be represented, but significant differences were identified in gene family composition, reflecting the different evolutionary history in the two lineages.

Most significantly, the analysis showed key differences in gene families for three key integrators of flowering responses in *Arabidopsis*: FT, FLOWERING LOCUS C (FLC), and

CONSTANS (CO). First, an expanded *FT* family was identified and shown to comprise five genes in three subfamilies, termed *FTa*, *FTb*, and *FTc* [3]. A third *FTa*-like gene (*FTa3*) has subsequently been identified in v4.0 of the *Medicago* genome [14]. Three of these genes (*FTa1*, *FTa2*, and *FTc*) are located in a cluster on chromosome 7, an arrangement that appears widely conserved across the two major clades of legume crops [15]. The *TFL1* gene subfamily, which is related to the *FT* family in sequence but has contrasting inhibitory roles, has also undergone expansion in *Medicago*, consistent with earlier observations in pea [3, 16].

Second, the FLC clade of MADS-domain transcription factors, which have a central role in vernalization and temperature control of flowering in Brassicaceae, are apparently absent from *Medicago* and other temperate legumes [1, 3, 17, 18].

Third, the *Arabidopsis* gene *CONSTANS*, which encodes a key activator of *FT* in LD, does not have a one-to-one orthologous relationship with any *M. truncatula* gene. In *Arabidopsis*, *CO* forms a small subclade (Group Ia) within the wider *CO*-like family, but neither of the other genes in this subclade (*COL1* and *COL2*) participates in control of flowering time or photoperiod responsiveness [19]. This group of genes is represented in *M. truncatula* by only a single gene, *COLa*. Subsequent more detailed analyses of *COL* gene families in *M. truncatula*, pea, and soybean have indicated the probable evolutionary loss of a second Group Ia *CO*-like gene from the *Medicago* lineage [20, 21].

Other more minor differences were identified in *M. truncatula* relative to *Arabidopsis*, and as more gene families have been analyzed in detail, more such differences have emerged. These include duplication of the blue-light photoreceptor CRY2, the circadian clock component ELF3 [22], the bZIP transcription factor FD [23], and several MADS-domain proteins including AP1, SOC1, FUL, and SVP [3]. Other families are smaller, including presence of only a single representative for each of the *Arabidopsis* circadian clock protein pairs LHY/CCA1 and LUX/NOX, only two proteins in the LOV-kelch blue-light receptor family, and only three phytochromes [3, 24].

Increasingly these differences can now be understood in the broader context of legume genome evolution, and many of the differences first observed in *M. truncatula* are also seen in other cool-temperate legume genomes, including chickpea [25].

4 Analysis of Natural Variation

Significant natural variation for flowering time in *M. truncatula* was first systematically documented by Julier et al. [26], who examined approximately 30 accessions grown under long-day conditions at variable temperature in a greenhouse following a one-week vernalization treatment. In this study, flowering time was assessed as degree days above 0 °C and varied from approximately 700 to 1500 across the accessions examined. Pierre et al. [27] evaluated variation for flowering response to photoperiod in eight accessions used as parents in the development of recombinant inbred mapping populations. All accessions were shown to have a strong response to photoperiod despite significant differences in inherent lateness.

The study of Julier et al. [26] also reported wide variation in flowering for a recombinant inbred progeny (LR4) derived from intermediate-flowering accessions Jemalong-6 and DZA315.16 that was grown several times under diverse conditions. QTL analysis on this population identified loci on chromosomes 1, 3, 5, 7, and 8, of which the QTL on chromosome 7 was stable across all environments/years/locations and explained the most variation. The location of this QTL in the center of chromosome 7 was shown to coincide with a genomic region containing the *FTa/c* gene cluster and a putative *CO* gene that has subsequently been shown to be a relatively distant homolog, *COLh* [21, 26]. This study was extended by Pierre et al. [27] who evaluated variation for flowering response to photoperiod in a selection of accessions and conducted QTL analysis for flowering time in three interrelated recombinant inbred populations, including LR4. The genetic analysis further supported the existence of a strong and stable QTL in the central region of chromosome 7 that was common to all three populations and identified one additional QTL on chromosome 4. A further study by Lagunes Espinoza [28] using the same populations showed that a major QTL for a number of shoot architecture traits including branching and main shoot growth co-located with the chromosome 7 flowering time QTL and suggested a similar co-location between architecture and flowering time QTL in other genomic regions. Pierre et al. [29] took advantage of residual heterozygosity in the LR4 population to develop a large population for fine-mapping of the major flowering time QTL on Chr 7. This study narrowed the position of the locus to an interval of approximately 2.4 cM, which contained the previously identified *FT*-like and *CO*-like candidates.

More recently, extensive whole-genome resequencing efforts in *M. truncatula* have opened up the possibility of genome-wide association analyses, and several recent studies have examined associations with flowering time. Stanton-Geddes et al. [8] identified genomic regions on all eight chromosomes that showed a statistical association with flowering time. Among the significantly associated SNPs, more than 75% clustered within the same approximate interval on chromosome 7 identified in the QTL analyses described by Pierre et al. [27]. This conclusion was reinforced in a more recent analysis examining association of variation in candidate flowering-related genes from the same resequencing dataset but with a more extensive flowering time data collected across several years [30].

This same region also emerged from studies examining adaptation to salinity [10, 31]. Phenotypic analysis of 39 accessions from populations growing in saline and nonsaline environments in Tunisia identified significant differentiation for flowering time with accessions from saline soils flowering earlier across several different conditions. Resequencing and association analysis identified significant associations between early flowering and "saline" alleles in a region of linkage disequilibrium that included the same *CO* and *FT* homolog candidates.

5 Induced Variation and Molecular Physiology: The FT Family

An initial investigation into the function of *Medicago FT* genes revealed differences in their regulation and activity in transgenic *Arabidopsis* [13]. Five out of the six were shown to promote flowering to some extent when expressed from the 35S promoter in the *Arabidopsis ft* mutant, with the most effective being *FTa1* and *FTc*. Expression analyses indicated that *FTb* genes were expressed mainly in leaf tissue, *FTc* mainly in apical buds and flowers, and *FTa* genes more broadly in expanded leaves, apical buds, and flowers [13]. These genes also showed distinct patterns of response to environmental factors. The *FTb* genes are induced by long days but not vernalization, whereas *FTa* genes are strongly induced by vernalization. *FTa1* also shows a qualitative induction by long-days. The induction of *FTa1* by vernalization was only seen following return to warm conditions, an evidence of cellular memory that is consistent with epigenetic regulation. Together with parallel information from the pea system, this information has suggested that three *FT* genes in particular, *FTa1*, *FTb1*, and *FTc*, are important for flowering time control and may have distinct physiological roles [1].

6 Response to Vernalization

Tnt1, a long terminal repeat (LTR) retrotransposon from tobacco, has been used to generate a large collection of insertion mutants [4]. These lines have been screened to identify phenotypes of interest and have also been used in reverse genetic screens, via PCR amplification of DNA from pools of different lines, to identify *Tnt1* insertional mutants in genes of interest (Chapter 07). Laurie et al. [13] reported the first reverse genetic analyses of flowering time in *Medicago*, taking advantage of the *Tnt1* insertion collection. Isolation of RNA-null *FTa1* insertion mutants showed that they are late flowering and have a dramatically reduced ability to respond to vernalization, consistent with the regulation of *FTa1* and its activity in *Arabidopsis*. However, *fta1* mutants retained an ability to respond to photoperiod, showing that although *FTa1* expression is promoted by long days, it is not essential for this response and implying that the strictly photoperiod-regulated *FTb1* gene may play a more dominant role [13]. This study also demonstrated the first use of *M. truncatula* transformation for functional analysis, reporting that *FTa1* overexpression conferred an early flowering phenotype in unvernalized plants, similar to that of vernalized WT plants [13]. In addition, the role and regulation of *FTc* was briefly examined. Somewhat surprisingly, given its strong overexpression phenotype in *Arabidopsis*, Tnt1 mutants for *FTc* showed no discernible flowering phenotype. However, expression of *FTc* was dramatically lower in the *Mtfta1* mutant, indicating a possible role for *FTc* downstream of *FTa1* [13], consistent with similar findings in pea [15].

In forward genetic screening for early flowering, Jaudal et al. [32] discovered that insertions in *FTa1* 3rd intron or in 3′ region confer a dominant early flowering phenotype under both SD and LD conditions, a phenotype similar to that found in plants overexpressing *FTa1*. These variants, referred to as *spring* mutants, eliminate the vernalization response, but retain a strong response to photoperiod. The *spring* insertions were found to correlate with elevated expression levels of *FTa1* itself, but were not associated with differences in DNA methylation at the *FTa1* locus [32]. Although the reason that *spring* mutations affect *FTa1* expression is not clear, the nature of their phenotype overall suggests that vernalization and photoperiod signals that converge on *FTa1* can be separated. Several MADS-domain transcription factor genes, including members of the *SOC1* and *FUL* family, also showed higher expression in the *spring* mutants, consistent with a role as targets of *FTa1* regulation [32].

Further evidence on how *FTa1* is regulated by vernalization has come from characterization of Tnt1 mutants for the *MtVRN2* gene, a homolog of the *Arabidopsis* polycomb repressive complex 2 (PRC2) component VRN2 [33]. In *Arabidopsis*, this complex is involved in epigenetic repression of the key vernalization target *FLC*, a repressor of *FT*. AtVRN2 acts to deposit the repressive epigenetic mark H3K27me3 across the *FLC* locus following exposure to cold, and *vrn2* mutants are late flowering with a reduced vernalization response [34]. However, in *M. truncatula*, a *vrn2* mutant shows early, vernalization independent flowering, like the *spring* mutants and *FTa1*-overexpressing plants. This phenotype is associated with elevated expression of *FTa1* and downstream MADS-domain genes and is dependent on *FTa1* [33], implying a role for MtVRN2 in epigenetic repression of *FTa1*. However, no effects on H3K27me3 at *FTa1* were detected, leaving open the question of how vernalization and VRN2 regulate *FTa1* expression.

In *Arabidopsis*, the MADS-domain protein SHORT VEGETATIVE PHASE has a role in mediation of temperature effects on flowering and acts in part through association with FLC and related MAF proteins [35, 36], raising the possibility that *SVP*-like genes might have a similar role in *Medicago*. To test this, Jaudal et al. [37] generated *Medicago* lines overexpressing *MtSVP1* and found that although flower development was abnormal in these lines, flowering time was not altered, indicating that a role of *SVP* in flowering time control is not conserved in *Medicago*.

7 Response to Photoperiod

A reverse genetic approach has also been used to explore the involvement of *Medicago* genes potentially involved in photoperiod regulation of *FT* gene expression and flowering. The involvement of *CO*-like genes has been of particular interest. In *Arabidopsis*, the *CO* gene encodes a protein with B-box and Zn finger domains that has a central role in the induction of *FT* expression under LD conditions. Long-day-specific accumulation of CO reflects two distinct mechanisms regulating *CO* transcription and CO protein stability. Long days relieve repression imposed on *CO* transcription by the CDF family of transcription factors, and this involves light-activated CDF degradation requiring the blue-light photoreceptor FKF1. In parallel, phytochrome and cryptochrome photoreceptors oppose the degradation of CO by the COP1/SPA ubiquitin ligase system [38]. Wong et al. [21] used the *Tnt1* collection to identify mutants for three of the four Group Ia *CO-like* genes in *M. truncatula*, including the co-ortholog of *CO* (*MtCOLa*) and two other closely related genes *MtCOLb* and *MtCOLc*. None of these mutants had a detectable effect on flowering time under LD, suggesting that the role of *CO* in photoperiod response might not be conserved. This observation was supported by a range of other evidence from *Medicago* and pea showing that *COLa* and other *COL* genes do not show characteristic patterns of regulation under different photoperiods and are unable to promote flowering when expressed in *Arabidopsis* [21]. One possibility is that the FKF1-CDF module may directly interact with *FT* promoters to confer photoperiod-responsive flowering. Consistent with this suggestion, the pea CDF protein LATE2 has been shown to inhibit transcription from the *MtFTb1* promoter [20].

Another possibility has been that other photoperiod-responsive regulators may act instead of CO to mediate photoperiod-responsive *FT* expression [1]. In soybean, the legume-specific transcription factor E1 plays a major role in the control of flowering and confers photoperiod-responsive flowering through inhibition of *FT* targets specifically under noninductive LD conditions [39]. To test the possibility that E1 might also be important in *M. truncatula* flowering,

Zhang et al. [40] characterized two Tnt1 mutants for *MtE1*, which both flowered slightly later than WT under LD conditions, suggesting that E1 might contribute to the LD induction of flowering and FT expression.

8 Comparative Genetics of Flowering in Legumes

In legumes, flowering time is an agronomically important trait, as it determines the suitability of cultivars to particular latitudinal zone, climatic regime, or cropping window [41]. The availability of *Medicago* transcriptome and genome sequences has provided a valuable initial means to identify potential flowering time genes in other legumes [42]. There is strong synteny between the genome of *Medicago* and of other legumes, and this has been used to identify likely candidate genes underlying mutants and QTLs in the related model species, pea [1]. Over the last decade, excellent progress has been made in determining how various genes from *Medicago*, as well as other legumes such as pea and soybean, function to control flowering time. In a few cases, variation within these genes has been shown to underlie variation in flowering. For example, allelic variation within the *ELF3* gene has been shown to underlie natural occurring variation in flowering in peas [22, 43], lentil [43], chickpea [25], and soybean [44]. Identifying the major alleles that influence flowering time will inform the rational design of new crop ideotypes that maximize yield for specific environments, production methods, and seasonal cycles. Of particular interest is the close relationship between *M. truncatula* and the tetraploid species *Medicago sativa* (alfalfa) and important crop globally for animal feed and forage. Unlike the annual *M. truncatula*, *M. sativa* is a perennial species, and understanding the underlying genetic differences between the two life cycles should provide ways to convert annual legume crops into perennials, which may offer a new range of possibilities and benefits associated with the perennial habit.

9 Conclusions

Medicago truncatula is clearly emerging as another useful model system for studying flowering time control—both in its own right and as a complementary system for gene discovery, validation, and functional analyses across a range of other legumes. Future studies will continue to help in the overall goal of understand how flowering time control mechanisms are conserved or have diverged within this group, how they may underlie adaptation in natural environments, and how they can be modified to improve legume crop production.

References

1. Weller JL, Ortega R (2015) Genetic control of flowering time in legumes. Front Plant Sci 6:207. https://doi.org/10.3389/fpls.2015.00207
2. Young ND, Cannon SB, Sato S, Kim D, Cook DR, Town CD, Roe BA, Tabata S (2005) Sequencing the genespaces of Medicago truncatula and Lotus japonicus. Plant Physiol 137(4):1174–1181. https://doi.org/10.1104/pp.104.057034
3. Hecht V, Foucher F, Ferrandiz C, Macknight R, Navarro C, Morin J, Vardy ME, Ellis N, Beltran JP, Rameau C, Weller JL (2005) Conservation of Arabidopsis flowering genes in model legumes. Plant Physiol 137(4):1420–1434
4. Tadege M, Wen J, He J, Tu H, Kwak Y, Eschstruth A, Cayrel A, Endre G, Zhao PX, Chabaud M, Ratet P, Mysore KS (2008) Large-scale insertional mutagenesis using the Tnt1 retrotransposon in the model legume Medicago truncatula. Plant J 54(2):335–347. https://doi.org/10.1111/j.1365-313X.2008.03418.x
5. Le Signor C, Savois V, Aubert G, Verdier J, Nicolas M, Pagny G, Moussy F, Sanchez M, Baker D, Clarke J, Thompson R (2009) Optimizing TILLING populations for reverse genetics in Medicago truncatula. Plant Biotechnol J 7(5):430–441. https://doi.org/10.1111/j.1467-7652.2009.00410.x
6. Rogers C, Wen J, Chen R, Oldroyd G (2009) Deletion-based reverse genetics in Medicago truncatula. Plant Physiol 151(3):1077–1086. https://doi.org/10.1104/pp.109.142919
7. Ronfort J, Bataillon T, Santoni S, Delalande M, David JL, Prosperi JM (2006) Microsatellite diversity and broad scale geographic structure in a model legume: building a set of nested core collection for studying naturally occurring variation in Medicago truncatula. BMC Plant Biol 6:28. https://doi.org/10.1186/1471-2229-6-28
8. Stanton-Geddes J, Paape T, Epstein B, Briskine R, Yoder J, Mudge J, Bharti AK, Farmer AD, Zhou P, Denny R, May GD, Erlandson S, Yakub M, Sugawara M, Sadowsky MJ, Young ND, Tiffin P (2013) Candidate genes and genetic architecture of symbiotic and agronomic traits revealed by whole-genome, sequence-based association genetics in Medicago truncatula. PLoS One 8(5):e65688. https://doi.org/10.1371/journal.pone.0065688
9. Yoder JB, Stanton-Geddes J, Zhou P, Briskine R, Young ND, Tiffin P (2014) Genomic signature of adaptation to climate in Medicago truncatula. Genetics 196(4):1263–1275. https://doi.org/10.1534/genetics.113.159319
10. Friesen ML, von Wettberg EJ, Badri M, Moriuchi KS, Barhoumi F, Chang PL, Cuellar-Ortiz S, Cordeiro MA, Vu WT, Arraouadi S, Djebali N, Zribi K, Badri Y, Porter SS, Aouani ME, Cook DR, Strauss SY, Nuzhdin SV (2014) The ecological genomic basis of salinity adaptation in Tunisian Medicago truncatula. BMC Genomics 15:1160. https://doi.org/10.1186/1471-2164-15-1160
11. Kang Y, Sakiroglu M, Krom N, Stanton-Geddes J, Wang M, Lee YC, Young ND, Udvardi M (2015) Genome-wide association of drought-related and biomass traits with HapMap SNPs in Medicago truncatula. Plant Cell Environ 38(10):1997–2011. https://doi.org/10.1111/pce.12520
12. Clarkson NM, Russell JS (1975) Flowering responses to vernalization and photoperiod in annual medics (Medicago spp.). Aust J Agric Res 26:831–838
13. Laurie RE, Diwadkar P, Jaudal M, Zhang L, Hecht V, Wen J, Tadege M, Mysore KS, Putterill J, Weller JL, Macknight RC (2011) The Medicago *FLOWERING LOCUS T* homolog, *MtFTa1*, is a key regulator of flowering time. Plant Physiol 156(4):2207–2224. https://doi.org/10.1104/pp.111.180182
14. Nelson MN, Ksiazkiewicz M, Rychel S, Besharat N, Taylor CM, Wyrwa K, Jost R, Erskine W, Cowling WA, Berger JD, Batley J, Weller JL, Naganowska B, Wolko B (2017) The loss of vernalization requirement in narrow-leafed lupin is associated with a deletion in the promoter and de-repressed expression of a Flowering Locus T (FT) homologue. New Phytol 213(1):220–232. https://doi.org/10.1111/nph.14094
15. Hecht V, Laurie RE, Vander Schoor JK, Ridge S, Knowles CL, Liew LC, Sussmilch FC, Murfet IC, Macknight RC, Weller JL (2011) The pea *GIGAS* gene is a *FLOWERING LOCUS T* homolog necessary for graft-transmissible specification of flowering but not for responsiveness to photoperiod. Plant Cell 23(1):147–161. https://doi.org/10.1105/tpc.110.081042
16. Foucher F, Morin J, Courtiade J, Cadioux S, Ellis N, Banfield MJ, Rameau C (2003) *DETERMINATE* and *LATE FLOWERING* are two *TERMINAL FLOWER1/CENTRORADIALIS* homologs that control two distinct phases of flowering initiation and development in pea. Plant Cell 15(11):2742–2754
17. Alves-Carvalho S, Aubert G, Carrere S, Cruaud C, Brochot AL, Jacquin F, Klein A, Martin C,

Boucherot K, Kreplak J, da Silva C, Moreau S, Gamas P, Wincker P, Gouzy J, Burstin J (2015) Full-length de novo assembly of RNA-seq data in pea (Pisum sativum L.) provides a gene expression atlas and gives insights into root nodulation in this species. Plant J 84(1):1–19. https://doi.org/10.1111/tpj.12967

18. Ruelens P, de Maagd RA, Proost S, Theissen G, Geuten K, Kaufmann K (2013) FLOWERING LOCUS C in monocots and the tandem origin of angiosperm-specific MADS-box genes. Nat Commun 4:2280. https://doi.org/10.1038/ncomms3280
19. Ledger S, Strayer C, Ashton F, Kay SA, Putterill J (2001) Analysis of the function of two circadian-regulated CONSTANS-LIKE genes. Plant J 26(1):15–22
20. Ridge S, Sussmilch FC, Hecht V, Vander Schoor JK, Lee R, Aubert G, Burstin J, Macknight RC, Weller JL (2016) Identification of LATE BLOOMER2 as a CYCLING DOF FACTOR homolog reveals conserved and divergent features of the flowering response to photoperiod in pea. Plant Cell 28(10):2545–2559. https://doi.org/10.1105/tpc.15.01011
21. Wong AC, Hecht VF, Picard K, Diwadkar P, Laurie RE, Wen J, Mysore K, Macknight RC, Weller JL (2014) Isolation and functional analysis of CONSTANS-LIKE genes suggests that a central role for CONSTANS in flowering time control is not evolutionarily conserved in Medicago truncatula. Front Plant Sci 5:486. https://doi.org/10.3389/fpls.2014.00486
22. Rubenach AJ, Hecht V, Vander Schoor JK, Liew LC, Aubert G, Burstin J, Weller JL (2017) EARLY FLOWERING3 redundancy fine-tunes photoperiod sensitivity. Plant Physiol 173(4):2253–2264. https://doi.org/10.1104/pp.16.01738
23. Sussmilch FC, Berbel A, Hecht V, Vander Schoor JK, Ferrandiz C, Madueno F, Weller JL (2015) Pea VEGETATIVE2 is an FD homolog that is essential for flowering and compound inflorescence development. Plant Cell 27(4):1046–1060. https://doi.org/10.1105/tpc.115.136150
24. Liew LC, Hecht V, Sussmilch FC, Weller JL (2014) The pea photoperiod response gene STERILE NODES is an Ortholog of LUX ARRHYTHMO. Plant Physiol 165(2):648–657. https://doi.org/10.1104/pp.114.237008
25. Ridge S, Deokar A, Lee R, Daba K, Macknight RC, Weller JL, Tar'an B (2017) The chickpea Early Flowering 1 (Efl1) locus is an ortholog of Arabidopsis ELF3. Plant Physiol 175(2):802–815. https://doi.org/10.1104/pp.17.00082
26. Julier B, Huguet T, Chardon F, Ayadi R, Pierre JB, Prosperi JM, Barre P, Huyghe C (2007) Identification of quantitative trait loci influencing aerial morphogenesis in the model legume Medicago truncatula. Theor Appl Genet 114(8):1391–1406. https://doi.org/10.1007/s00122-007-0525-1
27. Pierre JB, Huguet T, Barre P, Huyghe C, Julier B (2008) Detection of QTLs for flowering date in three mapping populations of the model legume species Medicago truncatula. Theor Appl Genet 117(4):609–620. https://doi.org/10.1007/S00122-008-0805-4
28. Espinoza Ldel C, Huguet T, Julier B (2012) Multi-population QTL detection for aerial morphogenetic traits in the model legume Medicago truncatula. Theor Appl Genet 124(4):739–754. https://doi.org/10.1007/s00122-011-1743-0
29. Pierre JB, Bogard M, Herrmann D, Huyghe C, Julier B (2011) A CONSTANS-like gene candidate that could explain most of the genetic variation for flowering date in Medicago truncatula. Mol Breeding 28(1):25–35. https://doi.org/10.1007/S11032-010-9457-6
30. Burgarella C, Chantret N, Gay L, Prosperi JM, Bonhomme M, Tiffin P, Young ND, Ronfort J (2016) Adaptation to climate through flowering phenology: a case study in Medicago truncatula. Mol Ecol 25(14):3397–3415. https://doi.org/10.1111/mec.13683
31. Friesen ML, Cordeiro MA, Penmetsa RV, Badri M, Huguet T, Aouani ME, Cook DR, Nuzhdin SV (2010) Population genomic analysis of Tunisian Medicago truncatula reveals candidates for local adaptation. Plant J 63(4):623–635. https://doi.org/10.1111/j.1365-313X.2010.04267.x
32. Jaudal M, Yeoh CC, Zhang L, Stockum C, Mysore KS, Ratet P, Putterill J (2013) Retroelement insertions at the Medicago FTa1 locus in spring mutants eliminate vernalisation but not long-day requirements for early flowering. Plant J 76(4):580–591. https://doi.org/10.1111/tpj.12315
33. Jaudal M, Zhang L, Che C, Hurley DG, Thomson G, Wen J, Mysore KS, Putterill J (2016) MtVRN2 is a Polycomb VRN2-like gene which represses the transition to flowering in the model legume Medicago truncatula. Plant J 86(2):145–160. https://doi.org/10.1111/tpj.13156
34. Bastow R, Mylne JS, Lister C, Lippman Z, Martienssen RA, Dean C (2004) Vernalization requires epigenetic silencing of FLC by histone methylation. Nature 427(6970):164–167
35. Lee JH, Yoo SJ, Park SH, Hwang I, Lee JS, Ahn JH (2007) Role of SVP in the control

of flowering time by ambient temperature in Arabidopsis. Genes Dev 21(4):397–402. https://doi.org/10.1101/gad.1518407

36. Li D, Liu C, Shen L, Wu Y, Chen H, Robertson M, Helliwell CA, Ito T, Meyerowitz E, Yu H (2008) A repressor complex governs the integration of flowering signals in *Arabidopsis*. Dev Cell 15:110–120
37. Jaudal M, Monash J, Zhang L, Wen J, Mysore KS, Macknight R, Putterill J (2014) Overexpression of Medicago SVP genes causes floral defects and delayed flowering in Arabidopsis but only affects floral development in Medicago. J Exp Bot 65(2):429–442. https://doi.org/10.1093/jxb/ert384
38. Andres F, Coupland G (2012) The genetic basis of flowering responses to seasonal cues. Nat Rev Genet 13(9):627–639. https://doi.org/10.1038/nrg3291
39. Xia Z, Watanabe S, Yamada T, Tsubokura Y, Nakashima H, Zhai H, Anai T, Sato S, Yamazaki T, Lu S, Wu H, Tabata S, Harada K (2012) Positional cloning and characterization reveal the molecular basis for soybean maturity locus E1 that regulates photoperiodic flowering. Proc Natl Acad Sci U S A 109(32):E2155–E2164. https://doi.org/10.1073/pnas.1117982109
40. Zhang X, Zhai H, Wang Y, Tian X, Zhang Y, Wu H, Lu S, Yang G, Li Y, Wang L, Hu B, Bu Q, Xia Z (2016) Functional conservation and diversification of the soybean maturity gene E1 and its homologs in legumes. Sci Rep 6:29548. https://doi.org/10.1038/srep29548
41. Nelson MN, Berger JD, Erskine W (2010) Flowering time control in annual legumes: prospects in a changing global climate. CAB Rev Perspect Agric Veterinary Sci Nutr Nat 5(017):1–14
42. Young ND, Debelle F, Oldroyd GE, Geurts R, Cannon SB, Udvardi MK, Benedito VA, Mayer KF, Gouzy J, Schoof H, Van de Peer Y, Proost S, Cook DR, Meyers BC, Spannagl M, Cheung F, De Mita S, Krishnakumar V, Gundlach H, Zhou S, Mudge J, Bharti AK, Murray JD, Naoumkina MA, Rosen B, Silverstein KA, Tang H, Rombauts S, Zhao PX, Zhou P, Barbe V, Bardou P, Bechner M, Bellec A, Berger A, Berges H, Bidwell S, Bisseling T, Choisne N, Couloux A, Denny R, Deshpande S, Dai X, Doyle JJ, Dudez AM, Farmer AD, Fouteau S, Franken C, Gibelin C, Gish J, Goldstein S, Gonzalez AJ, Green PJ, Hallab A, Hartog M, Hua A, Humphray SJ, Jeong DH, Jing Y, Jocker A, Kenton SM, Kim DJ, Klee K, Lai H, Lang C, Lin S, Macmil SL, Magdelenat G, Matthews L, McCorrison J, Monaghan EL, Mun JH, Najar FZ, Nicholson C, Noirot C, O'Bleness M, Paule CR, Poulain J, Prion F, Qin B, Qu C, Retzel EF, Riddle C, Sallet E, Samain S, Samson N, Sanders I, Saurat O, Scarpelli C, Schiex T, Segurens B, Severin AJ, Sherrier DJ, Shi R, Sims S, Singer SR, Sinharoy S, Sterck L, Viollet A, Wang BB, Wang K, Wang M, Wang X, Warfsmann J, Weissenbach J, White DD, White JD, Wiley GB, Wincker P, Xing Y, Yang L, Yao Z, Ying F, Zhai J, Zhou L, Zuber A, Denarie J, Dixon RA, May GD, Schwartz DC, Rogers J, Quetier F, Town CD, Roe BA (2011) The Medicago genome provides insight into the evolution of rhizobial symbioses. Nature 480:520. https://doi.org/10.1038/nature10625
43. Weller JL, Liew LC, Hecht VF, Rajandran V, Laurie RE, Ridge S, Wenden B, Vander Schoor JK, Jaminon O, Blassiau C, Dalmais M, Rameau C, Bendahmane A, Macknight RC, Lejeune-Henaut I (2012) A conserved molecular basis for photoperiod adaptation in two temperate legumes. Proc Natl Acad Sci U S A 109(51):21158–21163. https://doi.org/10.1073/pnas.1207943110
44. Lu S, Zhao X, Hu Y, Liu S, Nan H, Li X, Fang C, Cao D, Shi X, Kong L, Su T, Zhang F, Li S, Wang Z, Yuan X, Cober ER, Weller JL, Liu B, Hou X, Tian Z, Kong F (2017) Natural variation at the soybean J locus improves adaptation to the tropics and enhances yield. Nat Genet 49(5):773–779. https://doi.org/10.1038/ng.3819

Chapter 18

Functional Genomics and Genetic Control of Flower and Fruit Development in *Medicago truncatula*: An Overview

Edelín Roque, Concepción Gómez-Mena, Cristina Ferrándiz, José Pío Beltrán, and Luis A. Cañas

Abstract

A-, B-, and C-class genes code for MADS-box transcription factors required for floral organ identity in angiosperms. Other members of the family are also crucial to ensure proper carpel and fruit development. Development of genetic and genomic tools for *Medicago truncatula* has allowed its use as model system to study the genetic control of flower and fruit development in legumes. *M. truncatula* contains a single A-class gene, four B-function genes, and three C-class genes in its genome. This has made possible to do extensive functional characterization of these MADS-box transcription factors using gene expression analyses, protein-protein interactions, and forward and reverse genetic approaches. We have demonstrated the functions of these MADS-box transcription factors and the respective contributions of paralogous gene pairs to *M. truncatula* floral development. We have also defined the evolutionary outcomes of each duplicated pairs thus testing theoretical framework of several models about the evolution by gene duplication. Moreover, we have also studied the function of MADS-box fruit genes and how they may have contributed to the diversification of pod morphology within the *Medicago* genus. Our findings not only have contributed to increase knowledge in the field of the genetic control of flower and fruit development but also have provided a more complete understanding of the complexity of evolution by gene duplication and protein sequence diversification.

Key words *Medicago truncatula*, Floral development, MADS-box genes, Gene duplication, Paralog genes evolution

1 Flower Development: An Introduction

The flower is the most complex structure of plants. In evolutionary terms, flowers are the distinctive elements of a relatively recent group of land plants, the angiosperms. Floral structure is generally conserved in four concentric whorls of organs that are specified, from the outside to the center of the flower, in the sequence: sepals, petals, stamens, and carpels. Variants from this pattern, in which one type of floral organ develops at a wrong whorl, were known as

Luis A. Cañas and José Pío Beltrán (eds.), *Functional genomics in* Medicago truncatula*: Methods and Protocols*, Methods in Molecular Biology, vol. 1822, https://doi.org/10.1007/978-1-4939-8633-0_18,

homeotic mutants by the end of the nineteenth century [1]. Floral homeotic mutations affect organ identity, showing normal organs that develop in abnormal places. Although this type of mutants were available in many plant species, it was not until the end of 1980s when some floral homeotic mutants from *Antirrhinum majus* and *Arabidopsis thaliana* received more focused attention to understand the molecular mechanisms underlying floral organ identity [1]. In *Antirrhinum*, multiallelic series of spontaneous mutations induced by the insertion of transposable elements were available [2–5]. Also, the DNA sequences of the three active transposons in this species had been isolated and characterized [2, 3]. In *Arabidopsis*, the T-DNA insertional mutagenesis via transformation experiments began to be used as a powerful tool to study various aspects of plant growth and development. Such progresses allowed the discovery of the genes controlling key elements on the genetic control of floral development, leading to the proposal of the ABC model. The ABC model of floral development explained how each type of floral organ is specified by three gene functions (A, B, and C) that are expressed in overlapping regions of the flower [6–8]. A-class genes specify the formation of sepals in the first whorl and also repress the C-function in whorls 1 and 2. The combination of B- and A-function genes specifies petal identity in the second floral whorl, whereas a combination of B- and C-function genes controls stamen identity in the third whorl. C-class genes specify the formation of a carpel in the fourth whorl and also repress the A-function in the third and fourth whorls [6–8]. More recently, the ABC model has been expanded with the inclusion of the *Arabidopsis SEPALLATA* (*SEP*) genes (E-class genes) [9, 10] and the knowledge of the posttranscriptional regulation of *AP2* by the microRNA miR172 [11–13].

Cloning of the A, B, and C floral organ identity genes in these model species showed that all of them, except *AP2*, encoded MADS-box transcription factors [14, 15]. MADS proteins specify different floral organ identities in a combinatorial fashion, mediated by protein interactions and the formation of higher-order complexes (the quartet model) (e.g., [16]). They regulated their own expression and that of their target genes by binding to conserved sequences called CArG boxes present in the promoters of the regulated genes [14, 16–18]. The roles of MADS-box genes in flower development have been extensively studied in eudicot species being their core functions largely conserved across flowering plants. However, functional gene analyses are not possible in all plant species because it largely depend on the availability of genome resources and functional genomic tools [19].

2 *Medicago truncatula* as Model System to Study Floral Development in Legumes

Medicago truncatula (*Mt*) has been used as model system to study the genetic control of flower development in legumes. The *M. truncatula* flower (Fig. 1), as the majority of papilionoid flowers, is a typical example of zygomorphic symmetry [20]. It is pentamerous with members of each whorl alternating with those of adjacent whorls. It contains five fused sepals, five petals, ten stamens, and a single carpel. The corolla contains three types of yellow petals: a single standard petal (or vexillum), two wing petals, and two fused keel petals surrounding the reproductive organs, ten stamens and a central carpel (Fig. 1). Nine stamen filaments are joined in a staminal tube, and the tenth, the vexillary stamen filament at the adaxial region, is free standing [20, 21]. The study of *Medicago* floral ontogeny using scanning electron microscopy was an important contribution in this research area. Eight stages of flower development were defined by means of several previously chosen morphological markers [21]. Floral organ initiation in *Medicago truncatula* and the other papilionoid flowers is unidirectional in all whorls starting from the abaxial position of the flower with a high degree of overlap. Medicago floral meristems present distinctive features such as the existence of four common primordia from which petals and stamens differentiate and the early carpel primordium initiation [21, 22]. Medicago floral ontogeny study has been crucial for the systematic study of floral homeotic mutants, facilitating their subsequent characterization and the establishment of

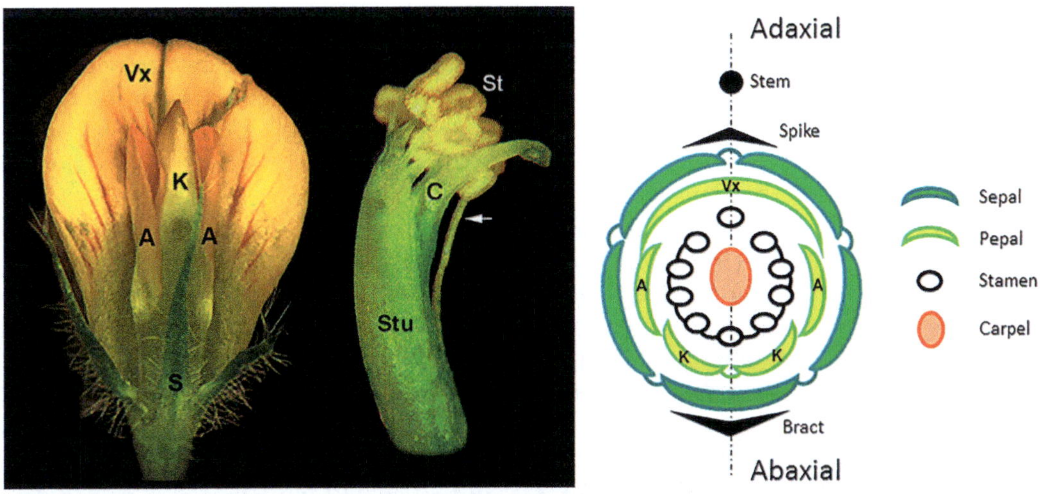

Fig. 1 *Medicago truncatula* flower: zygomorphic symmetry and floral formula. The pentamerous flower contains five fused sepals, five petals, ten stamens, and a single carpel. The corolla contains three types of yellow petals: a single standard petal (or vexillum), two wing petals, and two fused keel petals surrounding the reproductive organs, ten stamens and a central carpel. Nine stamen filaments are joined in a staminal tube and the tenth, the vexillary stamen filament at the adaxial region, is free standing. *S* sepal, *Vx* vexillum (standard), *A* allae (wing), *K* keel, *Stu* staminal tube, *St* stamen, *C* carpel

relationships with other homeotic mutations previously described in legume and non-legume plant species [21].

Legume floral development fits the classical ABC model with respect to organ identity specification and in the order in which whorls occur in the flower [21, 22]. It has been established that the ABC floral homeotic genes identified in *Arabidopsis* and *Antirrhinum* are conserved in the model legume *Medicago truncatula* [23–26, 27]. However, some variation has been observed: *M. truncatula* contains a single A-class gene [23], four B-function MADS-box genes [24, 26, 28], and three C-class genes in its genome [27], indicating that the B- and C-class lineages have been subjected to duplications and subsequent functional divergence during evolution. Functional genomics tools developed in *M. truncatula*, such as mutagenized populations (EMS, Tnt1) or transient and stable gene silencing (VIGS, RNAi), largely contributed in recent years to a comprehensive understanding of the genetic control of floral development in this model legume.

3 A-Class Genes in *Medicago truncatula*

According to the ABC model, the A-class genes are necessary for the specification of first whorl sepals (W1) and second whorl petals (W2) and for the repression of C-function activity in W1 and W2 [6]. The A-function is encoded by the homeotic genes *APETALA1* (*AP1*) and *APETALA2* (*AP2*) in *A. thaliana* [13, 29, 30, 31] and the redundant *AP2*-like genes *LIPLESS 1* and *2* (*LIP1* and *LIP2*) in *Antirrhinum majus* [32]. A-class mutants *ap1* and *ap2* exhibited organ identity defects primarily in W1 and W2. The organs of the W1 were replaced by carpels, and the W2 consisted of stamens in places normally occupied by the petals [13, 29, 33]. However, the phenotypical detailed study of several alleles of A-class mutants revealed some aspects that do not fit perfectly the model [13]. For example, *ap1* flowers have normal stamens and carpels, but petals are absent, and secondary floral meristems develop in the axils of the W1 organs. Thus, *APETALA1* (*AP1*) has been proposed to have a more general role in establishing floral meristem fate, which more accurately explains the phenotype of most *ap1* mutant alleles in *Arabidopsis* and those of orthologous genes in other plant species [34].

The *Medicago truncatula PROLIFERATING INFLORESCENCE MERISTEM* (*MtPIM*) gene, the homolog of *AP1* in this model legume, showed an expression pattern similar to that described for *AP1* and other *AP1*-like genes of different plant species. It was first strongly expressed in the secondary inflorescence meristem before the differentiation of the floral meristem. During early floral development, *MtPIM* expression was restricted to the periphery of the meristem, and in later stages expression is maintained in sepals and petals (Fig. 2), [23].

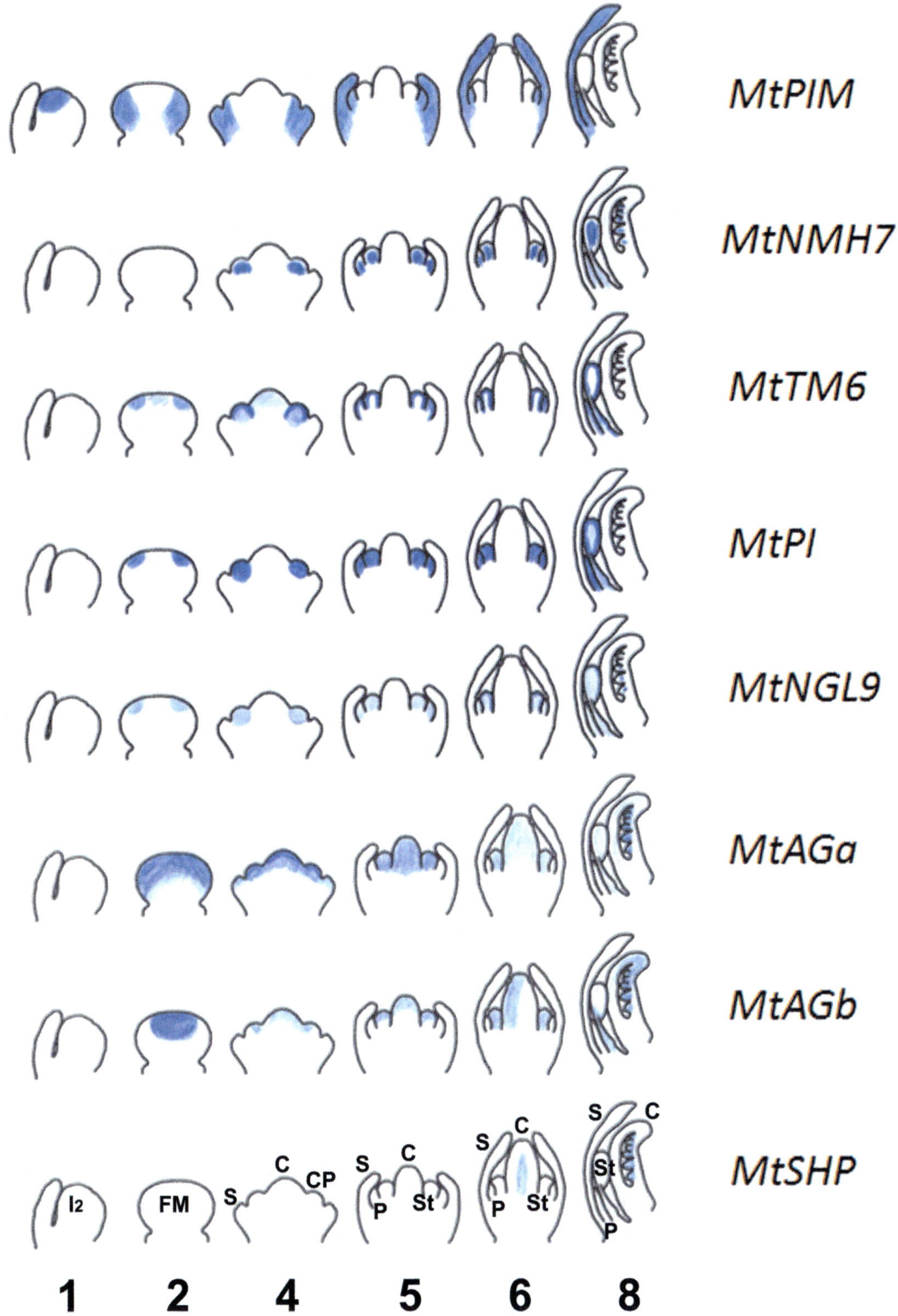

Fig. 2 Schematic representation of the expression patterns of *M. truncatula* A-, B-, and C-class floral homeotic genes during flower development. Numbers indicate the developmental stages of flowers defined in Benlloch et al., [21]. Stage 1: The secondary inflorescence meristems started producing floral meristems. Stage 2: Early floral meristem. Stage 4: The floral apex produced sepal primordia and the four common primordia that will rise to petal and stamen primordia. Stage 5: Common primordia differentiate into petal and stamen primordia. Stage 6: Flowers show elongated sepals that cover the rest of the floral organ primordia. Stage 8: Staminal tubes have been formed and the inflection of the carpel occurred. I_2 secondary inflorescence meristem, *FM* floral meristem, *S* sepal, *P* petal, *CP* common primordia, *St* stamen, *C* carpel

MtPIM was the first gene cloned and characterized based on reverse genetics screening of a *Tnt1* insertion population of *M. truncatula* [23, 35–37, 38]. The insertion of the retrotransposon *Tnt1* in the *MtPIM* gene caused dramatic phenotypical changes affecting inflorescence and flower architecture. The architecture of the inflorescence was dramatically altered in the *mtpim* mutant showing a high degree of ramification. *mtpim* flowers were replaced by complex structures with inflorescence characteristics (Fig. 3b) [23], being in this view equivalent to that of *ap1*, *squa*, *or pim* mutants from *Arabidopsis*, *Antirrhinum*, and pea, respectively [33, 39, 40, 41]. This flower-to-inflorescence conversion in *mtpim* is stronger than in the *Arabidopsis ap1* mutants. The flowers formed by the *mtpim* mutant showed defects in floral organ identity and number. A remarkable feature of *mtpim* flowers is that sepals were often converted into organs quite similar to leaves, indicating that *MtPIM* is involved in the specification of sepal identity [23].

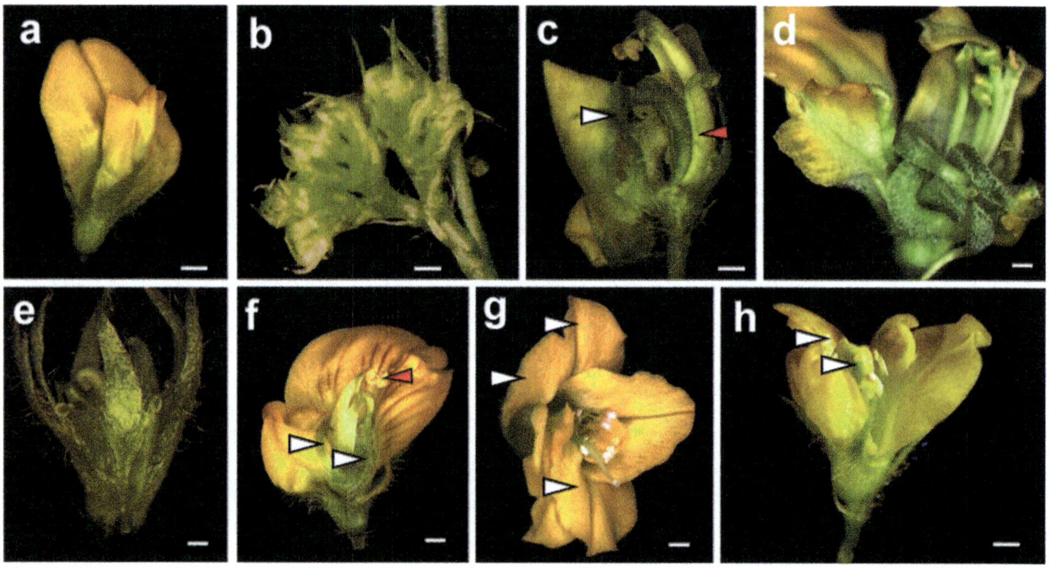

Fig. 3 Floral phenotypes of loss-of-function mutants of the *M. truncatula* A-, B-, and C-class floral homeotic genes. (**a**) *M. truncatula* wild-type flower. (**b**) Secondary inflorescence of a *mtpim* plant producing several proliferating meristems. (**c**) 35S::RNAi-*MtNMH7* flower. The second whorl contains petals with sepal-like (green) sectors (white arrowhead) and in whorl 3 several stamens show partial homeotic transformation into carpel-like structures (red arrowhead). (**d**) *mttm6–1* flower. The petals contain sepal tissues distributed in small areas, and the stamens were converted into carpel-like structures. (**e**) *mtpi-2* flower showing full conversion of petals into sepals and stamens into carpels. (**f**) Flower of the *mtaga* mutant with mild homeotic transformations of stamens into small petals (red arrow) and into two carpels (white arrowheads). (**g**) Flower of the *mtagb* mutant with homeotic transformations of stamens into petals resulting in a flower with more petals than the wild-type one (white arrowheads). (**h**) *MtAGab*-VIGS flower with severe homeotic transformation of stamens into petals (white arrows) and carpels into sepaloid structures. The *MtAGab*-VIGS flowers also showed indeterminacy of the floral meristem revealed by the presence of multiple petaloid or sepaloid concentric structures in the center. Bars indicate 1 mm

These studies indicated that the *MtPIM* gene is the *Medicago truncatula* homolog of *APETALA1/SQUAMOSA* (*AP1/SQUA*) and that works as a key controller of both floral meristem identity and floral development. Also, the isolation of *mtpim* represented a proof of concept demonstrating that the *Tnt1* populations can be efficiently used in reverse genetic screenings in *M. truncatula* [23].

4 B-Class Genes in *Medicago truncatula*

B-function MADS-box genes are involved in the specification of petal and stamen identity probably in all angiosperms [42]. B-function proteins form obligate heterodimers, which recruit E-function partners into tetrameric higher-order complexes [16]. A pair of MADS-box genes encodes the B-function activity in *A. thaliana*, *APETALA3* (*AP3*) and *PISTILLATA* (*PI*), and *A. majus*, *DEFICIENS* (*DEF*) and *GLOBOSA* (*GLO*) [6, 43, 44]. Mutations in either one of these genes produce transformation of petals into sepals and stamens into carpels [7, 43, 45, 46, 47, 48].

PI and AP3 represent lineages that arose from a duplication event that occurred after the split between extant gymnosperms and angiosperms, and subsequent duplication events of the AP3 and PI lineages have been documented throughout angiosperm evolution [49–52, 53, 54]. *M. truncatula* contains four B-function MADS-box genes: two *AP3*-like [26] and two *PI*-like paralogs [24, 28]. Thus, it is an ideal model system to study the effects of gene duplication and functional divergence within the B-function MADS-box lineages.

5 *APETALA3*-Like Genes in *Medicago truncatula*: *MtNMH7* and *MtTM6*

Two *M. truncatula AP3* homologs, *MtNMH7* and *MtTM6,* were isolated by screening of a flower cDNA library [26]. These two paralogous *MtAP3*-like genes arose from a duplication event that occurred early coinciding with the base of the core eudicot radiation and placed *MtNMH7* and *MtTM6* within their respective lineages, *euAP3* and *TM6,* of the clade of *AP3*-like genes [26].

The expression of *MtAP3*-like genes was detected in the regions of the floral meristem that give rise to petals and stamens and persists in these floral organs until later stages of development (Fig. 2), [26]. This was similar to that described for other *AP3*-like genes in agreement with their roles in specifying petals and stamens. However, *MtNMH7* and *MtTM6* appear to be expressed at different levels and with a mutually antagonistic expression pattern in different subpopulations of cells within petals and stamens (Fig. 2). The *MtTM6* transcript was strongly expressed in the outer cell layers that surround the second and third whorls, whereas

MtNMH7 was expressed inside them (Fig. 2). Moreover, *MtNMH7* is expressed in ovules at later stages of development (Fig. 2) [26].

Loss-of-*MtNMH7* function induced by RNAi resulted in defects predominantly in petal development, showing partial conversion of petals into sepaloid structures with patches of green tissue and sepal-like cells, whereas the stamens were weakly transformed into carpel-like organs (Fig. 3c). In contrast, RNAi-induced loss-of-*MtTM6* function plants produce flowers that display homeotic transformations predominantly in the third whorl [26]. Downregulation of *MtTM6* activity in *MtTM6*-RNAi plants did not result in homeotic defects in petal development [26]. However, flowers of the *mttm6* mutant, isolated from a *M. truncatula Tnt1*-mutagenized population, exhibited stronger conversion of stamens into carpels than *MtTM6*-RNAi lines (Fig. 3d), [26]. Also, *mttm6* flowers showed a weak homeotic conversion of petals into sepal-like organs (Fig. 3d), [26]. The lack of homeotic conversions in petals of the *MtTM6*-RNAi flowers is the result of low-level residual expression of *MtTM6* [26], because in the RNAi strategy, gene function is reduced but not completely eliminated. This suggests that it should not be excluded the lack of a homeotic phenotype without the analysis of the null allele.

Functional studies of the *MtAP3* homologs through a detailed analysis of their expression pattern and the use of both RNA silencing (RNAi) and *Tnt1* insertional mutagenesis have uncovered interesting aspects of the regulatory control of petal and stamen development in this legume species. The two *M. truncatula AP3*-like genes have slightly different B-class functions in agreement with this complementary expression pattern in petals and stamens. *MtNMH7* is more involved in petal identity whereas *MtTM6* plays a more important role in stamen identity [26]. Medicago *AP3*-like paralogs have experienced a loss or reduction of expression for different functions (subfunctionalization), being the combined action of both gene copies necessary to fulfill the requirements of the ancestral locus [26].

These studies provided information on the similarities and differences in petal and stamen development between core eudicots. *AP3*-like function is conserved in *M. truncatula* at the molecular level, but there are differences found at the level of the subfunctionalization/specialization between paralogous in the same gene lineage.

6 *PISTILLATA*-Like Genes in *Medicago truncatula: MtPI* and *MtNGL9*

M. truncatula contains two B-function MADS-box genes belonging to the PI/GLO subfamily (*MtPI* and *MtNGL9*) [24]. Paralogous *MtPI*-like genes arose from a duplication event that occurred prior to the speciation of legumes [28]. They may have originated during the whole-genome duplication (WGD) event that pre-dated speciation of *M. truncatula* and other legumes

~50–60 Mya [55]. Two copies of both *PI*-like genes are also found in other legume subfamilies [28].

MtPI and *MtNGL9* have similar temporal and spatial expression patterns in petals and stamens during floral development but differ in their expression levels [24, 28]. *MtPI* showed a high uniformly distributed expression in petals and stamens from early to late floral development (Fig. 2), [24, 28]. However, the *MtNGL9* signal seemed to be mainly confined to the epidermal cells [27]. Moreover, *MtNGL9* shows a differential expression in ovules (Fig. 2), [28].

MtPI function was assessed using RNAi-mediated gene silencing [24] and studying two *M. truncatula* floral mutants previously described [24, 28, 56, 57]. The *mtap* mutant was isolated from a *M. truncatula* EMS-mutagenized population [57]. It contains a nucleotide substitution in the *MtPI* coding sequence at position +100, causing an amino acid substitution (E34K) in the MADS domain of the MtPI protein [24]. *mtpi-2* was isolated as a case study to demonstrate the utility of the reverse genetic platform in the model legume *Medicago truncatula* [56]. Both *MtPI*-RNAi, *mtap*, and *mtpi*-2 flowers exhibited homeotic transformations of the second and third whorl organs [24, 28, 56, 57]. However, while *mtpl-2* and *MtPI*-RNAi flowers with strong phenotype exhibited a complete conversion of petals into sepals and stamens into carpels, the second whorl organs of *mtap* flowers were like wild-type sepals, but the third whorl organs showed partial homeotic transformations to carpels [24]. The *mtngl9-1 and mtngl9–2* mutants were isolated by a reverse genetics approach looking for retrotransposon *Tnt1* insertion mutants [28]. Although no *MtNGL9* transcripts were detected in homozygous *mtngl9-1* and *mtngl9-2* plants, homeotic changes or any obvious mutant phenotype were not observed in floral organs [28]. Knockdown of *MtPI* in *mtpi-2* flowers resulted in a marked reduction in the expression of all B-function MADS box genes, while the loss of *MtNGL9* caused a slight decrease in their expression [28]. The full loss-of-B-function phenotype in *mtpi-2* and the pronounced effects of the MtPI absence on the expression of the other B-function genes confirm *MtPI* as a master regulator in establishing the regulatory pathways for petal and stamen identity, while *MtNGL9* does not appear to have a role in this regard. Thus, *MtNGL9* could potentially be on the way to becoming a pseudogene. However, molecular evolution studies discarded pseudogenization as a plausible scenario for *MtNGL9* evolution [28]. *MtNGL9* has lost its ancestral expression level, and, as a consequence, its ancestral function seems to be decreased but not entirely eliminated. Indeed, in support of a functional role for *MtNGL9*, this gene encodes a protein that conserves this competence to interact with the MtAP3-like proteins, which it is an essential feature for the B-class MADS box proteins [28]. In line with this, MtNGL9 was able to partially replace PI in *A. thaliana* [28].

MtPI-like duplicated gene copies have been presumably preserved in the genome to achieve an optimal total dosage balance of their functional complex in which MtNGL9 would have a far low contribution to the total activity required to specify the B-function. Differential expression of *MtNGL9* in ovules makes neofunctionalization plausible as an alternative evolutionary fate for this gene. However, as the presence of transcripts in a given tissue does not necessarily mean that the gene activity is required for its proper development, the extent to which such divergences have led to a neofunctionalization of one of the duplicated *MtPI*-like copies and to a possible acquisition of other roles in different developmental processes requires further investigation [28].

7 C-Class Genes in *Medicago truncatula*: *MtAGa*, *MtAGb*, and *MtSHP*

C-function is required to promote stamen and carpel identity and to establish the determinate nature of the floral meristem. In *Arabidopsis* and *Antirrhinum*, the C-function is essentially represented by a single *AGAMOUS* (*AG*) gene [6, 43] and a single *PLENA* (*PLE*) gene, respectively [58]. Loss-of-function mutations in C-class genes have a similar phenotypic effect in all dicots studied to date. For example, the *agamous* (*ag*) mutant shows normal sepals and petals, whereas the stamens in W3 are homeotically transformed into petals, and in W4 a new *ag* flower develops instead of a pistil [43]. Additional C-function genes have been identified in these model species: *SHATTERPROOF1* (*SHP1*) and *SHATTERPROOF2* (*SHP2*) genes in *Arabidopsis* [59, 60] and *FARINELLI* (*FAR*) in *Antirrhinum* [61].

C-function MADS-box transcription factors belong to the *AGAMOUS* (*AG*) lineage. Phylogenetic studies using a large data set of *AG*-like sequences show that *AG* and *PLE* actually represent paralogous lineages derived from a duplication in a common ancestor early in the history of the core eudicots. This duplication gave rise to the *euAG* lineage which includes *AG* and *FAR*, and the *PLENA* lineage (*PLE*), where *SHP1/SHP2 and PLENA* are placed [62–64, 65, 66].

M. truncatula harbors two members of the *euAG* sub-clade (*MtAGa* and *MtAGb*) and one member of *PLENA* (*MtSHP*) [27, 67]. Paralogous *M. truncatula euAG* and *MtSHP* gene (*PLENA* clade) arose from a duplication event that occurred early coinciding with the base of the core eudicot radiation [27]. However, *MtAGa* and *MtAGb* are the products of a relatively recent duplication event, which probably occurred prior to the speciation of legumes around 50–60 Mya [27].

MtAGa and *MtAGb* showed nearly identical spatial expression patterns largely restricted to the developing male and female organs from early stages and throughout flower development. Differences were mainly observed at early stages of development, where *MtAGa* expression appears to be wider spread in the floral meristem than *MtAGb* (Fig. 2), [27]. In contrast, *MtSHP* was exclusively detected in the ovules at late stages of floral development (Fig. 2), [27].

The functions of *MteuAG* genes were studied using virus-induced gene silencing (VIGS) and RNAi. Also, two *M. truncatula* floral mutants (*mtaga* and *mtagb*), each containing a single *Tnt1* element inserted in the N-terminal part of the gene, were used [27, 56]. *mtaga* and *mtagb* mutants exhibited very similar floral phenotypes: flowers were wild type in appearance, showing only mild developmental defects on the third and fourth whorl organs (Fig. 3f, g), [27]. The carpels in whorl 4 presented stigmatic protuberances or multiple unfused carpels and exposed ovules. The effect of the *MtAG* loss-of-function was investigated by means of VIGS and stable *MtAG*::RNAi lines. *MtAG*-VIGS flowers showed an almost complete loss-of-C-function phenotype: stamens were replaced by petals or petal-like organs, and the central carpel was replaced by multiple petaloid or sepaloid concentric structures (Fig. 3h), [27]. The discrete mutant phenotypes of *mtaga* or *mtagb* were due to the presence of the remaining paralog that is sufficient to confer almost all the C-function. The expression analysis and phenotypic studies in *MtAG*::RNAi and *MtAG*-VIGS lines suggested that the functions of *MteuAG* genes fit in the quantitative model of *AG* activity [27].

C-function in *M. truncatula* is redundantly encoded by two *euAG*-genes. *MtAGa* and *MtAGb* are the unique genes showing a complete C-function activity, concomitant with their ancestral expression profile, early in the floral meristem and in the third and fourth floral whorls during floral development. In contrast, *MtSHP* expression appears late during floral development, suggesting it does not contribute significantly to the C-function in this legume species, while its expression in ovules would not exclude a possible function in ovule development [27]. *M. truncatula euAG*- and *PLENA*-like paralogs could have subfunctionalized concomitantly with the differential spatial and temporal expression pattern of the ancestral gene lineage [27]. Furthermore, *MtAG* paralogs are maintained in the genome consistent with the gene balance hypothesis, which predicts that the fate of duplicated genes largely depends on the maintenance of the stoichiometric balance among members of the macromolecular complex [68, 69, 70].

8 The Unusual Morphology of the *Medicago truncatula* Pod

Legumes take their name from their typical fruit. With very few exceptions, it is derived, after fertilization, from a pistil formed by a single superior carpel, fused at the margins to render a unilocular pod, where placentation is parietal, along the suture, and the number of ovules formed varies from two to many [71].

The legume fruit is most commonly a pod that shatters at maturity to free the seeds along two dehiscence zones: one placed at the adaxial suture, derived from the fusion of the carpel margins, and another one running along the central vein in the carpel, opposite to the marginal one [72]. While fruit morphology has diversified in many ways from this basic form to adapt to different seed dispersal agents such as wind, animals or water, the legume pod is the most frequent and basic fruit type in the huge Leguminosae family (Fig. 4a, b), including most legume crops.

Medicago truncatula belongs to the Papilionoideae subfamily, to which most legume crops belong, and shares with them the typical pistil morphology described previously. However, fruit morphology is quite different from those of closely related genera, such as pea, fava beans or lentils, and even many species within the *Medicago* genus. After fertilization, *M. truncatula* pods elongate asymmetrically, forming a coiled ball-shaped structure and developing spines in the pod surface; this morphological innovation implies a change in dispersal mechanisms, where seeds disseminate collectively inside the pod, and the spines, by attaching to passing animals, facilitate long-range distribution (Fig. 4c–f), [73, 74].

Comparative anatomical analysis of fruit morphology in several species of the genus *Medicago* with coiled fruits; the typical legume pod of, for example, *Pisum sativum*; and other *Medicago* species also with elongated pods indicates that the development of the presumptive dehiscence zones in all cases is similar, with small cells and lignification of adjacent zones at later stages of fruit development (Fig. 4g), [67]. Interestingly, differences in lignin deposition correlate perfectly with the coiled versus elongated pod structure: coiled pods have strong lignification at the carpel margins at the adaxial position and weak lignification at the abaxial zone along the main carpel vein, while elongated pods have symmetrical lignification at abaxial and adaxial positions (Fig. 4g). The extensive adaxial lignification at the margins of coiled pods may impose restricted growth at this position, and hence, it is likely the cause of the unequal spiral elongation of the fruit [67].

9 MADS-Box Genes Involved in Fruit Development in *Medicago*

The genetics of dehiscence zone formation and lignin deposition in fruits have been studied extensively in the model species *Arabidopsis thaliana* for almost two decades, and, currently, a wealth of information on the major factors controlling this process is available [75]. A basic set of four genes expressed in the valve margin directs the development of the lignified patches and of the domains of small cells with restricted growth at the dehiscence

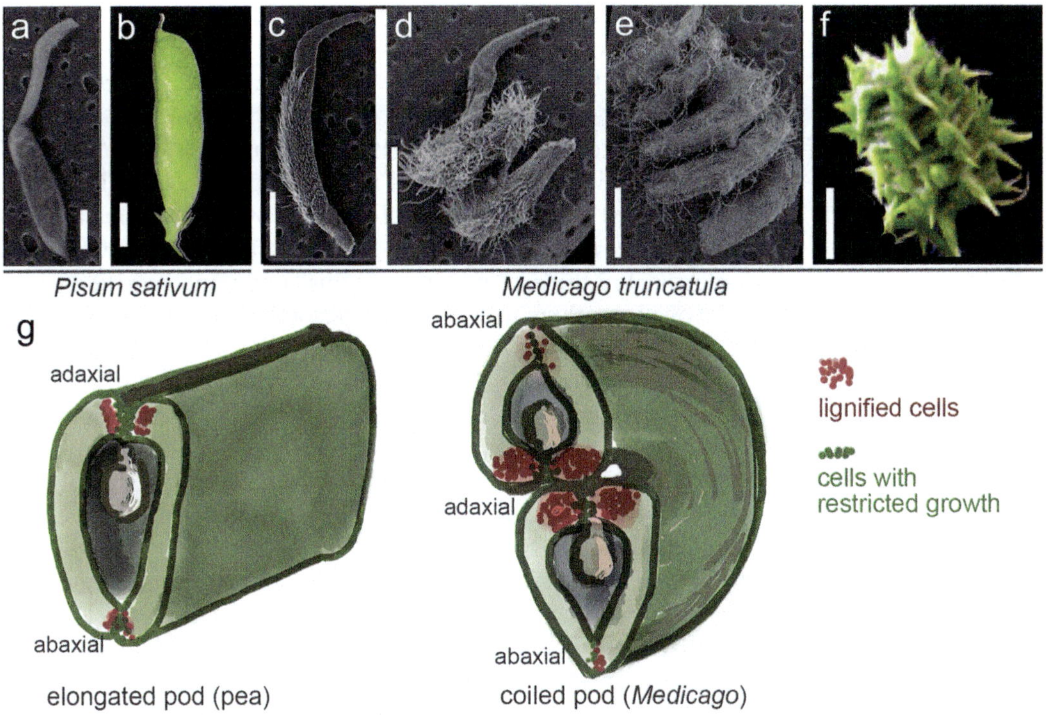

Fig. 4 Fruit morphology in *Medicago truncatula*. (**a**, **b**). Typical elongated legume pod of *Pisum sativum*. (**a**) SEM micrograph of a pistil at anthesis. (**b**) Fully elongated pod. (**c**–**f**) Coiled fruit development in *Medicago truncatula*. (**c**) SEM micrograph of a pistil at anthesis. (**d**) SEM micrograph of a fruit approximately 5 days after fertilization. (**e**) SEM micrograph of a fruit approximately 10 days after fertilization. (**f**) Fully grown fruit. Note how the number of coils increases with fruit elongation until the formation of a spiny ball from a pistil morphology that closely resembles that of a pea. (**g**) Cartoon representing the differential lignin deposition and the presence of cells with restricted growth in a typical legume pod (left) and in a coiled *Medicago* pod (right). Bars indicate 1 mm in **a**, **c**, **e** and **d**; 2 cm in **b**; 0.5 cm in **f**

zone, where the fruit opens at maturity. *SHP1* and *SHP2*, the pair of duplicated MADS-box genes belonging to the PLENA lineage, act redundantly to activate two genes of the bHLH family, *INDEHISCENT* (*IND*) and *ALCATRAZ* (*ALC*). *shp1 shp2* or *ind* mutants completely fail to develop the dehiscence zone, and, hence, cells at the valve margin appear very similar to those in the carpel wall, with no lignified patches or small cells forming, while *alc* mutants only show milder defects [76, 77]. An additional major factor in the network is another member of the MADS-box family, *FRUITFULL (FUL)*, which is related to the A-class gene *AP1* but has been shown to act in a wide range of developmental processes. *FUL* is expressed in the carpel wall where it restricts the expansion of *SHP* and *IND* expression. Thus, *ful* mutants show ectopic lignified tissue and restricted growth of the silique caused by ectopic formation of dehiscence zone characters in the carpel wall [78].

In contrast with the extensive functional studies in *Medicago truncatula* flower development described in previous sections, very little is known about the genetic pathways directing fruit development in this species. The *M. truncatula* genome, as already mentioned, contains a single *PLENA* gene, *MtSHP*, but *FUL* appears to have duplicated, and two close paralogs can be found, *MtFULa* and *MtFULb* [67]. *IND* and *ALC* are Brassicaceae-specific genes that arose recently by the whole genome duplication that pre-dated the diversification of this family, and have likely neo-functionalized thereafter, so no true orthologs can be found in legumes [75].

The basic functions of *SHP* and *FUL* orthologs in fruit development, and in particular dehiscence zone formation, appear to be conserved in a range of species [79]. However, no mutants have been identified through reverse genetic approaches for any of these genes in *Medicago*, and VIGS strategies have not been successful so far to characterize the effects of *MtSHP* or *MtFULa/b* loss of function. Interestingly, though, the intriguing pod morphology in *M. truncatula* and other species is the genus, and the correlation of pod shape with variations in the pattern of lignin deposition and cells with restricted growth at the carpel margins giving rise to the adaxial dehiscence zone, prompted a study where the molecular properties of *MedicagoFULa/b* and *MedicagoSHP* genes in the genus were analyzed [67]. In this study, a further correlation was found between *MedicagoSHP* sequences and lignification patterns. All the *MedicagoSHP* genes from the coiled pod species showed a sequence signature under strong positive selection in the C-terminal domain of the protein that altered SHP affinities for interacting protein partners and, thus, was likely affecting SHP activity to promote extensive lignification and restricted growth at the carpel margin, where they are expressed. The fact that this change in protein sequence appears to be evolutionary fixed suggests that the new pod morphology that evolved in the *Medicago* genus had an impact in fitness and seed dispersal strategies with adaptive value, constituting a nice example of modified protein activity as a driving force in evolution [67].

10 Conclusions and Perspectives

Studies of the *Medicago truncatula* ABC homeotic floral genes have contributed to increase knowledge in the field of the genetic control of floral development. Likewise they provided a more complete understanding of the complexity of evolution by gene duplication where different classes of duplicated MADS-box genes have been preserved in the genome with distinct evolutionary outcomes. Some paralog pairs have experienced a subfunctionalization process by the division of their ancestral expression domains, specializing in some aspects of their original function while having lost other functions.

However, in other cases, the duplicated genes have been retained because a mutation of one member would upset the correct stoichiometric relationships of the subunits of their complexes. Finally, certain paralogs have followed a complex model of evolution by gene duplication, being needed to combine molecular evolution studies and functional assays to give a parsimonious explanation for the preservation of both duplicated gene copies in the genome. Notwithstanding the complexity of the evolution by gene duplication, the A-, B-, and C-functions remain conserved in all eudicots, providing robustness to the specification of floral organ identity.

Despite functional specialization of the *Medicago truncatula* MADS-box genes as a result of gene duplication, control of floral organ identity is deeply conserved. However, such duplications increase complexity regarding MADS-box protein–protein interactions and their downstream and upstream regulations. In addition, the potential of protein modification at the base of evolutionary novelties such as the coiled pod morphology in the *Medicago* genus is also highlighted. Future studies need to elucidate a complex scenario where duplicated MADS-box transcription factors that form part of ternary protein complexes are competing for the same sequences in the regulatory regions of their target genes. Unraveling the transcriptional regulatory network integrating the MADS-box genes activity by the identification of upstream and downstream factors in *Medicago* will be also a challenge for future research. The use of the genetic and genomic tools developed for *M. truncatula* opens up a wide range of possibilities to better understand floral organogenesis in legumes, using this species as model system.

Acknowledgments

We gratefully acknowledge funding from the Spanish Ministry of Economy and Competitiveness (http://www.idi.mineco.gob.es/portal/site/MICINN) grant BIO2016-75485-R and BIO2015-64531-R.

References

1. Bowman JL, Smyth DR, Meyerowitz EM (2012) The ABC model of flower development: then and now. Development 139:4095–4098
2. Coen ES, Carpenter R, Martin C (1986) Transposable elements generate novel spatial patterns of gene expression in Antirrhinum majus. Cell 47:285–296
3. Sommer H, Bonas U, Saedler H (1988) Transposon-induced alterations in the promoter region affect transcription of the chalcone synthase gene of Antirrhinum majus. Mol Gen Genet 211:49–55
4. Carpenter R, Coen ES (1990) Floral homeotic mutations produced by transposon-mutagenesis in Antirrhinum majus. Genes Dev 4:1483–1493
5. Sommer H, Beltran JP, Huijser P, Pape H, Lonnig WE, Saedler H, Schwarz-Sommer Z (1990) Deficiens, a homeotic gene involved in the control of flower morphogenesis in Antirrhinum majus: the protein shows homology to transcription factors. EMBO J 9:605–613
6. Bowman JL, Smyth DR, Meyerowitz EM (1991) Genetic interactions among floral

homeotic genes of Arabidopsis. Development 112:1–20
7. Schwarz-Sommer Z, Huijser P, Nacken W, Saedler H, Sommer H (1990) Genetic control of flower development by homeotic genes in Antirrhinum majus. Science 250:931–936
8. Coen ES, Meyerowitz EM (1991) The war of the whorls: genetic interactions controlling flower development. Nature 353:31–37
9. Ditta G, Pinyopich A, Robles P, Pelaz S, Yanofsky MF (2004) The SEP4 gene of Arabidopsis thaliana functions in floral organ and meristem identity. Curr Biol 14:1935–1940
10. Pelaz S, Ditta GS, Baumann E, Wisman E, Yanofsky MF (2000) B and C floral organ identity functions require SEPALLATA MADS-box genes. Nature 405:200–203
11. Aukerman MJ, Sakai H (2003) Regulation of flowering time and floral organ identity by a MicroRNA and its APETALA2-like target genes. Plant Cell 15:2730–2741
12. Chen X (2004) A microRNA as a translational repressor of APETALA2 in Arabidopsis flower development. Science 303:2022–2025
13. Prunet N, Jack TP (2014) Flower development in Arabidopsis: there is more to it than learning your ABCs. Methods Mol Biol 1110:3–33
14. Smaczniak C, Immink RG, Muino JM, Blanvillain R, Busscher M, Busscher-Lange J, Dinh QD, Liu S, Westphal AH, Boeren S et al (2012) Characterization of MADS-domain transcription factor complexes in Arabidopsis flower development. Proc Natl Acad Sci U S A 109:1560–1565
15. Weigel D, Meyerowitz EM (1994) The ABCs of floral homeotic genes. Cell 78:203–209
16. Melzer R, Theissen G (2009) Reconstitution of 'floral quartets' in vitro involving class B and class E floral homeotic proteins. Nucleic Acids Res 37:2723–2736
17. Egea-Cortines M, Saedler H, Sommer H (1999) Ternary complex formation between the MADS-box proteins SQUAMOSA, DEFICIENS and GLOBOSA is involved in the control of floral architecture in Antirrhinum majus. EMBO J 18:5370–5379
18. Melzer R, Verelst W, Theissen G (2009) The class E floral homeotic protein SEPALLATA3 is sufficient to loop DNA in 'floral quartet'-like complexes in vitro. Nucleic Acids Res 37:144–157
19. Becker A, Lange M (2010) VIGS--genomics goes functional. Trends Plant Sci 15:1–4
20. Tucker SC (2003) Floral development in legumes. Plant Physiol 131:911–926
21. Benlloch R, Navarro C, Beltrán J, Cañas LA (2003) Floral development of the model legume Medicago truncatula: ontogeny studies as a tool to better characterize homeotic mutations. Sex Plant Reprod 15:231–241
22. Ferrandiz C, Navarro C, Gomez MD, Canas LA, Beltran JP (1999) Flower development in pisum sativum: from the war of the whorls to the battle of the common primordia. Dev Genet 25:280–290
23. Benlloch R, d'Erfurth I, Ferrandiz C, Cosson V, Beltran JP, Canas LA, Kondorosi A, Madueno F, Ratet P (2006) Isolation of mtpim proves Tnt1 a useful reverse genetics tool in Medicago truncatula and uncovers new aspects of AP1-like functions in legumes. Plant Physiol 142:972–983
24. Benlloch R, Roque E, Ferrandiz C, Cosson V, Caballero T, Penmetsa RV, Beltran JP, Canas LA, Ratet P, Madueno F (2009) Analysis of B function in legumes: PISTILLATA proteins do not require the PI motif for floral organ development in Medicago truncatula. Plant J 60:102–111
25. Hecht V, Foucher F, Ferrandiz C, Macknight R, Navarro C, Morin J, Vardy ME, Ellis N, Beltran JP, Rameau C et al (2005) Conservation of Arabidopsis flowering genes in model legumes. Plant Physiol 137:1420–1434
26. Roque E, Serwatowska J, Cruz Rochina M, Wen J, Mysore KS, Yenush L, Beltran JP, Canas LA (2013) Functional specialization of duplicated AP3-like genes in Medicago truncatula. Plant J 73:663–675
27. Serwatowska J, Roque E, Gomez-Mena C, Constantin GD, Wen J, Mysore KS, Lund OS, Johansen E, Beltran JP, Canas LA (2014) Two euAGAMOUS genes control C-function in Medicago truncatula. PLoS One 9:e103770
28. Roque E, Fares MA, Yenush L, Rochina MC, Wen J, Mysore KS, Gomez-Mena C, Beltran JP, Canas LA (2016) Evolution by gene duplication of Medicago truncatula PISTILLATA-like transcription factors. J Exp Bot 67:1805–1817
29. Irish VF, Sussex IM (1990) Function of the apetala-1 gene during Arabidopsis floral development. Plant Cell 2:741–753
30. Jofuku KD, den Boer BG, Van Montagu M, Okamuro JK (1994) Control of Arabidopsis flower and seed development by the homeotic gene APETALA2. Plant Cell 6:1211–1225
31. Mandel MA, Gustafson-Brown C, Savidge B, Yanofsky MF (1992) Molecular characterization of the Arabidopsis floral homeotic gene APETALA1. Nature 360:273–277
32. Keck E, McSteen P, Carpenter R, Coen E (2003) Separation of genetic functions controlling organ identity in flowers. EMBO J 22:1058–1066
33. Bowman JL, Alvarez J, Weigel D, Meyerowitz EM, Smyth DR (1993) Control of flower development in Arabidopsis thaliana by APETALA1

and interacting genes. Development 119:721–743

34. Causier B, Schwarz-Sommer Z, Davies B (2010) Floral organ identity: 20 years of ABCs. Semin Cell Dev Biol 21:73–79
35. Cheng X, Wen J, Tadege M, Ratet P, Mysore KS (2011) Reverse genetics in Medicago truncatula using Tnt1 insertion mutants. Methods Mol Biol 678:179–190
36. d'Erfurth I, Cosson V, Eschstruth A, Lucas H, Kondorosi A, Ratet P (2003) Efficient transposition of the Tnt1 tobacco retrotransposon in the model legume Medicago truncatula. Plant J 34:95–106
37. Tadege M, Ratet P, Mysore KS (2005) Insertional mutagenesis: a Swiss Army knife for functional genomics of Medicago truncatula. Trends Plant Sci 10:229–235
38. Tadege M, Wen J, He J, Tu H, Kwak Y, Eschstruth A, Cayrel A, Endre G, Zhao PX, Chabaud M et al (2008) Large-scale insertional mutagenesis using the Tnt1 retrotransposon in the model legume Medicago truncatula. Plant J 54:335–347
39. Huijser P, Klein J, Lonnig WE, Meijer H, Saedler H, Sommer H (1992) Bracteomania, an inflorescence anomaly, is caused by the loss of function of the MADS-box gene squamosa in Antirrhinum majus. EMBO J 11:1239–1249
40. Taylor SA, Hofer JM, Murfet IC, Sollinger JD, Singer SR, Knox MR, Ellis TH (2002) PROLIFERATING INFLORESCENCE MERISTEM, a MADS-box gene that regulates floral meristem identity in pea. Plant Physiol 129:1150–1159
41. Berbel A, Navarro C, Ferrandiz C, Canas LA, Madueno F, Beltran JP (2001) Analysis of PEAM4, the pea AP1 functional homologue, supports a model for AP1-like genes controlling both floral meristem and floral organ identity in different plant species. Plant J 25:441–451
42. Litt A, Kramer EM (2010) The ABC model and the diversification of floral organ identity. Semin Cell Dev Biol 21:129–137
43. Bowman JL, Smyth DR, Meyerowitz EM (1989) Genes directing flower development in Arabidopsis. Plant Cell 1:37–52
44. Jack T, Brockman LL, Meyerowitz EM (1992) The homeotic gene APETALA3 of Arabidopsis thaliana encodes a MADS box and is expressed in petals and stamens. Cell 68:683–697
45. Goto K, Meyerowitz EM (1994) Function and regulation of the Arabidopsis floral homeotic gene PISTILLATA. Genes Dev 8:1548–1560
46. Jack T, Fox GL, Meyerowitz EM (1994) Arabidopsis homeotic gene APETALA3 ectopic expression: transcriptional and posttranscriptional regulation determine floral organ identity. Cell 76:703–716
47. Schwarz-Sommer Z, Hue I, Huijser P, Flor PJ, Hansen R, Tetens F, Lonnig WE, Saedler H, Sommer H (1992) Characterization of the Antirrhinum floral homeotic MADS-box gene deficiens: evidence for DNA binding and autoregulation of its persistent expression throughout flower development. EMBO J 11:251–263
48. Trobner W, Ramirez L, Motte P, Hue I, Huijser P, Lonnig WE, Saedler H, Sommer H, Schwarz-Sommer Z (1992) GLOBOSA: a homeotic gene which interacts with DEFICIENS in the control of Antirrhinum floral organogenesis. EMBO J 11:4693–4704
49. Kim S, Yoo MJ, Albert VA, Farris JS, Soltis PS, Soltis DE (2004) Phylogeny and diversification of B-function MADS-box genes in angiosperms: evolutionary and functional implications of a 260-million-year-old duplication. Am J Bot 91:2102–2118
50. Kramer E, Di Stilio VS, Schlulter PM (2003) Complex patterns of gene duplication in the APETALA3 and PISTILLATA lineages of the Ranunculaceae. Int J Plant Sci 164:1–11
51. Kramer EM, Dorit RL, Irish VF (1998) Molecular evolution of genes controlling petal and stamen development: duplication and divergence within the APETALA3 and PISTILLATA MADS-box gene lineages. Genetics 149:765–783
52. Kramer EM, Irish VF (2000) Evolution of the petal and stamen developmental programs: evidence from comparative studies of the lower eudicots and basal angiosperms. Int J Plant Sci 161:S29–S40
53. Stellari GM, Jaramillo MA, Kramer EM (2004) Evolution of the APETALA3 and PISTILLATA lineages of MADS-box-containing genes in the basal angiosperms. Mol Biol Evol 21:506–519
54. Viaene T, Vekemans D, Irish VF, Geeraerts A, Huysmans S, Janssens S, Smets E, Geuten K (2009) Pistillata--duplications as a mode for floral diversification in (basal) asterids. Mol Biol Evol 26:2627–2645
55. Cannon SB, Sterck L, Rombauts S, Sato S, Cheung F, Gouzy J, Wang X, Mudge J, Vasdewani J, Schiex T et al (2006) Legume genome evolution viewed through the Medicago truncatula and Lotus japonicus genomes. Proc Natl Acad Sci U S A 103:14959–14964
56. Cheng X, Wang M, Lee HK, Tadege M, Ratet P, Udvardi M, Mysore KS, Wen J (2014) An efficient reverse genetics platform in the model legume Medicago truncatula. New Phytol 201:1065–1076

57. Penmetsa RV, Cook DR (2000) Production and characterization of diverse developmental mutants of Medicago truncatula. Plant Physiol 123:1387–1398
58. Bradley D, Carpenter R, Sommer H, Hartley N, Coen E (1993) Complementary floral homeotic phenotypes result from opposite orientations of a transposon at the plena locus of Antirrhinum. Cell 72:85–95
59. Liljegren SJ, Ditta GS, Eshed Y, Savidge B, Bowman JL, Yanofsky MF (2000) SHATTERPROOF MADS-box genes control seed dispersal in Arabidopsis. Nature 404:766–770
60. Pinyopich A, Ditta GS, Savidge B, Liljegren SJ, Baumann E, Wisman E, Yanofsky MF (2003) Assessing the redundancy of MADS-box genes during carpel and ovule development. Nature 424:85–88
61. Davies B, Motte P, Keck E, Saedler H, Sommer H, Schwarz-Sommer Z (1999) PLENA and FARINELLI: redundancy and regulatory interactions between two Antirrhinum MADS-box factors controlling flower development. EMBO J 18:4023–4034
62. Becker A, Theissen G (2003) The major clades of MADS-box genes and their role in the development and evolution of flowering plants. Mol Phylogenet Evol 29:464–489
63. Causier B, Castillo R, Zhou J, Ingram R, Xue Y, Schwarz-Sommer Z, Davies B (2005) Evolution in action: following function in duplicated floral homeotic genes. Curr Biol 15:1508–1512
64. Irish VF (2003) The evolution of floral homeotic gene function. BioEssays 25:637–646
65. Kramer EM, Jaramillo MA, Di Stilio VS (2004) Patterns of gene duplication and functional evolution during the diversification of the AGAMOUS subfamily of MADS box genes in angiosperms. Genetics 166:1011–1023
66. Zahn LM, Leebens-Mack JH, Arrington JM, Hu Y, Landherr LL, dePamphilis CW, Becker A, Theissen G, Ma H (2006) Conservation and divergence in the AGAMOUS subfamily of MADS-box genes: evidence of independent sub- and neofunctionalization events. Evol Dev 8:30–45
67. Fourquin C, del Cerro C, Victoria FC, Vialette-Guiraud A, de Oliveira AC, Ferrandiz C (2013) A change in SHATTERPROOF protein lies at the origin of a fruit morphological novelty and a new strategy for seed dispersal in Medicago genus. Plant Physiol 162:907–917
68. Birchler JA, Veitia RA (2007) The gene balance hypothesis: from classical genetics to modern genomics. Plant Cell 19:395–402
69. Edger PP, Pires JC (2009) Gene and genome duplications: the impact of dosage-sensitivity on the fate of nuclear genes. Chromosom Res 17:699–717
70. Freeling M, Thomas BC (2006) Gene-balanced duplications, like tetraploidy, provide predictable drive to increase morphological complexity. Genome Res 16:805–814
71. Polhill RM (1994) Classification of the Leguminosae and complete synopsis of legume genera. In: Bisby FA, Buckingham J, Harborne JB (eds) Phytochemical dictionary of the Leguminosae, vol 1. Chapman & Hall, London
72. Lewis GP, Schrire B, Mackinder B, Lock M (2005) Legumes of the world. Kew Publishing, Richmond, UK
73. Small E, Jomphe M (1989) A synopsis of the genus Medicago (Leguminosae). Can J Bot 67:3260–3294
74. Small E, Brookes B (1984) Coiling of alfalfa pods in relation to resistance against seed chalcids: additional observations. Canad J Plant Sci 64:659–665
75. Ballester P, Ferrándiz C (2017) Shattering fruits: variations in a dehiscent theme. Curr Opin Plant Biol 35:68–75
76. Liljegren SJ, Roeder AHK, Kempin SA, Gremski K, Østergaard L, Guimil S, Reyes DK, Yanofsky MF (2004) Control of fruit patterning in Arabidopsis by *INDEHISCENT*. Cell 116:843–853
77. Rajani S, Sundaresan V (2001) The Arabidopsis myc/bHLH gene *ALCATRAZ* enables cell separation in fruit dehiscence. Curr Biol 11:1914–1922
78. Ferrandiz C, Liljegren SJ, Yanofsky MF (2000) Negative regulation of the *SHATTERPROOF* genes by *FRUITFULL* during Arabidopsis fruit development. Science 289:436–438
79. Ferrandiz C, Fourquin C (2014) Role of the *FUL-SHP* network in the evolution of fruit morphology and function. J Exp Bot 65:4505–4513

Chapter 19

Toward Unravelling the Genetic Determinism of the Acquisition of Salt and Osmotic Stress Tolerance Through In Vitro Selection in *Medicago truncatula*

Adel M. Elmaghrabi, Hilary J. Rogers, Dennis Francis, and Sergio Ochatt

Abstract

Changes in global climate and the nonstop increase in demographic pressure have provoked a stronger demand for agronomic resources at a time where land suitable for agriculture is becoming a rare commodity. They have also generated a number of abiotic stresses which exacerbate effects of diseases and pests and result in physiological and metabolic disorders that ultimately impact on yield when and where it is most needed. Therefore, a major scientific and agronomic challenge today is that of understanding and countering the impact of stress on yield. In this respect, in vitro biotechnology would be an efficient and feasible breeding alternative, particularly now that the genetic and genomic tools needed to unravel the mechanisms underlying the acquisition of tolerance to stress have become available. Legumes in general play a central role in a sustainable agriculture due to their capacity to symbiotically fix the atmospheric nitrogen, thereby reducing the need for fertilizers. They also produce grains that are rich in protein and thus are important as food and feed. However, they also suffer from abiotic stresses in general and osmotic stress and salinity in particular. This chapter provides a detailed overview of the methods employed for in vitro selection in the model legume *Medicago truncatula* for the generation of novel germplasm capable of resisting NaCl- and PEG-induced osmotic stress. We also address the understanding of the genetic determinism in the acquisition of stress resistance, which differs between NaCl and PEG. Thus, the expression of genes linked to growth (*WEE1*), in vitro embryogenesis (*SERK*), salt tolerance (*SOS1*) proline synthesis (*P5CS*), and ploidy level and cell cycle (*CCS52* and *WEE1*) was upregulated under NaCl stress, while under PEG treatment the expression of *MtWEE1* and *MtCCS52* was significantly increased, but no significant differences were observed in the expression of genes *MtSERK1* and *MtP5CS*, and *MtSOS*1 was down-regulated. A number of morphological and physiological traits relevant to the acquisition of stress resistance were also assessed, and methods used to do so are also detailed.

Key words Callus, Cell cycle checkpoint, In vitro selection, *Medicago truncatula*, NaCl, PEG6000, Proline, Stress resistance, Endoreduplication

1 Introduction

Given its agroecological relevance [1], *Medicago truncatula* has been the object of many biotechnology studies [2–7], including on the generation of novel germplasm capable of tolerating salinity

Luis A. Cañas and José Pío Beltrán (eds.), *Functional genomics in* Medicago truncatula*: Methods and Protocols*, Methods in Molecular Biology, vol. 1822, https://doi.org/10.1007/978-1-4939-8633-0_19,

following in vitro selection with a step-up recurrent strategy [8] and of osmotic stress tolerance after selection on PEG 6000 [9].

M. truncatula grows in a wide range of habitats with varying environmental stress notably soil salinity and drought [10, 11], to which it is more tolerant than other legume crops such as pea, bean, and soybean [12–15]. Moreover, it includes a large diversity of ecotypes useful for exploring the molecular bases of environmental adaptation mechanisms. In addition, genetic tools are available to identify genes that could improve abiotic stress tolerance offering a range of approaches for exploiting such variability [16]. In spite of this, previous studies particularly on water stress resistance in *M. truncatula* mostly concerned gene transfer [5, 7, 17], while assessment of the physiological responses [18] and their genetic mechanisms [10] is more limited.

2 When Salinity Is the Problem

Salinity is typically caused by sodium chloride accumulated in the soil top layers [19, 20] where the electrical conductivity (EC) is 1.30 mM/cm^{-1} or more at 40 mM NaCl, and this is a salt level that dramatically decreases productivity of many crops [21–24].

Salt-tolerant ecotypes exist in different species that have developed mechanisms to control salt accumulation (e.g., exclusion, vacuolar sequestration, or control of sodium transport) by selecting other ions normally present in the soil at lower levels, such as potassium (K^+) and nitrate (NO_3^-) but compatible with plant growth [25]. Such mechanisms are not easily switched on by the plant, and high NaCl levels ($\geq$40 mM) may generate an imbalance of cellular ions provoking toxicity and/or osmotic stress. This results in increased reactive oxygen species coupled with reduced photosynthesis [26], which can in time result in whole plant death [27–29].

A typical feature of the acquisition of physiological tolerance to salt and osmotic stress is an increased endogenous content of osmolytes including soluble sugars, proline, and glycinebetaine [30, 31], which play an adaptive role in mediating osmotic adjustment to protect subcellular structures via the exclusion or vacuolar sequestration of Na^+ ions [20]. In many plants, a positive correlation has been observed between the accumulation of glycinebetaine and proline and stress tolerance [20, 32, 33]. Interestingly, using *M. truncatula* nodules, Aydi et al. [34] also found that salt stress tolerance was correlated to nodule osmotic adjustment due to Na^+ sequestration and an accumulation of soluble sugars and amino acids and coupled with an adequate nitrogen metabolism due to a high glutamine synthetase activity. Plant cells accumulate osmolytes including water soluble carbohydrates and proline to combat water loss in response to salt stress [35]. The increase in proline is driven by induction of the P5CS gene that encodes

Δ1-pyrroline-5-carboxylate synthetase, involved in proline biosynthesis [36]. P5CS is highly expressed in salt- and drought-tolerant plant species [37, 38] and also after in vitro selection of tissues on media with high NaCl contents [8, 39].

Phenotypic changes that can enable salt tolerance also include a modification of plasma membrane permeability through a differential lipid content whereby NaCl uptake is controlled, proteins stabilized, and cell antioxidant capacity enhanced [35, 40, 41]. In this context, plant ecotypes able to tolerate 100 mM NaCl or more can accumulate NaCl in vacuoles or exclude Na^+ [20, 42]. The *Salt Overly Sensitive* (*SOS*) gene family is important in maintaining ion homeostasis [43, 44], and conferring salt tolerance as recently reported in *Arabidopsis thaliana*, where *sos1*, *sos2*, and *sos3* mutants were hypersensitive to 60 or 120 mM NaCl with root growth, strongly suppressed compared to the wild type [45]. This was mainly due to the fact that *SOS1* encodes a membrane bound antiporter that is regulated by the *SOS2/3* kinase complex. The effect of overexpression of *P5CS* and *SOS1* genes on inducing tolerance seems however to be species-specific and, within the plant, related to a particular stage of development.

A recent transcriptional profiling of roots of *M. truncatula* under 180 mM NaCl using microarrays identified transcription factor *MtCBF4*, a member of the *AP2-EREBP* transcription factor family, as playing an important role between primary and secondary metabolism pathways in response to salt stress and may thus be a good candidate gene to obtain salt-tolerant plants [46].

3 When the Problem Is Water Shortage or Osmotic Stress

Osmotic stress, frequently equated to water deficit or drought, is defined as an absence of the adequate moisture needed for a plant to grow and complete its life cycle normally [47]. The three main mechanisms that enable plants to withstand osmotic stress are escape, avoidance, and tolerance [48–50], but water stress may also alter plant physiology, whereby plants acclimate to withstand drier conditions by mitigating osmotic stress through the production of osmolytes such as proline and soluble sugars to protect cells against osmotic perturbation [8, 37, 51–53]. On the other hand, water potential induces morphological variation in stressed tissues at the cellular level and particularly in terms of cell shape and size [54, 55] as recently shown in osmotic stress-resistant cells of *M. truncatula* [9].

Plants respond to abiotic stress by altering the expression of scores of genes resulting in many cellular and physiological modifications [5, 56, 57], but information about the effect on cell cycle genes for plants exposed to high NaCl or to osmotic stress remains more limited [8, 9, 58–61].

In *Arabidopsis*, *WEE1* which is a negative regulator of mitosis [62], is strongly expressed in response to abiotic stress [58], but not in the context of salinity or drought, while we recently showed that in *M. truncatula* the expression of *WEE1* is dramatically increased both by NaCl [8] and osmotic stress [9]. West et al. [63] had shown that the effects of salt stress on growth are mediated by effects on cell division, and, more recently, Skirycz et al. [64] showed that endoreduplication is also associated with osmotic stress in *Arabidopsis*. Taken together, there is increasing evidence for a central role of the *WEE1* gene in association with endoreduplication [65–67], as well as its strong expression when the DNA replication and DNA damage checkpoints are invoked [58], and in response to abiotic stresses such as salinity and osmolarity [8, 9]. Likewise, another key marker for endoreduplication is *CCS52* (cell cycle switch) [68], whose expression in *Medicago sativa* was enhanced in differentiating cells undergoing endoreduplication [69]. In this context, the studies of Elmaghrabi et al. on in vitro selection for salt [8] and osmotic stress tolerance [9] showed that the acquisition of tolerance was coupled with the onset of endoreduplication and a concomitant increased expression level of *CCS52* too.

4 Contribution of In Vitro Biotechnology Tools to Selection for NaCl and Osmotic Stress Tolerance

In vitro culture is a very powerful tool for the study of plant development. Given the potential regeneration of plants from cells, tissues, or organs cultured under suitable media, it can therefore be used also for in vitro selection to improve abiotic stress tolerance. From a fundamental perspective, in vitro selection can also be used to unravel the mechanism(s) underlying the acquisition of stress tolerance coupled to other tools as described above. The first in vitro plant regeneration strategy for *M. truncatula* was based on indirect somatic embryogenesis from callus tissues [70] and was thereafter followed by many protocols inducing indirect somatic embryogenesis from a range of explants including leaves, hypocotyls, cotyledons, petioles, and flower parts [71–74]. Direct somatic embryogenesis in *M. truncatula* has been achieved in both liquid and solid media [75–79].

It is well known today that during in vitro culture, some tissues may exhibit somaclonal variation (a random, nondirected, non-controlled variation of regenerating cells and/or callus sectors; [80]), even when the objective may have been the production of true-to-type regenerated plants. This apparent defect can be turned into an advantage by exploiting it in terms of in vitro selection by culturing tissues on media including abiotic stress-inducing factors. This can in turn be used for fundamental studies to probe mechanisms induced by abiotic stresses or for applied research in providing new salt- and drought-tolerant germplasm, provided the regeneration capacity is maintained. Somaclonal variation is characterized by changes in ploidy level and DNA sequence, coupled with movement of transposable elements to other chromosomal sites and also including epigenetic changes, such as DNA methylation and histone modification [81]. Abiotic factors (drought, salinity, extreme temperatures) can induce both somaclonal and epigenetic changes when applied over long culture periods as used for in vitro selection protocols [82–84].

Tissues and plants resistant to salinity have been obtained after in vitro culture on selection media for many years in species as wide apart as cherry [85, 86], potato [87], or *Zoysia matrella* L. [88] among many others, but the fundamental mechanisms underlying the acquisition of salinity tolerance often remained partially unexplained. This was likely due to the fact that the extent and range of DNA instability in vitro depend on many abiotic factors, among which the age of cell cultures, the genotype studied, the type of explants, the hormonal composition of the medium, and the type of exogenous abiotic factor added to induce (physical and/or chemical) stress. However, literature on in vitro selection for salt resistance has addressed the cellular mechanisms involved in salt tolerance [19, 85, 89, 90] and also the genetic determinism of acquisition of resistance to salinity [8, 38, 44, 62, 63] by using selected NaCl-tolerant cell lines as study systems. Several reports also examined alternative ways to exploit stress in vitro to characterize the biology and genetic diversity of the early stages of plantlet growth [33, 87, 91–97], and the effects of salinity stress on plant morphology have also been extensively studied [26, 31, 33, 98, 99]. However, for example, cellular studies mainly focused on the effects of abiotic stresses on cell size [100] until recent studies on the effects on cell morphometry of NaCl and osmotic stress in resistant cells of *M. truncatula* [8, 9]. In this context and as a direct consequence of this lack of detailed analysis, some of the apparently resistant tissues obtained were only acclimated [101] followed by physiological adaptation to the stress induced, but not truly resistant. Thus the trait would not be heritable unless a recurrent selection strategy, encompassing back and forth passages from a stress to a nonstress condition, was applied [102]. Such a strategy results in a progressive enrichment of the selected cell population in stress-resistant cells facilitated by a reprogramming of gene expression pattern responses to the stress, e.g., receptor kinases [103, 104], transcription factors [105], enzymes, and structural proteins with a central physiological and metabolic role under stress [80]. Such recurrent selection strategies can be direct (where the maximum resistance level aimed for is added to the culture medium from the start [102]), or indirect where tissues are successively transferred to increasing concentrations of the stress-inducing factor until the level of resistance sought is reached [8]. As mentioned above, exposure to increasing NaCl levels through acclimation may result in long-term tolerance of cells/tissues to salinity [88, 106, 107] and water stress [108]. Increases in both embryogenic and organogenic calli have also been achieved by acclimation [87, 89, 109, 110], even if best conditions for in vitro embryogenesis are highly species specific in terms of both medium and explant [87, 111]. Therefore, a useful marker for embryogenic callus is the *SERK* (*SOMATIC EMBRYOGENESIS RECEPTOR KINASE*) gene [103, 112], a key gene required for somatic embryogenesis and whose expression in rice was upregulated in embryogenic calli [113]. *SERK* was also recently found to be upregulated following selection for NaCl tolerance in vitro in *M. truncatula* [8].

Likewise, water stress-tolerant cell lines have been selected in vitro, and resistant plants were recovered in crops like alfalfa, tomato, soybean, and wheat [114–116]. Most frequently, the stress agent used was PEG 6000 [85, 115, 117, 118] whose adsorbent features imposes on cells and tissues the same or comparable effects to those obtained by drying soil at the same φw but generally without any other associated detrimental effects [119]. Note however that PEG 6000 was also reported to induce DNA damage on in vitro plantlets [120] and seeds [121] of *M. truncatula*. Thus, PEG 6000 imitates soil water shortage [122] and also increases total soluble sugars [9, 123], while stimulating somatic embryogenesis in vitro [124, 125], including in *M. truncatula* [9].

5 Protocols for In Vitro Plant Regeneration in *M. truncatula*

In *M. truncatula*, the most frequently used and most successful medium components for the induction and maintenance of somatic embryogenesis are various combinations of auxins such as 2,4-D (2,4-D, dichlorophenoxyacetic acid), NAA (α-naphthaleneacetic acid), or picloram with either BAP (6-benzylaminopurine) or kinetin as cytokinins [70, 73, 126, 127] as summarized in Table 1.

The synthetic auxin, 2,4-D is known to generate callus for indirect embryogenesis in legumes and particularly in Medicago species [73, 134, 135], with a clear dose-response mechanism in action whereby the optimum concentration for somatic embryo induction is highly species- and genotype-specific [2, 136]. As far as NAA is concerned, its importance is mostly for the induction of callus formation rather than for the differentiation of somatic embryos, although it also proved successful in that respect with both *M. truncatula* [70] and *M. polymorpha* [137]. Production of callus tissue followed by embryo formation has been achieved in *M. truncatula* on solid medium supplemented with the cytokinin, BAP (6-benzylaminopurine) [73]. However, induction of somatic embryogenesis by a single cytokinin is relatively rare with legumes, and sequences of different cytokinins (e.g., BAP, kinetin, thidiazuron) are often required [111, 138]. The molecular processes

Table 1
Culture media used to induce somatic embryogenesis in *Medicago truncatula*

Medium code	Basal medium	Supplements[a]	References
B5E1	B5 [128] + 3% sucrose + 7 g/L agar	1 mg/L 2,4-D + 0.2 mg/L kin + 1 mg/L adenine + 500 mg/L CEH + 500 mg/L myoinositol	[129]
EID	MS [128] + 3% sucrose + 7 g/L agar	1 mg/L 2,4-D + 0.2 mg/L BAP + 1 g/L CEH + 100 mg/L myoinositol	[129]
B5,4	B5 [130] + 3% sucrose + 7 g/L agar	4 mg/L 2,4-D + 0.2 mg/L kin + 1 mg/L adenine + 500 mg/L CEH	[76]
MS4	MS [128] + 3% sucrose + 7 g/L agar	4 mg/L 2,4-D + 0.2 mg/L BAP + 100 mg/L myoinositol	[77]
PCI-4	SH [131] + 3% sucrose + 7 g/L agar	4 mg/L 2,4-D + 0.2 mg/L BAP	[72]
R2B5	B5 [130] + 3% sucrose + 7 g/L agar	1 mg/L IAA + 5 mg/L kin+1 g/L CEH	[91]
CIM-D	MS [128] + 3% sucrose + 8 g/L agar	1 mg/L 2,4-D + 2 mg/l Z + 100 mg/L myoinositol	[132]
CIM-C	UM (Uchimiya and Murashige; 133) + 3% sucrose + 8 g/L agar	1 mg/L 2,4-D + 2 mg/L Z + 100 mg/L myoinositol + 0.2 mg/L bacto-tryptone	[71]
MANA	MS [128] + 3% sucrose + 7 g/L agar	2 mg/L NAA + 0.5 mg/L BAP + 100 mg/L myoinositol	[8, 9, 111]
MPIC	MS [128] + 3% sucrose + 7 g/L agar	0.2 mg/L picloram + 0.52 mg/L kin + 100 mg/L myoinositol	[8, 9, 111]

[a]*2,4-D* 2,4-dichlorphenoxyacetic acid, *BAP* 6-benzylaminopurine, *CEH* casein enzymatic hydrolysate, *IAA* 3-indoleacetic acid, *Kin* kinetin (6-furfurylaminopurine), *NAA* alpha-naphthalene acetic acid, *Pic* picloram (4-amino-3,5,6-trichloro-2-pyridinecarboxylic acid)

activated during the induction and development of embryogenesis are still poorly understood. However, recently, the *SERK* gene from *M. truncatula* (*MtSERK1*) has been cloned and its expression evaluated on defined culture media [74, 104]. Its expression is strongly linked to somatic embryogenesis.

Lowering or a complete removal of auxin is a key for embryo development and maturation beyond the globular stage in *M. truncatula* [73, 75–79]. In addition, moderate reductions in cytokinin levels are sometimes also useful for a normal progress of embryogenesis in vitro [71, 76]. Moreover, a total removal of phytohormones from the culture medium is often the only way to enable somatic embryos to convert into a robust plantlet [2, 79], and the stage of embryo maturation at which auxins are removed is also critical for its subsequent development [71, 79]. Last but not least, in vitro cultures in general, and those of recalcitrant species like legumes in particular, tend to lose regeneration competence as they age [2, 88, 111], and this may have dramatic effects on the efficiency of long-term in vitro selection studies on callus and cell suspension cultures, as the subsequent regeneration of plants from the tolerant tissues might be impaired.

6 Protocols for In Vitro Selection for Abiotic Stress in *M. truncatula*

6.1 Callus and Cell Suspension Culture Initiation and Maintenance

In two recent articles, Elmaghrabi et al. reported protocols useful for the in vitro selection of NaCl-induced salinity [8] and PEG 6000-induced osmotic stress [9] with *M. truncatula*, and the procedures employed are described and discussed below.

For both protocols, a recurrent selection strategy was adopted, where *M. truncatula* cv. Jemalong line (A17) ($2n = 2x = 16$, 1C value = 0.48 pg) was used as the source of material.

As in any in vitro selection study, the first step was the optimization of the composition of culture conditions required for the induction of callus tissues and also for the initiation and maintenance of cell suspension cultures from such callus. This entailed a number of preliminary assessments to identify the best original explant source, basal salts mixture, and hormonal composition of the culture medium and light and temperature regimes. During such tests, hypocotyls, cotyledons, and green leaves from aseptically germinated seedlings were compared as explant sources, all being wounded with a scalpel in order to ensure contact of the explant tissue with the culture medium, and they were cultured on four different media all based on Murashige and Skoog [128] salts formula, with supplements as detailed in Table 2.

Table 2
The four different media used to optimize the ideal conditions of in vitro growth of callus derived from explants of *Medicago truncatula*

	Growth regulators and other supplements (mg L^{-1})		
Medium codes[a]	Auxin[b]	Cytokinin[c]	Other[d]
MANA	NAA (2.0)	BAP (0.5)	Nil
EID	2,4-D (1.0)	BAP (0.2)	CEH (1000)
MS4	2,4-D (4.0)	BAP (0.2)	Nil
MPIC	Pic (0.2)	Kin (0.5)	Nil

See also Fig. 1

[a]MANA and MPIC media, as in Ochatt et al. [111], *EID* Embryo Induction Development, modified from Iantcheva et al. [129]. *See* also Table 1

[b]*NAA* alpha-naphthalene acetic acid, *2,4-D* 2,4-dichlorphenoxyacetic acid, *Pic* picloram (4-amino-3,5,6-trichloro-2-pyridinecarboxylic acid). *See* also Table 1

[c]*BAP* 6-benzylaminopurine, *Kin* kinetin (6-furfurylaminopurine). *See* also Table 1

[d]*CEH* casein enzymatic hydrolysate. *See* also Table 1

For culture, multi-well 5 × 5 dishes with 2 mL of medium per well were used, providing 25 replicates for each treatment (medium and explant), and all experiments were repeated twice. Dishes were kept at 24/22 °C, in the dark or under a 16/8 h (light/dark) photoperiodic regime of 90 μmol m^{-2} s^{-1} from warm white fluorescent tubes for initiation of callus proliferation. After 4 weeks, explants were subcultured on the four different media and the frequency of callus initiation assessed as reported elsewhere [127]. The optimum medium was MANA, and the most responsive explants were leaves, where mature callus cultures were established within 5–6 months with monthly subculturing.

6.2 Full Experimental Protocol for In Vitro Acclimation, Recurrent Selection, Growth Assessments, and Characterization of Stress-Resistant Material Produced

During establishment and once no remaining material from the initial explants was left within the culture tissues, the callus was observed by light microscopy for signs of embryogenesis and organogenesis. Then, calli were transferred to Petri dishes (10 cm diameter × 2 cm) with 25 mL MANA medium supplemented with or without 50 mM NaCl. One month into the 50 mM treatment, embryogenic callus only was stepped up to 100 mM NaCl for 1 month, when half of such calli were kept at 100 mM NaCl and the other half was stepped up to 150 mM NaCl. Likewise, 1 month into this treatment, calli were split into two halves where one was subcultured onto fresh medium with 150 mM NaCl and the other half stepped up onto either 250 or 350 mM NaCl for 3 months (Fig. 1). Thus, by 5 months from the start of the acclimation strategy, calli were available that had remained continuously on 50 mM NaCl (6 months), 100 mM NaCl (5 months), 150 mM NaCl (4 months), and 250 or 350 mM NaCl (3 months), in addition to the controls on 0 mM NaCl (10 months). This first phase of acclimation was termed Phase I. At this point in time, the first phase of indirect recurrent selection (Phase II) was initiated by transferring all calli onto MANA medium minus NaCl for 2 months and then back to the same Phase I NaCl concentration they came from. Phase III (second cycle of recurrent selection) was a repeat of Phase II, and by the end of Phase III, calli were 24 months old.

In parallel to these treatments, 5-month-old control calli were also tested for their growth potential on medium containing PEG 6000 (1%, w/v), and they were submitted to the same strategy only that there were no step-up stages, thereby constituting a direct recurrent selection methodology when it came to the induction of tolerance to osmotic stress.

For the quantification of growth, data were first examined for linearity following log transformation of y axis coordinates, and exponentially growing calli were then measured in terms of absolute (g $month^{-1}$) and relative growth rates (RGR g g^{-1} day^{-1}), as follows:

For the relative growth rate calculation the formula used was

$$\mathrm{RGR} = \frac{lnB - \ln A}{t}$$

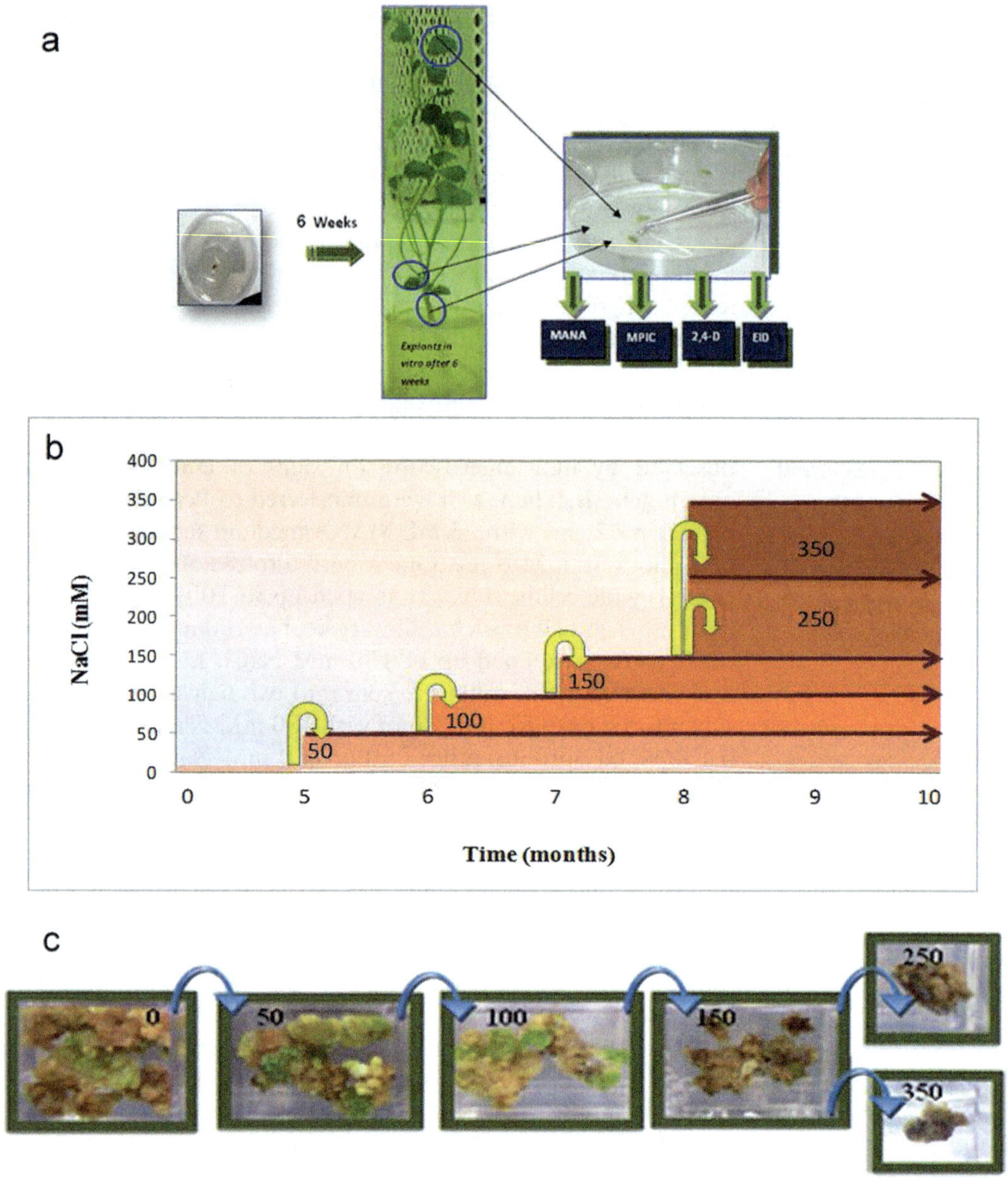

Fig. 1 Full in vitro culture protocol from callus initiation to selection for NaCl resistance. (**a**) Leaf explants from in vitro germinated plantlets are grown on various media for callus initiation, which is best on MANA medium where they are kept by either continuous culture on 0 mM NaCl (control treatment) or after 5 months entered in the acclimation protocol. (**b**) Acclimation protocol (until month 10 of culture, Phase I) begins by subculturing calli onto 50 mM NaCl for 1 month, and thereafter, monthly, calli are also stepped up to 100 mM NaCl, then to 150 mM NaCl, and finally to 250 mM or 350 mM NaCl where they were kept for 4, 3, and 2 months, respectively. This marks completion of Phase I, which is followed by recurrent selection consisting of two back and forth passages between stress-free and stress-inducing medium, repeated twice, i.e., Phases II and III (the same protocol is applied for PEG6000 only that direct recurrent selection is applied). (**c**) Callus morphology by the end of Phase I as calli are acclimated to increasing NaCl concentrations from 0 to 350 mM NaCl

where *A* and *B* are the sampling times within the exponential phase of growth for each treatment and t is the time interval between successive samplings. Thereafter, when results conformed to linear growth, they were analyzed by regression and growth (g month^{-1}) determined according to the gradient of the straight line (y = mx + c), and finally statistically significant regressions were then used to calculate growth (g m^{-1}).

When growth was neither exponential nor linear, absolute callus growth per month was estimated as

$$\mathrm{GR} = \frac{\mathrm{FW2} - \mathrm{FW1}}{t}$$

where FW1 and FW2 are the fresh weight of callus at the beginning and end of treatments, respectively, and t spans this time interval.

6.3 Assessment Morphogenetic Potential

Once Phase III above was completed, stress-tolerant callus tissues were subcultured onto an embryo development medium (EDM), derived from that reported by Iantcheva et al. [129] and supplemented with the same NaCl concentration as used during Phase III, for 4–6 weeks. Then, such calli were transferred for a further 2 months onto an embryo conversion and rooting medium (ECR) [129] supplemented again with the same NaCl concentration as in the previous passage (Fig. 2). At this stage, callus tissues were examined under a binocular loupe, and the extent of embryo and/or organ formation at each concentration of NaCl was recorded.

6.4 Endogenous K^+ and Na^+ Measurements

For the determination of the sodium accumulation in the callus, 100 mg fresh weight pieces were collected from three independent replicates following Phases I, II, and III, i.e., with calli that were 11, 17, and 23 months old, respectively. The dry weight of each replicated fresh callus piece was determined by incubation in a glass beaker for 24 h in an oven at 80 °C. Then, 10 mL of 15.8 M HNO_3 (Fisher-MOS) was added to each beaker, and beakers were placed on a sand bath at 240 °C in a fume cupboard for 2–3 h until the digests started to clear (more acid should be added as necessary to prevent the digest from drying up), and once samples had cooled (15 min) at room temperature, they were filtered and made up to 25 mL with ultrapure water. Na^+ accumulation for each treatment was measured using an atomic absorption spectrophotometer (Varian Spectr AA-100, version 2.00 software) and [Na^+] calculated as

$$\frac{\mu g\ \text{ion}}{\text{g sample material}} = \frac{\text{ppm value}\left(\dfrac{\mu g}{\text{cm}^3}\right) \times \text{volume dilution factor}}{\text{dry weight}}$$

The same methodology was used to measure K^+ accumulation in 23-month-old calli.

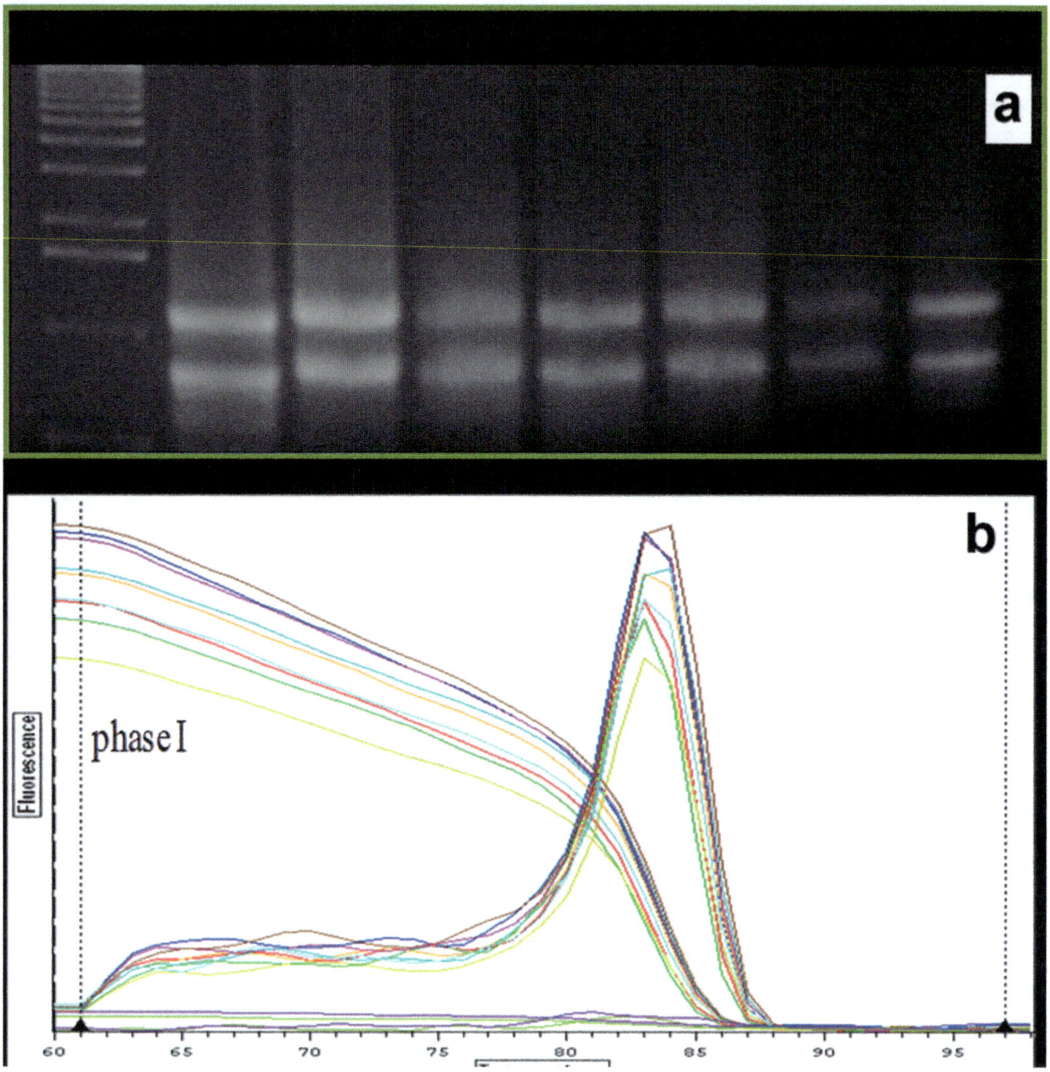

Fig. 2 Quality checks for RNA quality and primer specificity. (*a*) Gel electrophoresis of RNA extracted from leaf and callus from *Medicago truncatula* grown with different levels of abiotic stress. Lanes: 1, 1 kb ladder; 2, leaf (0 mM NaCl); 3, callus (0 mM NaCl); 4, callus (50 mM NaCl); 5, callus (100 mM NaCl); 6, callus (150 mM NaCl); 7, callus (250 mM NaCl); 8, callus (10 v/v PEG) showing sharp bands for both major rRNAs. (*b*) Melting curve of Mt18S gene as control, following Phase I of *M. truncatula* calli on NaCL stress treatments (*see* Fig. 1 for further details) showing a sharp single peak indicating that there are no primer dimers or other non-specific products in the real-time PCR reaction

6.5 Proline Measurement After Selection for NaCl and Osmotic Stress Tolerance

For all experiments on in vitro selection for salt and osmotic stress tolerance with *M. truncatula*, we evaluated the proline content [8, 9] following methods reported by Troll and Lindsley [139] and Boukel and Houassine [140]. Briefly, 100 mg of callus tissue per sample per treatment was examined over the concentration range 0, 50, 100, 150, 250, and 350 mM NaCl, repeated three times. Samples were collected in 30 mL tubes, and 2 mL of 40% ethanol was added per tube prior to heating in a water bath at 85 ° C for 60 min. After cooling samples (2 h, room temperature), 1 mL extract per sample was transferred to a tube with 1 mL acetic acid and 1 mL ninhydrin reagent (120 mL distilled water +30 mL acetic acid +80 mL orthophosphoric acid (density 1.7) + 25 mg ninhydrin). This mixture (sample and reagent) was then boiled (30 min), and after cooling at room temperature, 5 mL benzene was added. At this time, an upper layer yellow-colored band indicated the presence of proline [140], whereby determination of the sample optical densities was performed using a spectrophotometer (*Unicam*, UV-Vis spectrometry) at a wavelength of 528 nm. A standard proline solution was prepared from a stock solution (125 mg proline in 100 mL 0.3 M HCl) that was diluted with 0.3 M HCl to obtain concentrations of 0.1, 0.2, and 0.4 mg mL^{-1} for the calibration used to determine the proline concentration of different samples.

6.6 Water Soluble Carbohydrate Measurements

Water soluble sugars were measured with the anthrone method of Plummer [141] by extraction from three 100 mg callus samples from each NaCl treatment (0, 50, 100, 150, 250, and 350 mM NaCl) that were immersed for 48 h in 3 mL of 80% ethanol and then were heated in a water bath at 70 °C (30 min). Then, 2 mL were taken from the extract solution from each sample and were diluted 10 times with 80% ethanol plus 4 mL of anthrone (0.2 g dissolved in 100 mL of 18 M H_2SO_4, prepared at least 4 h in advance). Tubes were shaken, placed in a water bath at 92 °C for 8 min, and finally cooled (30 min on ice in the dark). The water soluble sugar content was read spectrophotometrically (*Unicam*, UV-Vis spectrometry) at a 585 nm wavelength, and the data were converted to mg L^{-1} using preestablished calibrations.

6.7 Osmolarity Measurements

All measurements of osmolarity were carried out using a Wescor (model VAPRO 5520, USA) vapor pressure micro-osmometer on a minimum of three 10-μL samples for each treatment, and the intracellular osmolarity and that of MANA medium were also evaluated at the end of step-up acclimation, i.e., Phase I.

For assessments at the intracellular level, 2 g fresh weight of callus were collected from all salt stress treatments of 0, 50, 100, 150, 250, and 350 mM NaCl after callus terminated Phase I and also for callus tissues selected for resistance to PEG-induced osmotic stress. They were crushed in an Eppendorf tube with a pestle and then centrifuged (1000 g, 10 min, 4 °C), and, finally, the supernatant was employed for measurements of osmolarity after calibrating the micro-osmometer. For calibration, three standard solutions are used: 100 ± 0.3 mM/Kg (equivalent to osmosed deionized water osmolarity), 290 ± 0.5 mM/Kg (equivalent to milliQ water osmolarity), and 1000 ± 1.4 mM/Kg (equivalent to sea water osmolarity) [54].

Calibration is a delicate and time-consuming operation, as a minimum of five measurements of each standard solution are performed in order to determine precisely the range of technical error to be found in the subsequent measurements. Thus, the variations from the standard solution data evoked above are those corresponding to the actual measurements performed at the time of experiments, but they are not necessarily identical to figures that may be obtained every time the experiments are repeated. In general, an admissible margin of error would be of ≤1.5 mM/Kg against the 100 mM standard, ≤3.5 mM/Kg against the 290 mM standard, and up to 6 mM/Kg against the 1000 mM standard.

Results, are the mean ± S.E of measurements performed at least three times and are expressed in terms of mM/kg and, for all parameters assessed, should be analyzed statistically ($P \geq 0.05$).

6.8 Determination of the Water Content of Callus Tissues

For determinations of the water (according to 142), about 0.3 g fresh weight of 11-month-old calli, at the end of Phase I, was gathered in three replicates from each NaCl treatment: 0, 50, 100, 150, 250, and 350 mM. The fresh weight (FW) was determined, and they were oven-dried (70 °C, for 72 h) to determine the dry weight (DW). Then, the water content (WC) was simply calculated according to the formula: WC(%) = FW−DW/FW × 100 [143].

6.9 Determination of Mitotic Index, Cell Viability, and Cell Morphology

C value of calli was determined following in vitro selection for several months as compared to leaf tissues from the original plants in order to confirm their cytogenetic stability. Thus, nuclei were isolated from abiotic stress-resistant calli and also from leaves of greenhouse-grown *M. truncatula* A17 plants, by chopping tissues in 400 mL of nuclear extraction buffer to which 1.6 mL of staining buffer containing 4, 6 diamidino-2-phenylindole (DAPI; Sysmex), an A-T binding specific fluorochrome, were added [144]. The resulting nucleus suspension was filtered through a 20 mm nylon mesh, and DNA contents of the isolated nuclei were analyzed using a Partec PAS-II flow cytometer equipped with an HBO-100 W mercury lamp and a dichroic mirror (TK420). Ten replicated calli for each treatment were analyzed, with a minimum of 3000 to 10,000 nuclei per run. The mitotic index was calculated according to the formula: MI D 4 × 4C/6 2C C 4C, where 2C and 4C correspond to the mean integrated value of nuclei in G1 phase and G2, respectively [144].

Cell viability was estimated by dual staining with fluorescein diacetate (FDA; Sigma, Poole, UK) and propidium iodide (PI; Sigma, Poole, UK). Cell suspensions (75 mL) from each treatment were mixed with 75 mL of dual staining solution containing FDA (200 mg mL^{-1}; 145) and propidium iodide (PI at 120 mg mL^{-1}) on ice and incubated for 20 min. FDA is cleaved by esterases in the cytoplasm into acetate and fluorescein which is hydrophilic and accumulates in the cytoplasm of metabolically active cells.

Thus, after excitation under UV light, live cells fluoresce yellow-green, while dead cells appear red using a fluorescent microscope. A minimum of 300 cells are counted, and results are expressed as the percentage of fluorescing cells referred to the total number of cells in the field.

Following application of a long-term in vitro selection strategy, the cell morphology should also be characterized [1]. For this, FDA stained slides of control and stress-treated cells are observed using a microscope under UV light. The surface area of cells and nuclei is determined after variable times of culture (generally 2, 4, and 6 months), using image acquisition programs such as ArchimedPlus and Histolab (Microvision, France) as reported by Ochatt et al. [54] or with ImageJ software, and it is also advisable to apply a shape coefficient [55] once selected cells are stabilized under the stress factors tested, i.e., at about 6 months of culture. Briefly, this shape coefficient (SC) is calculated based on the half length of the cell along its longest (a) and shortest (b) axes, as:

$$SC = \frac{\sqrt{a^2 - b^2}}{a}$$

For each treatment, nucleus and cell size should be measured on 10 to 20 cells, and results are expressed as the mean ± SE. The SC distinguishes round from elongated shapes, since SC values close to 1.0 correspond to elongated cells while SC values close to 0.5 correspond to rounder cell shapes [55].

For vacuolar measurement, calli having completed Phase III (23 months ± acclimation) on 0, 50, 100, 150, 250, and 350 mM NaCl were used. Vacuoles were measured by placing 50 mg callus in small vials containing a 1% (w/v) aqueous solution of Lucifer yellow dipotassium salt (Sigma) for 2 h. Lucifer yellow is a trace that stains the apoplast and accumulates in the pro-vacuole; it cannot penetrate membranes, but it can penetrate the tonoplast [146]. Size of stained vacuoles was measured by image analysis using SigmaScan Pro (objective, DPlan Apo 20 UV, 0.70, 160/0.17).

6.10 Real-Time PCR

In the in vitro selection studies of Elmaghrabi et al. for salt [8] and osmotic stress resistance [9], all procedures were according to Spadafora et al. [147, 148]. Thus, RNA was extracted using TRI Reagent (Sigma Aldrich, UK) with a ratio of about 200 mg of plant material to 2 mL of TRI Reagent. Keeping this ratio is important to ensure sufficient concentration of the resulting RNA without contamination from proteins and other cellular components. Leaf or callus material was ground to a fine powder using liquid nitrogen in a sterile pestle and mortar (pre-chilled to −20 °C) ensuring that the plant material remained frozen throughout. Then, 2 mL of TRI Reagent were added to the mix and ground to form a homogeneous paste, and the paste was transferred to a

1.5 mL Eppendorf tube. Following a brief (15 s) vortex, the extract was left at room temperature for 5 min, before being centrifuged at 12,000 × *g* for 10 min at 4 °C in a microcentrifuge (BECKMAN COULTERTM, AllegraTM 12R Centrifuge) to pellet the cellular debris. The resulting supernatant was added to 0.2 mL of chloroform, vortexed (10 s) and left to stand at room temperature for 5 min. This was followed by centrifugation at 12,000 × *g* in a microcentrifuge (as above) for 15 min at 4 °C. The supernatant produced contained an aqueous layer of RNA which was carefully removed, avoiding the interface between the two phases and mixed with 0.5 mL of isopropanol to precipitate the nucleic acids at room temperature for 10 min. The mix was then centrifuged for 10 min at 12,000 × *g* in a microcentrifuge at 4 °C (as above). The supernatant was removed and the pellet washed in 1 mL of 75% ethanol to remove residual salts. A quick vortex followed before another round of centrifugation for 10 min at 12,000 × *g* in a microcentrifuge at 4 °C (as above). The pellet was air-dried thoroughly (in a flow cabinet or fume hood to increase air flow and accelerate the process) and resuspended in sterile distilled water. The RNA was kept on ice until used or stored at −80 °C. Concentration and purity were tested using a Nanodrop spectrophotometer (Nano Drop® ND-1000 spectrophotometer) and by agarose gel electrophoresis (Fig. 2) to visualize the rRNA bands. This is important to assess integrity of the RNA. To ensure that the RNA was not degraded during electrophoresis, the tank, comb, and tray were soaked in 0.1 M NaOH for at least 10 min and then rinsed thoroughly with distilled water before use.

To ensure that real-time PCR results are a true reflection of changes in gene expression, it is very important to remove residual genomic DNA. DNase treatment was therefore used to remove the residual genomic DNA. Depending on the RNA concentration, from 2 to 16 μL of RNA (2 μg) were added to a solution of 2 μL of RQ1 DNAase 10× buffer (Promega), 2 μL of RQ1 DNAse and sterile water to a 20 μL final volume and were incubated at 37 °C for 10 min. To stop the reaction, 2 μL of RQ1 DNAse stop solution were added to the mix and incubated at 65 °C for 10 min. To test whether the DNAse treatment was successful, the absence of contaminating genomic DNA was verified by PCR amplification with primers to 18S rRNA (PUV2F 5`-TTCCATGCTAATGTATTCAGAG-3` and PUV4R 5`-ATGGTGGTGACGGGTGAC-3`) using appropriate positive and negative controls (plant genomic DNA and sterile distilled water), and products were analyzed by agarose gel electrophoresis. If residual genomic DNA was present, the DNAse treatment was repeated, checking the integrity and concentration of the RNA again after the second round of DNAse treatment.

Purified RNA (2 μg of in a total volume of 19 μL) was retrotranscribed using an Ambion kit (RETROscript® Reverse transcription for RT-PCR), and cDNA was stored at −80 °C until future processing. cDNA quality was tested by PCR using 18S rRNA primers and the primers for the target genes of interest. PCR using 18S rRNA and limited cycle number was used to equalize the cDNA concentrations prior to real-time PCR.

An ABsoluteTM QPCR SYBR® Green Mix (Thermo Scientific) kit was used in the real-time PCR process, where each reaction consisted of a total volume of 25 μL: 5 μL cDNA (diluted 1:20), 12.5 μL ABsolute™ QPCR SYBR® Green Mix, 1.75 μL of each primer (10 μM), and 4 μL H_2O. Reactions were cycled in an MJ Research OPTICONTM 2, in triplicate, at 95 °C for 10 min, followed by 40 cycles of 95 °C for 15 s, 60 °C for 30 s, and 72 °C for 30 s and finally 1 cycle of 72 °C for 30 s. A melting curve analysis (60–98 °C with an increasing heat rate of 0.5 °C s^{-1}) was performed to test primer specificity after amplification. Mt18S rRNA primers were used to normalize the results in Elmaghrabi et al. [8, 9] with *M. truncatula* samples. 18S rRNA is a reliable reference gene that has been used widely for developmental and stress-response studies with many species [149, 150]. To ensure correct quantification, the real-time PCR was repeated adjusting the dilution of the cDNA if cycle number variability among the samples was >2 cycles. At least two technical and three biological replicates are recommended for real-time PCR. The relative quantification of gene expression data was carried out with the 2-DDCT or comparative CT method [151].

7 Conclusions

Use of in vitro selection for the development of salt- and drought-tolerant germplasm is underpinned by a range of techniques described above that assess the physiological, biochemical, and gene expression status of the cells as they progress through the selection process. This ensures that enhancement of the tolerance can be followed and optimized throughout these long experiments. Using the three-phase step protocol in *M. truncatula* resulted in a change in these parameters over the treatments (Fig. 3) [8, 9]. For example, mean callus growth and *SOS1* gene expression changed from moderate decrease to moderate increase compared to controls during the course of the experiment. In contrast many other parameters showed a consistent but moderate increase even from the earliest time points measured at 6 months after the start of the selection. As techniques for the analysis of gene expression become cheaper and faster, future analyses may incorporate approaches such as RNAseq, proteomics, metabolomics, and whole genome methylation experiments to assess the overall status of the germplasm, offering new insights into the cellular changes that are taking place and the mechanisms that underlie those changes [152, 153].

	Treatments															
NaCl (mM)	50			100			150			250			350			PEG 6000 (1%, w/v)
Phases	I	II	III	I	II	III	I	II	III	I	II	III	I	II	III	
Mean Callus Growth				*		*	*		*	**	**	**	**	**	**	
Na⁺ accumulation	*	*	*	*	*	*	*	*	*	**	**	**	**	**	**	
K⁺ accumulation	ND	ND		ND	ND	*	ND	ND	*	ND	ND	**	ND	ND	**	
Proline	ND	*	*	ND	*	**	ND	*	**		*	*		*	*	*
Soluble sugars	ND	ND		ND	ND		ND	ND	*			*				*
Water content		ND	ND		ND	ND		ND	ND	*	ND	ND	*	ND	ND	
Osmolarity	*	ND	ND	*	ND	ND	*	ND	ND	**	ND	ND	**	ND	ND	*
Vacuolar size	ND	ND		ND	ND	*	ND	ND	*	ND	ND	*	ND	ND	*	
SOS1 expression	*	*		*	*	**	**	**	*	ND	ND	ND	ND	ND	ND	**
P5CS		**	*	*	**	**	**	**	**	ND	ND	ND	ND	ND	ND	
SERK1	*	*	*	*	**	**	**	*	*	ND	ND	ND	ND	ND	ND	
WEE1	*	*	*	*	**	**	**	*	**	ND	ND	ND	ND	ND	ND	*
CCS52	*	*	*	*	**	**	**	**	**	ND	ND	ND	ND	ND	ND	*

Fig. 3 Summary of physiological responses and molecular results following in vitro selection for NaCl and PEG resistance with *M. truncatula* cultures compared with nonstressed control tissues. Blue color indicates increase/upregulation (moderate, light; high, dark), red color indicates decrease/downregulation (moderate, light; high, dark), and when non-significant compared to controls, white is used. *P = <0.05; **P = <0.01; *ND* not determined

References

1. Ochatt SJ (2015) Agroecological impact of an *in vitro* biotechnology approach of embryo development and seed filling in legumes. Agron Sustain Dev 35:535–552
2. Atif RM, Patat-Ochatt EM, Svabova L et al (2013) Gene transfer in legumes. In: Lüttge U, Beyschlag W, Francis D, Cushman J (eds) Progress in botany, vol 74. Springer, Berlin, Heidelberg, pp 37–100
3. Gatti I, Guindón F, Bermejo C et al (2016) *In vitro* tissue culture in breeding programs of leguminous pulses: use and current status. Plant Cell Tissue Organ Cult 127:543–559
4. Jacob C, Carrasco B, Schwember AR (2016) Advances in breeding and biotechnology of legume crops. Plant Cell Tissue Organ Cult 127:561–584
5. Araújo SS, Beebe S, Crespi M et al (2015) Abiotic stress responses in legumes: strategies used to cope with environmental challenges. Crit Rev Plant Sci 34:237–280
6. Araújo S, Balestrazzi A, Faè M et al (2016) MtTdp2α-overexpression boosts the growth phase of Medicago truncatula cell suspension and increases the expression of key genes involved in the antioxidant response and genome stability. Plant Cell Tissue Organ Cult 127:675–680
7. Duque AS, López-Gómez M, Kráčmarová J et al (2016) Genetic engineering of polyamine metabolism changes Medicago truncatula responses to water deficit. Plant Cell Tissue Organ Cult 127:681–690
8. Elmaghrabi AM, Ochatt S, Rogers HJ, Francis D (2013) Enhanced tolerance to salinity following cellular acclimation to increasing NaCl levels in Medicago truncatula. Plant Cell Tissue Org Cult 114:61–70
9. Elmaghrabi AM, Rogers HJ, Francis D, Ochatt SJ (2017) PEG induces high expression of the cell cycle checkpoint gene WEE1 in embryogenic callus of Medicago truncatula: potential

link between cell cycle checkpoint regulation and osmotic stress. Front Plant Sci 8:1479. https://doi.org/10.3389/fpls.2017.01479
10. Badri M, Ilahi H, Huguet T et al (2007) Quantitative and molecular genetic variation in sympatric populations of Medicago laciniata and M. truncatula (Fabaceae): relationships with eco-geographical factors. Genet Res 89:107–122
11. www.noble.org/medicago/ecotypes.html
12. Motan JF, Becana M, Iturbeormaetxe I et al (1994) Drought induces oxidative stress in pea plants. Planta 194:346–352
13. Gonzalez EM, Aparicio-Tejo PM, Gondon AJ et al (1998) Water-deficit effects on carbon and nitrogen metabolism of pea nodules. J Exp Bot 49:1705–1714
14. Costa França MG, Pham Thi AT, Pimentel C et al (2000) Differences in growth and water relations among Phaseolus vulgaris cultivars in response to induced drought stress. Environ Exp Bot 43:227–237
15. Galvez L, Gonzalez EM, Arrese-Igor C (2005) Evidence for carbon flux shortage and strong carbon/nitrogen interaction in pea nodules at early stages of water stress. J Exp Bot 56:2551–2561
16. Zahaf O, Blanchet S, de Zélicourt A et al (2012) Comparative transcriptomic analysis of salt adaptation in roots of contrasting Medicago truncatula genotypes. Mol Plant 5:1068–1081
17. Alcântara A, Morgado RS, Silvestre S et al (2015) A method to identify early-stage transgenic Medicago truncatula with improved physiological response to water deficit. Plant Cell Tissue Organ Cult 122:605–616
18. Nunes CMJ, Araújo SS, Marques da Silva J et al (2008) Physiological responses of the legume model Medicago truncatula cv. Jemalong to water deficit. Environ Exp Bot 63:289–296
19. Queiros F, Fidalgo F, Santos I, Salema R (2007) *In vitro* selection of salt tolerant cell lines in Solanum tuberosum L. Biol Plant 51:728–734
20. Feki K, Quintero FJ, Pardo JM, Masmoudi K (2011) Regulation of durum wheat NaC/HC exchanger TdSOS1by phosphorylation. Plant Mol Biol 76:545–556
21. Tester M, Leigh RA (2001) Partitioning of transport processes in roots. J Exp Bot 52:442–457
22. Tester M, Davenport R (2003) Na^+ tolerance and Na^+ transport in higher plants. Ann Bot 91:503–507
23. Kim DY, Jin J-Y, Alejandro S et al (2010) Overexpression of AtABCG36 improves drought and salt stress resistance in Arabidopsis. Physiol Plant 139:170–180
24. Alet IA, Sánchez SH, Cuevas CJ et al (2012) New insights into the role of spermine in Arabidopsis thaliana under long-term salt stress. Plant Sci 182:94–100
25. Munns R (2002) Comparative physiology of salt and water stress. Plant Cell Environ 25:239–250
26. Hasegawa PM, Bressan RA, Zhu J-K, Bohnert HJ (2000) Plant cellular and molecular responses to high salinity. Annu Rev Plant Physiol Plant Mol Biol 51:463–499
27. Hernandez JA, Campillo A, Jimenez A et al (1999) Response of antioxidant system and leaf water relation to NaCl stress in pea plants. New Phytol 141:241–251
28. Abebe T, Guanzi AC, Martin B, Cushman JC (2003) Tolerance of mannitol accumulating transgenic wheat to water stress and salinity. Plant Physiol 131:1748–1755
29. Trapp S, Feificova D, Rasmussen FN, Bauer-Gottein P (2008) Plant uptake of NaCl in relation to enzyme kinetics and toxic effects. Environ Exp Bot 64:1–7
30. Trinchant JC, Boscari A, Spennato G et al (2004) Proline betaine accumulation and metabolism in alfalfa plants under sodium chloride stress. Exploring its compartmentalization in nodules. Plant Physiol 135:1583–1594
31. Negrão S, Schmöckel SM, Tester M (2017) Evaluating physiological responses of plants to salinity stress. Ann Bot 119:1–11. https://doi.org/10.1093/aob/mcw191
32. Yang W-J, Rich PJ, Axtell JD et al (2003) Genotypic variation for glycinebetaine in sorghum. Crop Sci 43:162–169
33. Parida AK, Das AB (2005) Salt tolerance and salinity effects on plants: a review. Ecotoxicol Environ Saf 60:324–349
34. Aydi S, Sassi S, Debouba M et al (2010) Resistance of Medicago truncatula to salt stress is related to glutamine synthetase activity and sodium sequestration. J Plant Nutr Soil Sci 173:892–899
35. Surjus A, Durand M (1996) Lipid changes in soybean root membranes in response to salt treatment. J Exp Bot 47:17–23
36. Chen J-B, Wang S-M, Jing R-L, Mao X-G (2009) Cloning the PvP5CS gene from common bean (Phaseolus vulgaris) and its expression patterns under abiotic stresses. J Plant Physiol 166:12–19
37. Choudhary NL, Sairam RK, Tyagi A (2005) Expression of 1-pyrroline–5 carboxylate synthetase gene during drought in rice (Oryza sativa L.). Indian J Biochem Biophys 42:366–370
38. Silva-Ortega CO, Ochoa-Alfaro AE, Reyes-Agüero JA et al (2008) Salt stress increases the expression of p5cs gene and induces

proline accumulation in cactus pear. Plant Physiol Biochem 46:82–92
39. Somboonwatthanaku I, Dorling S, Leung S et al (2010) Proline biosynthetic gene expression in tissue cultures of rice (Oryza sativa L.) in response to saline treatment. Plant Cell Tissue Organ Cult 103:369–376
40. Sairam RK, Tyagi A (2004) Physiology and molecular biology of salinity stress tolerance in plants. Curr Sci 86:407–421
41. Vinocur B, Altman A (2005) Recent advances in Engineering plant tolerance to abiotic stress: achievements and limitations. Curr Opin Biotechnol 16:1913–1923
42. Zsigmond L, Szepesib A, Tari I et al (2012) Overexpression of the mitochondrial PPR40 gene improves salt tolerance in Arabidopsis. Plant Sci 182:87–93
43. Shi HZ, Lee B-H, Wu S-J, Zhu J-K (2003) Overexpression of a plasma membrane Na?/H? Antiporter gene improves salt tolerance in Arabidopsis thaliana. Nat Biotechnol 21:81–85
44. Türkan I, Demiral T (2009) Recent developments in understanding salinity tolerance. Environ Exp Bot 67:2–9
45. Tang R-J, Liu H, Bao Y et al (2010) The woody plant poplar has a functionally conserved salt overly sensitive pathway in response to salinity stress. Plant Mol Biol 74:367–380
46. Li D, Zhang Y, Hu X et al (2011) Transcriptional profiling of Medicago truncatula under salt stress identified a novel CBF transcription factor MtCBF4 that plays an important role in abiotic stress responses. BMC Plant Biol 11:109. https://doi.org/10.1186/1471-2229-11-109
47. Cabuslay SG, Ito O, Alejar AA (2002) Physiological evaluation of responses of rice (Oryza sativa L.) to water deficit. Plant Sci 163:815–827
48. Levitt J (1972) Responses of plants to environmental stresses. Academic Press, New York, NY
49. Zhu J-K (2002) Salt and drought stress signal transduction in plants. Annu Rev Plant Biol 53:2273–2467
50. Golldack D, Li C, Mohan H et al (2014) Tolerance to drought and salt stress in plants: unraveling the signaling networks. Front Plant Sci 5:151. https://doi.org/10.3389/fpls.2014.00151
51. Valliyodan B, Nguyen HT (2006) Understanding regulatory networks and engineering for enhanced drought tolerance in plants. Curr Opin Plant Biol 9:189–195
52. Fulda S, Mikkat S, Stegmann H, Horn R (2011) Physiology and proteomics of drought stress acclimation in sunflower (Helianthus annuus L.). Plant Biol 13:632–642
53. Deinlein U, Stephan AB, Horie T et al (2014) Plant salt-tolerance mechanisms. Trends Plant Sci 19:371–379
54. Ochatt S, Muilu R, Ribalta F (2008) Cell morphometry and osmolarity as early indicators of the onset of embryogenesis from cell suspension cultures of grain legumes and model systems. Plant Biosyst 142:480–486
55. Ochatt SJ, Moessner A (2010) Rounding up plant cells. Int J Plant Biol 1:e8. https://doi.org/10.4081/pb.2010.e8
56. Cushman JC, Bohnert HJ (2000) Genomic approaches to plant stress tolerance. Curr Opin Plant Biol 3:117–124
57. Sreenivasulu N, Varshney RK, Kavi Kishor PB et al (2004) Functional genomics for tolerance to abiotic stress in cereals: a functional genomics approach. In: Gupta PK, Varshney RK (eds) Cereal genomic. Springer, Dordrecht, pp 483–514. https://doi.org/10.1007/1-4020-2359-6_16
58. De Schutter K, Joubes J, Cools T et al (2007) Arabidopsis WEE1 kinase controls cell cycle arrest in response to activation of the DNA integrity checkpoint. Plant Cell 19:211–225
59. Gill SS, Tuteja N (2010) Reactive oxygen species and antioxidant machinery in abiotic stress tolerance in crop plants. Plant Physiol Biochem 48:909–930
60. Zhao L, Wang P, Hou H et al (2014) Transcriptional regulation of cell cycle genes in response to abiotic stresses correlates with dynamic changes in histone modifications in maize. PLoS One 9:e106070. https://doi.org/10.1371/journal.pone.0106070
61. Roy S (2016) Function of MYB domain transcription factors in abiotic stress and epigenetic control of stress response in plant genome. Plant Signal Behav 11:e1117723. https://doi.org/10.1080/15592324.2015.1117723
62. Sorrell DA, Marchbank A, McMahon K et al (2002) A WEE1 homologue from Arabidopsis thaliana. Planta 215:518–522
63. West G, Inzé D, Beemster GTS (2004) Cell cycle modulation in the response of the primary root of Arabidopsis to salt stress. Plant Physiol 135:1050–1058
64. Skirycz A, Claeys H, De Bodt S et al (2011) Pause-and-stop: the effects of osmotic stress on cell proliferation during early leaf development in Arabidopsis and a role for ethylene signaling in cell cycle arrest. Plant Cell 23:1876–1888
65. Gonzalez N, Hernould M, Delmas F et al (2004) Molecular characterization of a WEE1 gene homologue in tomato (Lycopersicon esculentum mill.). Plant Mol Biol 56:849–861
66. Gonzalez N, Gevaudant F, Hernould M, Chevalier C, Mouras A (2007) The cell cycle associated protein kinase WEE1 regulates cell

size in relation to endoreduplication in developing tomato fruit. Plant J 51:642–655
67. Sun Y, Dilkes BP, Zhang C et al (1999) Characterization of maize (Zea mays L.) Weel and its activity in developing endosperm. Proc Natl Acad Sci U S A 96:4180–4185
68. Cebolla A, Vinardell JM, Kiss E et al (1999) The mitotic inhibitor ccs52 is required for endoreduplication and ploidy-dependent cell enlargement in plants. EMBO J 18:4476–4484
69. Larson-Rabin Z, Li Z, Masson PH, Day CD (2009) FZR2/CCS52A1 expression is a determinant of endoreduplication and cell expansion in Arabidopsis. Plant Physiol 149:874–884
70. Nolan KE, Rose RJ, Gorst JG (1989) Regeneration Medicago truncatula from tissue culture: increased somatic embryogenesis using explants from regenerated plants. Plant Cell Rep 8:278–281
71. Chabaud M, Larsonneau C, Marmouget C et al (1996) Transformation of barrel medic (Medicago truncatula Gaertn.) by agrobacterium tumefaciens and regeneration via somatic embryogenesis of transgenic plants with the MtENOD12 nodulin promoter fused to the gus reporter gene. Plant Cell Rep 15:305–310
72. Hoffmann B, Trinh TH, Leung J et al (1997) A new Medicago truncatula line with superior *in vitro* regeneration, transformation and symbiotic properties isolated through cell culture selection. Mol Plant-Microbe Interact 10:307–315
73. Trinh TH, Ratet P, Kondorosi E et al (1998) Rapid and efficient transformation of diploid Medicago truncatula and Medicago sativa ssp. Falcata lines improved in somatic embryogenesis. Plant Cell Rep 17:345–355
74. Wang X-D, Nolan KE, Irwanto RR et al (2011) Ontogeny of embryogenic callus in Medicago truncatula: the fate of the pluripotent and totipotent stem cells. Ann Bot 107:599–609
75. Iantcheva A, Vlahova M, Bakalova E et al (1999) Regeneration of diploid annual medics via direct somatic embryogenesis promoted by thidiazuron and benzylaminopurine. Plant Cell Rep 18:904–910
76. Iantcheva A, Vlahova M, Trinh TH et al (2001) Assessment of polysomaty, embryo formation and regeneration in liquid media for various species of diploid annual Medicago. Plant Sci 160:621–627
77. Svetoslavova G, Vlahova M, Iantcheva A et al (2005) High frequency plant regeneration of diploid Medicago coerulea through somatic embryogenesis. Biotech Biotech Equip 19:57–61
78. Duque AS, Pires AS, Santos DM et al (2006) Efficient somatic embryogenesis and plant regeneration from long-term cell suspension cultures of Medicago truncatula cv. Jemalong. In Vitro Cell Dev Biol Plant 42:270–273
79. Ochatt S, Jacas L, Patat-Ochatt EM, Djennane S (2013) Flow cytometric analysis and molecular characterization of agrobacterium tumefaciens-mediated transformants of Medicago truncatula. Plant Cell Tissue Organ Cult 113:237–244
80. Anjanasree K, Neelakandan WK (2012) Recent progress in the understanding of tissue culture-induced genome level changes in plants and potential applications. Plant Cell Rep 31:597–620
81. Kaeppler SM, Kaeppler HF, Rhee Y (2000) Epigenetic aspects of somaclonal variation in plants. Plant Mol Biol 43:179–188
82. Jain SM (2001) Tissue culture-derived variation in crop improvement. Euphytica 118:153–166
83. Sotirova V, Shtereva L, Zagorska N et al (1999) Resistance responses of plants regenerated from tomato anther and somatic tissue cultures to Clavibacter michiganense. Israel J Plant Sci 47:237–243
84. Lu S, Peng X, Guo Z et al (2007) *In vitro* selection of salinity tolerant variants from triploid bermudagrass (Cynodon transvaalensis x C. dactylon) and their physiological responses to salt and drought stress. Plant Cell Rep 26:1413–1420
85. Ochatt SJ, Power JB (1989) Selection for salt/drought tolerance using protoplast and explant-derived tissue cultures of Colt cherry (Prunus avium x pseudocerasus). Tree Physiol 5:259–266
86. Ochatt SJ, Power JB (1989) Cell wall synthesis and salt (saline) sensitivity of Colt cherry (Prunus avium x pseudocerasus) protoplasts. Plant Cell Rep 8:365–367
87. Ochatt SJ, Marconi PL, Radice S et al (1999) *In vitro* recurrent selection of potato: production and characterization of salt tolerant cell lines and plants. Plant Cell Tissue Org Cult 55:1–8
88. Chen S, Chai M, Jia Y et al (2011) *In vitro* selection of salt tolerant variants following 60Co gamma irradiation of longterm callus cultures of Zoysia matrella [L.] Merr. Plant Cell Tissue Organ Cult 107:493–500
89. Davenport SB, Gallego SM, Benavides MP et al (2003) Behaviour of antioxidant defense system in the adaptive response to salt stress in Helianthus annuus L. cells. Plant Growth Reg 40:81–88
90. Gu R, Liu Q, Pei D, Jiang X (2004) Understanding saline and osmotic tolerance of Populus euphratica suspended cells. Plant Cell Tissue Organ Cult 78:261–265

91. Basu S, Gangopadhyay G, Mukherjee BB, Gupta S (1997) Plant regeneration of salt adapted callus of indica rice (var.Basmati 370) in saline condition. Plant Cell Tissue Organ Cult 50:153–159
92. Tao L, Van Staden SJ (2000) Selection and characterization of sodium chloride-tolerant callus of Glycine max (L) Merr cv. Acme. Plant Growth Reg 31:195–207
93. Zhu J-K (2000) Genetic analysis of plant salt tolerance using Arabidopsis. Plant Physiol 124:941–948
94. Shi H, Quintero FJ, Pardo JM, Zhu JK (2002) The putative plasma membrane Na^+/H^+ antiporter SOS1 controls long-distance Na^+ transport in plants. Plant Cell 14:465–477
95. Merchan F, Breda C, Hormaeche JP et al (2003) A Kruppel-like transcription factor gene is involved in salt stress responses in Medicago spp. Plant Soil 257:1–9
96. Veatch ME, Smith SE, Vandemark G (2004) Shoot biomass production among accession of Medicago truncatula exposed to NaCl. Crop Sci 44:1008–1013
97. Machuka J, Rasha AO, Magiri E et al (2008) *In vitro* selection and characterization of drought tolerant somaclones of tropical Maize (Zea mays L.). Biotechnology 7:641–650
98. Claeys H, Van Landeghem S, Dubois M et al (2014) What is stress? Dose-response effects in commonly used *in vitro* stress assays. Plant Physiol 165:519–527
99. Zhang L, Ma H, Chen T et al (2014) Morphological and physiological responses of cotton (Gossypium hirsutum L.) plants to salinity. PLoS One 9(11):e112807
100. Kurth E, Cramer GR, Lauchli A, Epstein E (1986) Effects of NaCI and $CaCl_2$ on cell enlargement and cell production in cotton roots. Plant Physiol 82:1102–1106
101. Farooq M, Wahid A, Kobayashi N (2009) Plant drought effects, mechanisms and management. Agron Sustain Dev 29:185–212
102. Ochatt SJ (1994) *In vitro* selection for salt/drought tolerance in Colt cherry. In: Bajaj YPS (ed) Biotechnology in agriculture and forestry, Somaclonal variation in crop improvement II, vol 36. Springer-Verlag, Heidelberg, pp 223–238
103. Nolan KE, Irwanto RR, Rose RJ (2003) Auxin up-regulates MtSERK1 expression in both Medicago truncatula root-forming and embryogenic cultures. Plant Physiol 133:218–230
104. Nolan KE, Kurdyukov S, Rose RJ (2009) Expression of the somatic embryogenesis receptor like kinase1 (SERK1) gene is associated with developmental change in the life cycle of model legume Medicago truncatula. J Exp Bot 60:1759–1771
105. Liu L, White MJ, MacRae TH (1999) Transcription factors and their genes in higher plants. Functional domains, evolution and regulation. Eur J Biochem 262:247–257
106. Miki Y, Hashiba M, Hisajima S (2001) Establishment of salt stress tolerant rice plant through set up NaCl treatment *in vitro*. Plant Biol 44:391–395
107. Yamaguchi T, Blumwald E (2005) Developing salt-tolerant crop plants: challenges and opportunities. Trends Plant Sci 10(12):615–620
108. Leone A, Costa A, Tucci M, Grillo S (1994) Adaptation versus shock response to polyethylene glycol-induced low water potential in cultured potato cells. Physiol Plant 92:21–30
109. Gossett DR, Banks SW, Millhollon EP, Lucas MC (1996) Antioxidant response to NaCl stress in a control and a NaCl-tolerant cotton cell line grown in the presence of paraquat, buthionine sulfoximine and exogenous glutathione. Plant Physiol 112:803–809
110. Shankhdhar D, Shankhdhar SC, Mani SC (2000) *In vitro* selection for salt tolerance in rice. Plant Biol 43:477–480
111. Ochatt SJ, Pontecaille C, Rancillac M (2000) The growth regulators used for bud regeneration and shoot rooting affect the competence for flowering and seed set in regenerated plants of protein pea. In Vitro Cell Dev Biol Plants 36:188–193
112. Rubio MC, González EM, Minchin FR et al (2002) Effects of water stress on antioxidant enzymes of leaves and nodules of transgenic alfalfa overexpressing superoxide dismutases. Plant Physiol 115:531–540
113. Hu H, Xiong L, Yang Y (2005) Rice SERK1 gene positively regulates somatic embryogenesis of cultured cell and host defense response against fungal infection. Planta 222:107–117
114. Sakthivelu G, Akitha Devi MK, Giridhar P et al (2008) Drought-induced alterations in growth, osmotic potential and *in vitro* regeneration of soybean cultivars. Gen Appl Plant Physiol 34:103–112
115. Guóth A, Benyo D, Csiszar J et al (2010) Relationship between osmotic stress-induced abscisic acid accumulation, biomass production and plant growth in drought-tolerant and -sensitive wheat cultivars. Acta Physiol Plant 32:719–727
116. Mahmood I, Razzaq A, Hafiz AI et al (2012) Interaction of callus selection media and stress duration for *in vitro* selection of drought tolerant callus of wheat. Afr J Biotechnol 11:4000–4006
117. Rai MK, Kaliaa RK, Singh R et al (2011) Developing stress tolerant plants through *in vitro* selection—an overview of the recent progress. Environ Exp Bot 71:89–98

118. Yang L, Li Y, Shen H (2012) Somatic embryogenesis and plant regeneration from immature zygotic embryo cultures of mountain ash (Sorbus pohuashanensis). Plant Cell Tissue Organ Cult 109:547–556
119. Michel BE, Kaufmann MR (1973) The osmotic potential of polyethylene glycol 6000. Plant Physiol 51:914–916
120. Lu Z, Neumann PM (1998) Water-stressed maize, barley and rice seedlings show species diversity in mechanisms of leaf growth inhibition. J Exp Bot 49:1945–1952
121. Srivastava DK, Gupta VK, Sharma DR (1995) *In vitro* selection and characterization of water stress tolerant callus cultures of tomato (Lycopersicon esculentum L.). Indian J Plant Physiol 38:99–104
122. Attree SM, Pomeroy MK, Fowke LC (1995) Development of white spruce (Picea glauca (Moench.) Voss) somatic embryos during culture with abscisic acid and osmoticum, and their tolerance to drying and frozen storage. J Exp Bot 46:433–439
123. Igasaki T, Sato T, Akashi N et al (2003) Somatic embryogenesis and plant regeneration from immature zygotic embryos of Cryptomeria japonica D. Don. Plant Cell Rep 22:239–243
124. Macovei A, Balestrazzi A, Confalonieri M, Carbonera D (2010) The Tyrosyl-DNA phosphodiesterase gene family in Medicago truncatula Gaertn.: bioinformatic investigation and expression profiles in response to copper- and PEG-mediated stress. Planta 232:393–407
125. Balestrazzi A, Confalonieri M, Macovei A, Carbonera D (2011) Seed imbibition in Medicago truncatula Gaertn.: expression profiles of DNA repair genes in relation to PEG-mediated stress. J Plant Physiol 168:706–713
126. Pintos B, Martin JP, Centeno ML et al (2002) Endogenous cytokinin levels in embryogenic and non- embryogenic calli of Medicago arborea L. Plant Sci 163:955–960
127. Elmaghrabi AM, Ochatt SJ (2006) Isoenzymes and flow cytometry for the assessment of true-to-typeness of calluses and cell suspension of barrel medic prior to regeneration. Plant Cell Tissue Organ Cult 85:31–43
128. Murashige T, Skoog F (1962) A revised medium for rapid growth and bioassays with tobacco tissue cultures. Physiol Plant 15:473–497
129. Iantcheva A, Slavov S, Prinsen E et al (2005) Somatic embryogenesis of the model legume- Medicago truncatula and other diploid Medics. Plant Cell Tissue Organ Cult 81:37–43
130. Gamborg OL, Miller RA, Ojima K (1968) Nutrient requirements of suspension cultures of soybean root cells. Exp Cell Res 50:151–158
131. Schenk RU, Hildebrandt AC (1972) Medium and techniques for induction and growth of monocotyledonous and dicotyledonous plant cell cultures. Can J Bot 50:199–204
132. das Neves L, Duque S, de Almeida J et al (1999) Repetitive somatic embryogenesis in Medicago truncatula ssp. Narbonensis and M. truncatula Gaertn cv. Jemalong. Plant Cell Rep 18:398–405
133. Uchimiya T, Murashige M (1974) Evaluation of parameters in the isolation of viable protoplasts from cultured tobacco cells. Plant Physiol 54:936–944
134. Zafar Y, Nenz E, Damiani F et al (1995) Plant regeneration from explant and protoplast derived calluses of Medicago littoralis. Plant Cell Tissue Organ Cult 41:41–48
135. Denchev P, Velcheva M, Atanassov A (1991) A new approach to direct somatic embryogenesis in Medicago. Plant Cell Rep 10:338–341
136. Vergana R, Verbc F, Pitto L et al (1990) Reversible variation in the methylation pattern of carrot DNA during somatic embryogenesis. Plant Cell Tissue Organ Cult 8:697–701
137. Scarpa GM, Pupilli F, Damiani F, Arcioni S (1993) Plant regeneration from callus and protoplasts in Medicago polymorpha. Plant Cell Tissue Organ Cult 35:49–57
138. Chabaud M, de Carvalho-Niebel F, Barker DG (2003) Efficient transformation of Medicago truncatula cv. Jemalong using the hypervirulent Agrobacterium tumefaciens strain AGL1. Plant Cell Rep 22:46–51
139. Troll W, Lindsley J (1955) A photometric method for the determination of proline. J Biol Chem 215:655–660
140. Boukel M, Houassine D (1997) Adaptation au stress hydrique de quelques variétés de blé dur (Triticum durum). Thèse Magistère, INA, Algérie 90 pp
141. Plummer DT (1987) Introduction to practical biochemistry, 3rd edn. McGraw Hill Book Company Ltd, London, pp 179–180
142. Yazici I, Türkan I, Sekmen SH et al (2007) Salinity tolerance of purslane (Portulaca oleracea L.) is achieved by enhanced antioxidative system, lower level of lipid peroxidation and proline accumulation. Environ Exp Bot 61:49–57
143. Smart RE, Bingham GE (1974) Rapid estimates of relative water content. Plant Physiol 53:258–260
144. Ochatt SJ (2008) Flow cytometry in plant breeding. Cytometry A 73:581–598
145. Widholm JM (1972) The use of fluorescein diacetate and phenosafranine for determining viability of cultured plant cells. Stain Technol 47:189–194

146. Oparka KJ (1991) Uptake and compartmentation of fluorescent probes by plant cells. J Exp Bot 42:565–579
147. Spadafora ND, Doonan JH, Herbert RJ et al (2011) Arabidopsis T-DNA insertional lines for CDC25 are hypersensitive to hydroxyurea but not to zeocin or salt stress. Ann Bot 107:1183–1192
148. Spadafora ND, Parfitt D, Li S et al (2012) Perturbation of cytokinin and ethylene-signalling pathways explain the strong rooting phenotype exhibited by Arabidopsis expressing the Schizosaccharomyces pombe mitotic inducer, cdc25. BMC Plant Biol 12:45
149. Price A, Orellana D, Salleh F et al (2008) A comparison of leaf and petal senescence in wallflower reveals common and distinct patterns of gene expression and physiology. Plant Physiol 147:1898–1912
150. Livak KJ, Schmittgen TD (2001) Analysis of relative gene expression data using real-time quantitative PCR and the 2-DDC method. Methods 25:402–408
151. Wagstaff C, Bramke I, Breeze E et al (2010) A specific group of genes respond to cold dehydration stress in cut Alstroemeria flowers whereas ambient dehydration stress accelerates developmental senescence expression patterns. J Exp Bot 61:2905–2921
152. Shinozaki K, Yamaguchi-Shinozaki K (2007) Gene networks involved in drought stress response and tolerance. J Exp Bot 58:221–227
153. Young DY, Udvardi M (2009) Translating Medicago truncatula genomics to crop legumes. Curr Opin Plant Biol 12:193–201

Chapter 20

Functional Genomics in the Study of Metabolic Pathways in *Medicago truncatula*: An Overview

Chenggang Liu, Chan Man Ha, and Richard A. Dixon

Abstract

In addition to its value as a model system for studies on symbiotic nitrogen fixation, *Medicago truncatula* has recently become an organism of choice for dissection of complex pathways of secondary metabolism. This work has been driven by two main reasons, both with practical implications. First *Medicago* species possess a wide range of flavonoid and terpenoid natural products, many of which, for example, the isoflavonoids and triterpene saponins, have important biological activities impacting both plant and animal (including human) health. Second, *M. truncatula* serves as an excellent model for alfalfa, the world's major forage legume, and forage quality is determined in large part by the concentrations of products of secondary metabolism, particularly lignin and condensed tannins. We here review recent progress in understanding the pathways leading to flavonoids, lignin, and triterpene saponins through utilization of genetic resources in *M. truncatula*.

Key words Flavonoid, Lignin, Metabolite transport, Proanthocyanidin, Saponin, Transcriptional regulation, Transposon insertion mutant, Triterpene

1 Introduction

Plant metabolites can be conveniently classified into primary metabolites, which are essential for plant growth and development, and secondary or specialized metabolites which are not required for growth and development but rather enhance the fitness of plants to their environment. However, in many cases the boundaries between primary and secondary metabolism are not obvious. Secondary metabolites, unlike primary metabolites, are usually specific to certain taxonomic groups of plants, and, as such, are extremely diverse. Based on their structures, plant secondary metabolites can be grouped into three major classes: phenolic and polyphenolic compounds, terpenoids, and nitrogen-containing compounds [1]. Phenolic and polyphenolic compounds contain one or more phenolic hydroxyl groups and are synthesized in whole or part through the shikimic acid pathway. This group of

Luis A. Cañas and José Pío Beltrán (eds.), *Functional genomics in* Medicago truncatula*: Methods and Protocols*, Methods in Molecular Biology, vol. 1822, https://doi.org/10.1007/978-1-4939-8633-0_20,

compounds includes lignins, lignans, phenolic acids, and flavonoids [2, 3]. Terpenoids are compounds containing various numbers of 5-carbon isoprene units, which are synthesized from the mevalonic acid (MVA) or methylerythritol phosphate (MEP) pathways [4]. This group of compounds includes carotenoids, gibberellic acid and abscisic acid (two fundamental plant hormones), and triterpene saponins. Nitrogen-containing secondary metabolites usually originate from amino acids, and this group includes alkaloids, cyanogenic glycosides, and nonprotein amino acids [5].

Medicago truncatula, initially developed as a model plant for studies of symbiosis between nitrogen-fixing rhizobacteria and legumes, has now been developed as a model for studies on all aspects of plant biology [6], including compound leaf development [7]; flower development and flowering time regulation [8–10]; interactions with bacteria, fungi, and insects [11–14]; and drought tolerance [15–17]. Not surprisingly, *M. truncatula* has also been used as a model plant for studies of plant secondary metabolism.

M. truncatula produces many secondary metabolites that are primarily distributed in leguminous plants, making it an indispensable model plant for studies on such metabolites. For example, isoflavonoids are flavonoid derivatives prevalent in legumes but lacking in the model plants *Arabidopsis* and rice [18]. In addition to their roles as antimicrobial compounds [19, 20] and inducers of rhizobial nodulation genes [21, 22], there has been a long history of interest in the biosynthesis of isoflavonoids due to their phytoestrogenic activity [18, 23]. *M. truncatula* is also rich in triterpenoidal saponins, many of which may be specific to this species [24–27], and it has now become the model plant of choice for studies on saponin biosynthesis.

Alfalfa (*M. sativa*), a close relative of *M. truncatula*, is the world's major forage legume, and there is considerable interest in engineering the metabolic composition of alfalfa to improve forage quality [28]. However, alfalfa is a tetraploid, outcrossing species, hindering direct genetic studies on this species. In contrast, *M. truncatula* is diploid and self-pollinated and possesses a relative small genome, making it an excellent model plant to address metabolic pathways in alfalfa. The flavan-3-ol polymers known as proanthocyanidins (PAs) are of particular importance for forage quality because a moderate content of PAs can prevent potentially lethal pasture bloat in ruminant animals [29]. The goal of enhancing PA content in alfalfa has driven studies on the PA biosynthesis pathway in *M. truncatula*. These studies have revealed several features of the pathway different from that in *Arabidopsis* [30–32]. Similarly, the push to develop low lignin alfalfa with improved forage digestibility has led to *M. truncatula* also being used as a model plant to study lignin biosynthesis [33–36]. These studies have indicated that there are features of the lignin pathway in *Medicago* that differ from the commonly accepted paradigm arising from studies in *Arabidopsis*.

With completion of the *Medicago* genome [37], research with *M. truncatula* entered the post-genome era. To study the function of every gene, it is essential to generate mutants in each gene. In the case of *M. truncatula*, chemical mutagenesis has been used to generate an EMS (ethyl methanesulfonate) mutant population [38]. Fast neutron bombardment (FNB) has also been successfully utilized to generate a mutant population harboring large fragment deletions [39]. EMS and FNB mutant populations can be used for both forward and reverse genetic studies. For forward genetics, the causative mutation underlying a phenotype can be identified by gene mapping technology. For reverse genetics, a mutation in a gene of interest can be identified by screening the mutant population with TILLING (Targeting Induced Local Lesions IN Genomes) technology [40, 41]. However, the most powerful mutagenesis approach in *M. truncatula* has involved the construction of a *Tnt1* (tobacco retro-transposon) transposon-tagged population [42]. Currently, this population contains about 22,000 lines harboring *Tnt1* insertions in about 90% of the genes (https://medicago-mutant.noble.org/mutant/index.php), close to saturation. Unlike the EMS and FNB populations, the positions at which the mutations have occurred (*Tnt1* insertion sites) can be identified or screened by regular PCR-based approaches, greatly facilitating both forward and reverse genetics approaches [43, 44].

In this chapter, we will mainly focus on the progress in secondary metabolism studies made through accessing the *Tnt1* transposon-tagged population. The review is not intended to be comprehensive but instead addresses the most recent progress in studies on flavonoids, lignin, and triterpene saponins in *M. truncatula*, with special emphasis on how mutant phenotypes, pathways, and regulatory mechanisms can differ from those reported in *Arabidopsis*.

2 Flavonoids

Flavonoids are a subgroup of polyphenolic molecules that exist ubiquitously in the plant kingdom. They contain a 15-carbon skeleton composed of two aromatic rings connected via a heterocyclic pyran ring [45]. Various functional groups are attached to the 15-carbon skeleton to generate the structural diversity of the different flavonoids [46]. *M. truncatula* produces members of all classes of flavonoids found in plants, i.e., isoflavones (e.g., formononetin, medicarpin, genistein, and biochanin A), anthocyanidins (e.g., cyanidin), proanthocyanidins (e.g., epicatechin polymer), flavanones (e.g., naringenin), flavonols (e.g., quercetin and myricetin), and flavones (e.g., apigenin, chrysoeriol, tricin, and luteolin)

[47–51]. These flavonoids are usually conjugated with sugars or other functional groups to change their solubility or activity [52].

Flavonoids are synthesized in part through the general phenylpropanoid pathway. The first committed step of flavonoid biosynthesis (pathway summarized schematically in Fig. 1) is the condensation of three units of malonyl-CoA and one unit of 4-coumaroyl-CoA, catalyzed by chalcone synthase (CHS). The resulting chalcone is further isomerized to flavanone, the precursor of most other flavonoids, by chalcone isomerase (CHI). Flavanones can be converted into isoflavones via 2-hydroxyisoflavanone synthase (2-HIS, which catalyzes the migration of the B ring from C2 to C3) and a specific dehydratase. Alternatively, flavanone can be converted into flavone by flavone synthase (FNS), which introduces a double bond between C2 and C3, with subsequent loss of chirality at these positions. The most common pathway beyond flavanones is their conversion into dihydroflavonol by flavanone 3-hydroxylase (F3H). Dihydroflavonol is the common precursor of both flavonols and anthocyanidins/flavan-3-ols. Dihydroflavonol is converted into flavonol by flavonol synthase (FLS), or into leucoanthocyanidin by dihydroflavonol reductase (DFR). Leucoanthocyanidin can be further converted into anthocyanidin by anthocyanidin synthase (ANS), or converted into catechin (a C2-C3 *trans*-flavan-3-ol) by leucoanthocyanidin reductase (LAR). Anthocyanidins are further glycosylated and acylated to form anthocyanins. Anthocyanidin can also be reduced to (−)-epicatechin (a C2-C3 *cis*-flavan-3-ol) by anthocyanidin reductase (ANR) (Fig. 1). Flavan-3-ols are then condensed to form polymeric proanthocyanidins (PAs). Several aspects of this pathway and its regulation have been illuminated through analysis of *M. truncatula* mutants (Table 1).

2.1 A Novel Function for LAR

The activity of LAR to convert leucoanthocyanidin to catechin (Fig. 1) has been well established by in vitro enzymatic assay [49, 53–55]. *M. truncatula* possesses a highly expressed *LAR* gene, but the PAs in *Medicago* are almost exclusively epicatechin-based [49]. Moreover, overexpression of *LAR* genes in plants often produces either no catechin, or both catechin and epicatechin, depending on the species [55, 56]. This apparent paradox raised the question of whether LAR might have an additional function *in planta*. To address this, we screened the *Tnt1* transposon-tagged *Medicago* population and obtained *lar* mutants. The most striking phenotype of the *lar* mutants was the almost complete loss of extractable PAs in seeds, concurrent with the increase of non-extractable, insoluble PAs [57], suggesting a role for LAR in regulating PA condensation in *Medicago*. Further analyses revealed that LAR converts the extension unit, 4-β-cysteinyl epicatechin (epicatechin-cysteine, a molecule that accumulates in seed coats of *lar* mutants), to the terminal unit, (−)-epicatechin [57] (Fig. 1), with chain

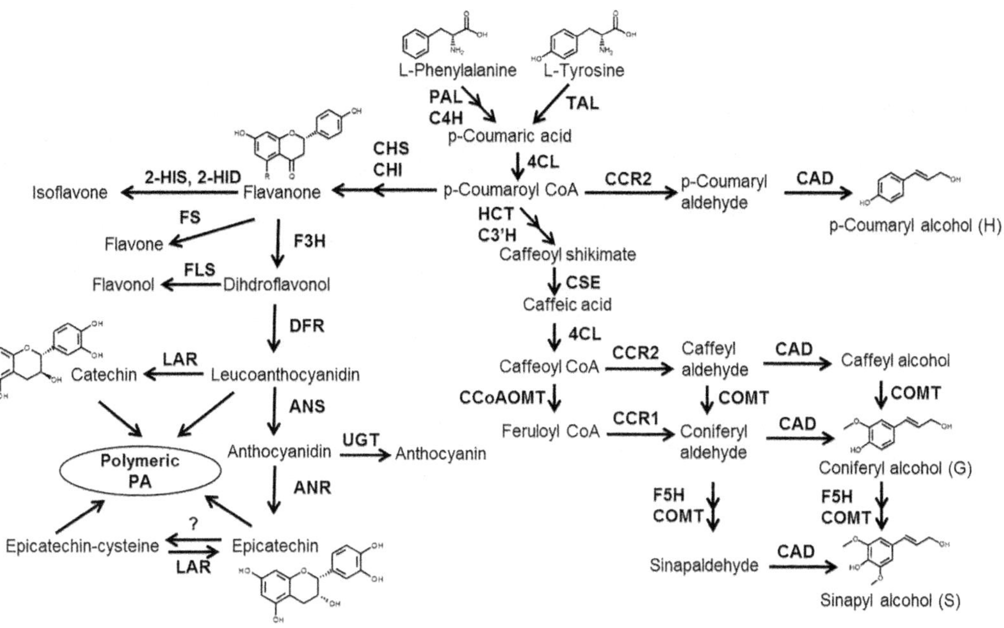

Fig. 1 The metabolic pathways to flavonoids and lignin in *M. truncatula*. Arrows with bold letters indicate the enzymatic steps. *PA* proanthocyanidin, *CHS* chalcone synthase, *CHI* chalcone isomerase, *2-HIS* 2-hydroxyisoflavanone synthase, *2-HID* 2-hydroxyisoflavanone dehydratase, *FS* flavone synthase, *F3H* flavanone 3-hydroxylase, *FLS* flavonol synthase, *DFR* dihydroflavonol reductase, *LAR* leucoanthocyanidin reductase, *ANS* anthocyanidin synthase, *UGT* UDP-glucosyltransferase, *ANR* anthocyanidin reductase, *CSE* caffeoyl shikimate esterase, *CCR* cinnamoyl-CoA reductase, *CAD* cinnamyl-alcohol dehydrogenase, *PAL* L-phenylalanine ammonia-lyase, *C4H* cinnamate 4-hydroxylase, *TAL* L-tyrosine ammonia-lyase (probaly not in *M. truncatula*, mainly found in grasses), *4CL* 4-coumarate:CoA ligase, *C3H* p-coumarate 3-hydroxylase, *HCT* hydroxycinnamoyl-CoA quinate/shikimate hydroxycinnamoyltransferase, *CCoAOMT* caffeoyl-CoA 3-*O*-methyltransferase, *F5H* ferulate/5-hydroxyconiferaldehyde 5-hydroxylase, *COMT* caffeic acid 3-*O*-methyltransferase

length being determined by the relative proportions of terminal and extension units.

It is generally accepted that the catechin-based PAs are formed by condensation between leucoanthocyanidin, serving as extension units, and terminal catechin/or epicatechin [58]. In the case of catechin-based PA condensation, LAR not only produces the terminal catechin unit but also removes the extension units, leucoanthocyanidin (by conversion to catechin). Thus, in this model, LAR functions to control PA chain length in the same manner during condensation of both epicatechin- and catechin-based PAs (Fig. 1).

2.2 Anthocyanin and PA Transport

Anthocyanins and PAs accumulate in vacuoles, but their precursors need to be glycosylated before they can serve as substrates for vacuolar transporters. The *M. truncatula* glucosyltransferase UGT78G1 most efficiently glycosylates isoflavone(s) to their 7-*O*-glycosides in vitro, with poorer efficiency with anthocyanidins [59, 60], suggesting a role in isoflavone glycosylation in vivo. However, overexpression of UGT78G1 in alfalfa resulted in

Table 1
***M. truncatula* mutants described in this chapter**

Metabolic pathway	Target gene name	Gene ID	Line number[a]	Phenotype	Refs.
Flavonoid	*WD40–1*	Medtr3g092840	NF0977 NF2745	Loss of accumulation of PA and mucilage in seeds. Loss of anthocyanin accumulation in leaf. White seed coat.	[30]
	MtTT8	Medtr1g072320	NF2225	Loss of accumulation of PA in seeds. Loss of anthocyanin accumulation in leaf and flower. White seed coat.	[75]
	MtPAR	Medtr8g020490	NF2466 NF4419 NF1358 NF3308	Reduced PA content in seeds. Slightly lighter seed coat.	[80]
	MtMYB5	Medtr3g083540	NF11932 (*Mtmyb5-1*) NF12338 (*Mtmyb5-2*) NF13006 (*Mtmyb5-3*)	Reduced PA content in seeds. Dark-red seed coat.	[31]
	MtMYB14	Medtr4g125520	NF0084 (*Mtmyb14-2*) NF14565 (*Mtmyb14-1*)	Reduced PA content in seeds. Slightly darker seed coat.	[31]
	MtMYB2	Medtr5g079670	NF10162 NF19285	Increased PA content in seeds Accumulation of more anthocyanin in seedling and flower.	[32]
	MtLAR	Medtr1g093720	NF9870 (*Mtlar-1*) NF18997 (*Mtlar-2*)	Loss of soluble PA. Accumulation of more insoluble PA.	[57]
	MtANR	Medtr4g092080	NF9161 (*Mtanr-1*) NF18737 (*Mtanr-2*)	Loss of soluble and insoluble PA. Dark-red seed coat.	[57]
	MtMATE1	Medtr5g090290	NF2629	Reduced PA accumulation in seeds. White seed coat.	[63]
	MtMATE2	Medtr1g100180	NF5697 (*Mtmate2-1*) NF5176 (*Mtmate2-2*) NF0737 (*Mtmate2-3*) NF3103 (*Mtmate2-4*)	Reduced anthocyanin accumulation in leaf and flower.	[65]
	UGT78G1	Medtr4g128690	NF1507	Embryo lethal.	[30]
	UGT72L1	Medtr8g009063	NF1263 NF2826	Partially reduced soluble PA content.	[62]

(continued)

Table 1 (continued)

Metabolic pathway	Target gene name	Gene ID	Line number[a]	Phenotype	Refs.
Lignin	*MtCSE*	Medtr4g127220	NF13103 (*Mtcse-1*)	Dwarf.	[36]
			NF17462 (*Mtcse-2*)	Dwarf.	
	MtCCR1	Medtr2g104960	NF5145	Dwarf.	[35]
	MtCAD	Medtr1g107425	NF1807	Semi-dwarf.	[111]
	MtWRKY	Medtr8g007270	NF3788 (*Mtstp-1*)	Normal.	[33]
			NF1715 (*Mtstp-2*)	Normal.	
	MtNST1	Medtr8g024480	NF3056 (*Mtnst1-1*)	Bending stem.	[112]
			NF2911 (*Mtnst1-2*)	Bending stem.	
			NF4041 (*Mtnst1-3*)	Bending stem.	
			NF3687 (*Mtnst1-4*)	Bending stem.	
Saponin	*MtLHA* (*CYP716A12*)	Medtr8g100135	E25-10 (*Mtlha-1*; Activation tagging)	Growth retardation.	[131]
			CYP2768 (*Mtlha-2*; EMS)	Normal.	
			CYP3201 (*Mtlha-3*; EMS)	Growth retardation.	
	CYP72A67	Medtr2g023680	Mt1903 (EMS)	Normal shoot, decreased number of nodules.	[26]
			Mt2961 (EMS)	Normal.	
	UGT73F3	Medtr2g035020	NF5746	Dwarf.	[132]
			NF8981	Dwarf.	

[a] *NF* TNT1 mutant line number, *EMS* ethyl methanesulfonate, *PA* proanthocyanidin

enhanced anthocyanin accumulation when the plants were exposed to high light intensity, indicating that UGT78G1 can function as an anthocyanidin glycosyltransferase *in planta* [60]. A *Tnt1* insertional mutant was obtained, but the homozygous mutant is embryo lethal [60], suggesting a critical role of anthocyanidin/isoflavone glycosylation in plant development.

M. truncatula UGT72L1 specifically catalyzes conversion of epicatechin to epicatechin-3′-*O*-glucoside [61, 62]. Both anthocyanidin 3-*O*-glucoside and epicatechin-3′-*O*-glucoside can be transported into vacuoles by MATE1 in *M. truncatula* [63]. MATE1 is the homolog of *Arabidopsis* TT12 (Transparent Testa 12) and is a flavonoid/H^+ antiporter located in the vacuolar membrane. Both *Mtmate1* and *Atttl2* mutants exhibit a transparent testa phenotype (clear seed coat), consistent with the role of their encoded proteins in anthocyanin and epicatechin-3′-*O*-glucoside transport [63, 64]. Another MATE transporter in *M. truncatula*, MATE2, appears to be a broad-spectrum transporter for flavonoid glucosides [65].

Since PAs accumulate in the vacuole, it is tempting to propose that PA condensation also occurs in vacuoles, which have an acidic pH. However, the condensation between epicatechin and epicatechin-cysteine or between catechin and leucoanthocyanidin occurs optimally at neutral pH [57, 66], suggesting that PA condensation occurs in the cytosol or another neutral compartment such as the ER lumen. This raises the question of the function of epicatechin-3′-*O*-glucoside transport. Theoretically, transport of epicatechin-3-O′-glucoside into the vacuole could provide building blocks for PA condensation or be a mechanism for detoxification of epicatechin [61]. Because of the neutral pH optimum of PA condensation, the latter may be the more likely. To reconcile the discrepancy between PA condensation and accumulation sites, we speculate that PA condensation and transport might be coupled processes. Alternatively, PA extension units (epicatechin-cysteine) and terminal units (epicatechin) could be transported into the ER independently and then condense into PA polymers in the ER lumen.

2.3 Regulation of Anthocyanin and PA Biosynthesis

The biosynthesis of anthocyanins and PAs appears to be mainly regulated at the transcriptional level. Three kinds of transcription factors, i.e., MYB, bHLH, and WD40-repeat proteins, form the so-called MBW ternary complex [67], a highly conserved transcription factor complex that regulates the biosynthesis of not only anthocyanidins and PAs but also epidermal differentiation and mucilage biosynthesis [68]. A WD40 repeat transcription factor is required for all the cellular processes regulated by the MBW complex, and loss of function mutants of the WD40 transcription factor usually show pleiotropic phenotypes. Loss of function of *ttg1*, encoding the WD40 repeat transcription factor in *Arabidopsis*, results in a transparent testa phenotype due to loss of anthocyanidin/PA biosynthesis, loss of trichome and root hair growth (glabrous phenotype), and loss of mucilage biosynthesis [69, 70]. In *M. truncatula*, loss of function of *Mtwd40–1*, the homolog of TTG1, also results in the loss of anthocyanin and PA biosynthesis and mucilage accumulation [30]. However, in contrast to *ttg1*, trichome and root hair growth in the *Mtwd40–1* mutant are normal [30], suggesting that an another WD40 repeat transcription factor may control these phenotypes in *M. truncatula* or that trichome and root hair growth do not require involvement of a WD40 repeat transcription factor in this species.

The bHLH transcription factor involved in anthocyanidin/PA biosynthesis is called TT8 (Transparent Testa 8). In *Arabidopsis*, the *tt8* mutant exhibits loss of anthocyanin/PA phenotypes but has normal trichome and root hair growth and mucilage biosynthesis [71], suggesting that TT8 is specific to anthocyanin/PA biosynthesis. Other *Arabidopsis* bHLH transcription factors, GL3 and EGL3, are involved in trichome and root hair growth [72, 73], and there are partially redundant functions between TT8, GL3, and EGL3 [74]. The homolog of TT8 in *M. truncatula* has been characterized [75]. As with *Arabidopsis* TT8, *M. truncatula* TT8 can interact with WD40 and MYB transcription factors forming a MBW complex to activate anthocyanin and PA biosynthetic genes [75], and the *Tnt1* insertional mutant of *Mttt8* exhibits a transparent testa phenotype [75].

The specificity of the MBW complex is mainly determined by the MYB transcription factor component, and different MYBs are required for the activation of the anthocyanin or PA biosynthesis pathways [76]. Overexpression of LAP1 (*L*egume *A*nthocyanin *P*roduction *1*) can induce massive anthocyanin accumulation in alfalfa and clover without PA accumulation, suggesting that here LAP1 is the major MYB transcription factor regulating the biosynthesis of anthocyanins [60]. Overexpressing LAP1 cannot induce the expression of *LAR* and *ANR*, two structural genes specific for PA biosynthesis [60]. The ability of LAP1 to induce anthocyanin accumulation seems specific to leguminous plants, since the homolog of LAP1 in *Arabidopsis*, PAP1 (*P*roduction of *A*nthocyanin *P*igment *1*), can induce anthocyanin accumulation in *Arabidopsis* but fails to induce it in alfalfa [60].

Two MYB transcription factors, TT2 and MYB5, are involved in PA biosynthesis in *Arabidopsis* [77]. TT2 is the major regulator, with the null mutant *tt2* exhibiting a transparent testa phenotype, indicating loss of PA accumulation [78]. The null *myb5* mutant exhibits normal seed color [77, 79], but the *tt2/myb5* double mutant exhibits slightly less pigmentation than the *tt2* single mutant, indicating that MYB5 has a minor role in PA biosynthesis [77]. The main function of MYB5 in *Arabidopsis* is to control mucilage biosynthesis and trichome branching. In *M. truncatula*, three MYB transcription factors, PAR (*pro*anthocyanidin *r*egulator), MYB14, and MYB5, are involved in PA biosynthesis [31].

Overexpression of each of these MYBs in *M. truncatula* hairy roots induces massive accumulation of PAs [31]. The *Tnt1* insertion mutants of these three MYBs exhibit reduced PA accumulation in the seed coat, but the accumulation of anthocyanin in these mutants is normal, indicating that the three MYBs are specific for PA biosynthesis [31, 80]. The seed coat color changes caused by these mutations are different from those observed in *Arabidopsis*. Although both PAR and MYB14 are homologs of *Arabidopsis* TT2, as demonstrated by their abilities to complement the transparent testa phenotype of *Arabidopsis tt2* [31], the *par* and *myb14* mutants exhibit a slightly lighter or darker pigmentation than wild-type seeds, respectively [31, 80]. The most striking phenotype of these three MYBs is the seed coat color of *myb5*, which is a conspicuous dark red color, presumably resulting from anthocyanin accumulation [31]. This is in sharp contrast to the *Arabidopsis myb5* mutant, which has a normal seed coat color. These results suggest that MYB5 plays a much more important role in PA biosynthesis in *Medicago* than its homolog in *Arabidopsis*.

In addition to the positive regulators described above, there are also MYB repressors involved in anthocyanin and PA biosynthesis. In *M. truncatula*, MYB2 is an R2R3 MYB transcription factor that represses the accumulation of anthocyanins and PAs [32]. MYB2 belongs to a conserved family of MYB repressors found in strawberry (FaMYB1) [81, 82], petunia (PhMYB27) [83], and poplar (PtMYB182) [84]. The common feature of this MYB group is the conserved EAR motif (LxLxL). Besides the conserved EAR motif (pdLNLD/ELxiG/S), another semiconserved EAR domain exists at the C-terminal of MYB2 [32]. MYB2 functions as a co-suppressor by joining the ternary activating MBW complex to form a quaternary complex [32, 83]. This mode of suppression is different from that of R3 small MYBs, which suppress anthocyanin and PA biosynthesis by competing with activating MYBs for bHLH and WD40 transcription factors [85]. The expression pattern of MYB2 in the seed coat is opposite to that of ANR [32]. MYB2 is expressed in epidermal cells excluding the hilum region, whereas ANR is expressed strongly in the hilum region and weaker in epidermal cells. This suggests that MYB2 progressively fine-tunes PA accumulation which is initiated at the hilum region and expands progressively to epidermal cells [32].

3 The Monolignol Biosynthesis Pathway

The plant cell wall is a heterogeneous complex consisting mainly of cellulose, hemicellulose, and lignin and is the major component of the lignocellulosic biomass for the production of biofuels through chemical or biological processing [86]. Lignin is an abundant aromatic heteropolymer present primarily in secondary cell walls, where it plays crucial roles in imparting mechanical strength and hydrophobicity, water transport, and pathogen defense [87].

The content of lignin varies widely between different types of biomass. Softwood lignin accounts for approximately 25–32% of dry matter, whereas this amount is lower (18–25%) in hardwoods [88]. Lower amounts of lignin are present in grasses, and lignin is almost completely absent in mosses and green algae [89]. High lignin content is correlated with the recalcitrance of biomass to enzymatic saccharification during biofuel production and reduced in-rumen digestibility of forages [90–94]. Many studies have reported transgenic or mutational approaches to reduce lignin amounts for improving forage digestibility in both grasses [92, 95, 96] and forage legumes such as alfalfa [90, 97, 98] and facilitating processing of lignocellulosic biomass for biofuels [91, 92, 99–104].

There are three major monomeric building blocks in the lignin polymer, abbreviated as H (*p*-hydroxyphenyl), G (guaiacyl), and S (syringyl), differing in the number of methoxyl groups on the aromatic ring: 0, 1, or 2, respectively [87]. These monomers are derived from the aromatic amino acid L-phenylalanine (Phe), and also from tyrosine in grasses [105], via the activity of 11 currently known enzymes [87, 106, 107].

Phe is first deaminated by L-phenylalanine ammonia-lyase (PAL) to produce cinnamic acid, the substrate of the cytochrome P450 enzyme cinnamate 4-hydroxylase (C4H) for production of *p*-coumaric acid (Fig. 1). There are three different cytochrome P450-dependent hydroxylases involved in monolignol biosynthesis which are thought to reside on the cytosolic side of the endoplasmic reticulum: C4H, *p*-coumaroyl-shikimate-3′-hydroxylase (C3′H), and ferulate/5-hydroxyconiferaldehyde 5-hydroxylase (F5H) [87, 89]. 4-Coumarate:CoA ligase (4CL) catalyzes the formation of the CoA thiol ester of *p*-coumarate which serves as substrate for the BAHD family acyltransferase hydroxycinnamoyl-CoA quinate/shikimate hydroxycinnamoyltransferase (HCT). After hydroxylation by C3′H, caffeoyl shikimate is either converted back to its corresponding CoA ester by HCT operating in the reverse direction or, according to a recently discovered pathway, is hydrolyzed by caffeoyl shikimate esterase (CSE) to produce caffeic acid which is reactivated to caffeoyl CoA by 4CL activity (Fig. 1).

Introduction of the methoxyl groups into monolignols occurs via the activities of caffeoyl-CoA-3-*O*-methyltransferase (CCoAOMT) and caffeic acid/5-hydroxyferulic acid 3/5-*O*-methyltransferase (COMT). The cinnamoyl-CoA esters are converted into monolignols by two enzymes that successively reduce the monolignol side chain: cinnamoyl-CoA reductase (CCR) and cinnamyl alcohol dehydrogenase (CAD). Polymerization of the monolignols in the cell wall occurs by free radical coupling initiated by peroxidases and laccases.

M. truncatula has become an important model system for the study of monolignol biosynthesis for several reasons. Firstly, although many of the studies on the molecular identification of lignin biosynthetic enzymes and their functions have been performed in *Arabidopsis thaliana*, this species is not the best for studying lignification in stems as it is a rosette plant, and the stem structure studied is the inflorescence. Alfalfa is the major forage species for genetic manipulation of lignin to improve digestibility, and having *M. truncatula* as a genetically tractable model for this species has facilitated much of the work leading to the commercialization of low lignin alfalfa [108, 109]. Finally, recent reports studying lignin pathway mutants in *M. truncatula* have revealed differences from *Arabidopsis*, particularly as regards gene redundancy. These reports, and the approaches taken to characterize lignin mutants in *M. truncatula*, are outlined below.

3.1 Regulation of Anthocyanin and PA Biosynthesis

Studies on lignin biosynthesis in *M. truncatula* have utilized both forward and reverse genetic approaches. High-throughput screening of the transposon insertion population by simple examination of lignin autofluorescence of cross sections under UV light, using appropriate filters (UG1 excitation filter (300–400 nm) and 41 barrier filter (0–400 nm) [110]), has led to the discovery of mutants with less lignin and/or ectopic deposition of lignin. The genes affected were in cinnamyl alcohol dehydrogenase [111], a NAC master regulator [112], and a WRKY transcription factor [33] (*see* below). Reverse genetics has been utilized to isolate knockout lines in most of the remainder of the currently accepted monolignol pathway genes. These lines are summarized in Table 1.

Lignin content and composition are routinely measured by means of the Klason, acetyl bromide (AcBr), and thioacidolysis methods [113]. Thioacidolysis measures monomeric breakdown products of lignin by GC-MS, revealing the proportions of H, G, and S monomers that are β–O–4-linked [114], whereas the other methods provide a grosser estimate of lignin quantity [113, 115, 116]. Lignin can also be stained in situ with Wiesner (phloroglucinol) reagent, which gives a red-pink product with the cinnamaldehyde groups present in lignins, or the Mäule reagent, specific for detection of the S moieties [110, 117].

3.2 Approaches for Analysis of Lignin Pathway Mutants

We will here discuss the phenotypes of a selection of the *M. truncatula* lignin pathway transposon insertion mutants, with particular reference to comparisons to the corresponding *Arabidopsis* mutants. Further details of additional mutants will appear elsewhere.

The *Arabidopsis* T-DNA insertional *cse* mutant (*Atcse*) exhibits a semi-dwarf phenotype, whereas the *M. truncatula Mtcse* Tnt1 retrotransposon insertion lines obtained by reverse genetic screening has a very strong dwarf phenotype (Fig. 2a, e) [36, 107].

Fig. 2 Phenotypes of lignin pathway mutants in *Medicago truncatula*. Photographs of *cse* (**a**), *ccr1* (**b**), *cad1* at 22 °C (**c**), and *cad1* at 30 °C (**d**). Photographs were taken from the following references (**a**) [36]; (**b**) [35]; and (**c**) and (**d**) [111]; with permission. *WT* wild type

In the *Atcse* mutant, UV autofluorescence and phloroglucinol and Mäule staining were all decreased [107], whereas, paradoxically, the *Mtcse* mutant exhibited increased UV autofluorescence and enhanced phloroglucinol and Mäule staining.

The total amount of lignin in the *Atcse* mutant was reduced by 36% as compared to control plants, whereas thioacidolysis analysis indicated a much greater decrease in total lignin in *Mtcse* mutant stems, by 78–86% as compared to control plants, in spite of the higher phloroglucinol staining. The staining therefore suggests accumulation of coniferaldehyde end groups in the lignins rather than increased total lignin content [36]. The level of H monomer is highly increased in both *Atcse* (27%) and *Mtcse* (52–55%) mutants. NMR analysis showed a greater increase in H units in both *Atcse* (44%) and *Mtcse* (77–84%) compared with the amount determined by thioacidolysis, suggesting that a large proportion of the H units are not joined by β-O-4 linkages. Although more time-consuming, NMR analysis can give a better estimation of lignin composition [118].

CCR catalyzes the conversion of cinnamoyl-CoA esters to their respective cinnamaldehydes. *M. truncatula* possesses two *CCR* genes. MtCCR1 protein exhibits preference for feruloyl CoA as substrate, whereas MtCCR2 prefers caffeoyl and 4-coumaroyl CoAs (Fig. 1) [35]. It is not clear whether *MtCCR2* expression is related with plant defense, as has been suggested for *AtCCR2* [119].

The *Mtccr1* mutant obtained by reverse genetic screening shows a strong dwarf phenotype (Fig. 2b), whereas the *Atccr1* mutant had a relatively mild semi-dwarf phenotype [35, 120]. The *Mtccr1* mutant showed reduced lignin autofluorescence and decreased staining by phloroglucinol and Mäule reagent [35], similar to the *Atccr1* mutant [120]. Consistent with the staining, total lignin content was strongly decreased in *Mtccr1* mutant stems, whereas *Atccr1* mutant stems had only a mild decrease in total lignin content. Both G- and S-lignin were decreased in *Mtccr1* mutant stems, whereas S-lignin was slightly increased in *Atccr1* stems.

CAD catalyzes the final step of monolignol biosynthesis, reducing the hydroxycinnamaldehydes to their corresponding hydroxycinnamyl alcohols, before their transport to the wall for lignification. An *Arabidopsis CAD* double mutant (*cad4 cad5*) resulted in a phenotype with a limp floral stem at maturity, whereas the *Mtcad* mutant obtained by forward genetic screening showed a temperature-sensitive dwarf phenotype under high-temperature growth conditions (Fig. 2c, d) [111]. Thioacidolysis indicated that H, G, and S units were strongly decreased in the *Mtcad* mutant. In *Arabidopsis*, the *Atcad4 Atcad5* double mutant exhibited a similar decrease in the level of total lignin and each monolignol as seen in the *Mtcad* mutant, but did not show a dwarf phenotype [121]. NMR analysis of lignin extracted from the *Mtcad* mutant showed that it is derived almost exclusively from coniferaldehyde and sinapaldehyde and is therefore strikingly different from traditional lignin. Therefore, in *M. truncatula* a single *CAD* gene appears to be sufficient for the bulk of G and S units synthesized.

3.3 Upstream Regulators of the Lignin Biosynthesis Pathway in *M. truncatula*

The lignin biosynthetic pathway is under tight and complex transcriptional control, including both positive and negative feedback/forward regulation [122–124]. Many MYB transcription factors (TFs) bind to the AC elements in promoters of lignin biosynthetic pathway genes and thereby regulate the coordinated expression of the pathway [123, 124]. MYB46 and MYB83 function as central regulators for biosynthesis of secondary cell wall polymers including lignin [123]. Secondary wall NAC (NAM, ATAT1/2, and CUC2) TFs, such as NST1, NST2, NST3/ANAC012/SND1, VND6, and VND7, are direct upstream regulators of the MYB46/MYB83 central switch [123–126].

The *Medicago NAC SECONDARY WALL THICKENING PROMOTING FACTOR 1* (*MtNST1*) likewise functions as an upstream positive regulator of the MYB46/MYB83 switch [112]. The *Mtnst1* mutant exhibits a semi-dwarf phenotype with a decrease in total lignin thioacidolysis yield and also shows secondary cell wall formation defects in stems, anther dehiscence defects, and reduced ferulic acid fluorescence in stomatal guard cell walls [112]. The mutant has greatly improved enzymatic saccharification efficiency of cell walls with or without acid pretreatment. These results highlight a major difference in lignin pathway regulation between *M. truncatula* and *Arabidopsis*; whereas different pairs of NAC transcription factors separately control lignification in the stems and anthers in *Arabidopsis*, a single TF (MtNST1) controls both in *M. truncatula*.

A WRKY TF functions as a negative regulator of *NST2* in *Arabidopsis* [33]. Loss of function of *MtWRKY* and *AtWRKY* genes results in secondary cell wall thickening in pith cells associated with ectopic deposition of lignin, xylan, and cellulose [33]. The *Mtwrky* mutant showed normal stature with increased thioacidolysis lignin yield only in the pith area of stem tissue, but not in total stem tissue. This pith area in the *Mtwrky* mutant displayed enhanced UV autofluorescence and phloroglucinol and Mäule staining, resulting in increased stem density.

4 Biosynthesis of Triterpene Saponins

Saponins are plant glycosides that derive their name from their soap-like properties and consist of an isoprenoidal-derived aglycone, designated genin or sapogenin, covalently linked to one or more sugar moieties [127, 128]. Plant-derived saponins are considered to be part of defense systems against pathogens and herbivores. Saponins also have potential health benefits for humans, and plants containing saponins have long been used in traditional medicine in East Asian culture.

Saponin biosynthesis starts with the condensation of acetyl-CoA in the cytosolic mevalonate pathway, ultimately leading to 2,3-oxidosqualene (Fig. 3) [127–129]. Linear precursors are cyclized by a variety of oxidosqualene cyclases (OSCs) to polycyclic structures, e.g., β-amyrin, which in higher plants marks the branch point between primary metabolism and specialized triterpene metabolism [128]. Then, these common precursors undergo various modifications prior to glycosylation. Most sapogenin modifications are believed to be introduced by enzymes belonging to the class of cytochromes P450 (CYP P450) that are membrane-associated hemoprotein monooxygenases [127, 128]. CYP P450 dependent-oxidations at different positions of the triterpene backbone increase the polarity of the scaffold by introducing reactive functional hydroxyl groups. Then, the triterpene backbone is subsequently tailored by an array of transferases, including UDP-dependent glycosyltransferases (UGTs) (Fig. 3), acyltransferases, malonyltransferases and methyltransferases, resulting in enormous structural and potentially functional diversity [128, 129].

Saponins which show hemolytic activity due to their affinities for membrane sterols and nonhemolytic saponins are synthesized through different pathways branching immediately downstream of β-amyrin formation [26, 130]. A *lacking hemolytic activity* (*lha*) mutant of *M. truncatula* was isolated by forward genetic screening of activation-tagged lines and TILLING analysis of EMS-treated *Medicago* seed pools [131]. The *lha* mutant has a defect in CYP716A12, the enzyme that catalyzes the first step of the hemolytic sapogenin pathway; as a result, the plants cannot synthesize

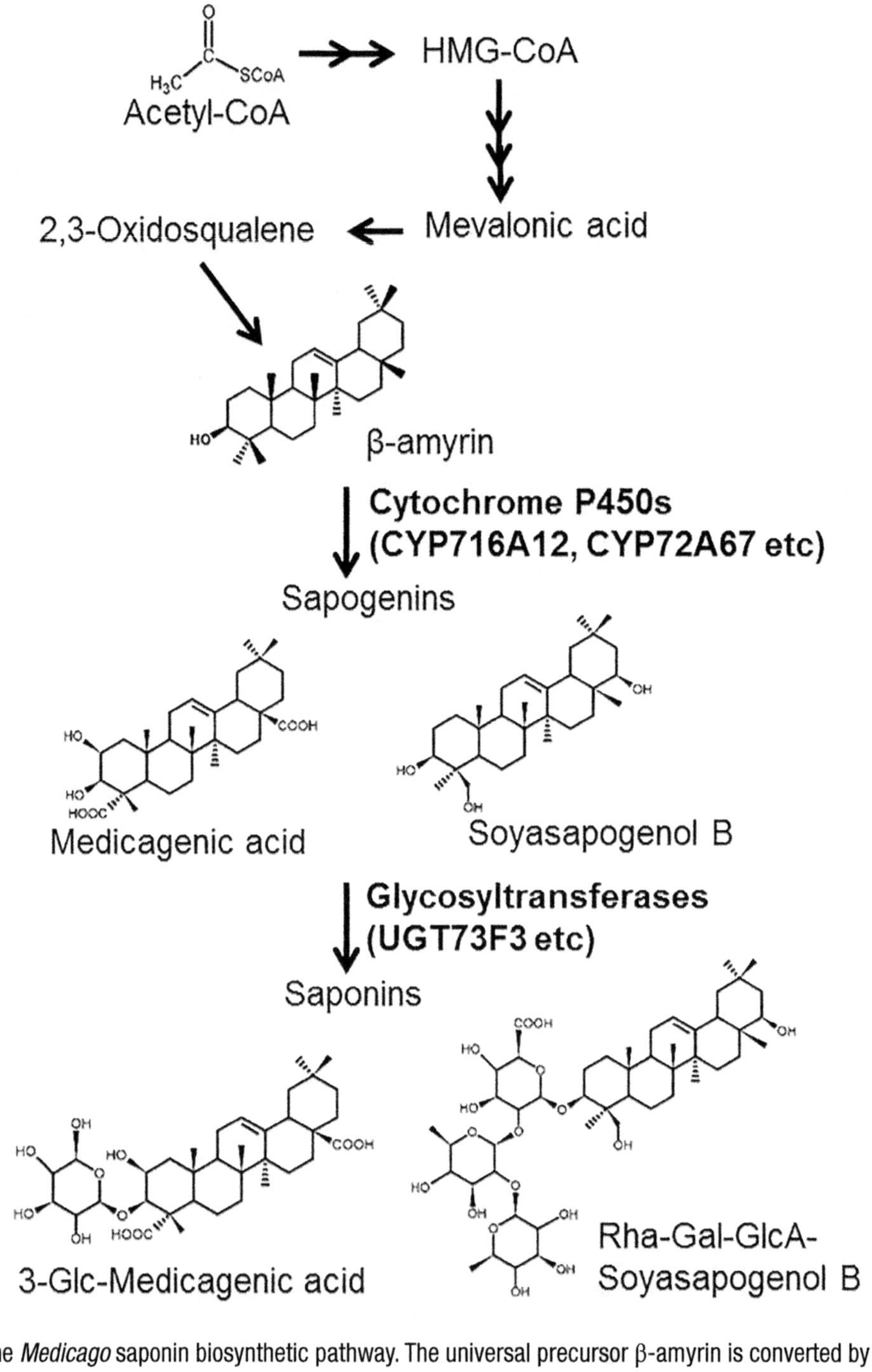

Fig. 3 The *Medicago* saponin biosynthetic pathway. The universal precursor β-amyrin is converted by a series of oxidative reactions to sapogenin aglycones, which are converted by glycosyltransferases to saponins

hemolytic sapogenins such as hederagenin, bayogenin, and medicagenic or zanhic acid, although synthesis of nonhemolytic soyasapogenol is not affected. The *lha* mutants were dwarf plants. *M. truncatula* mutants with loss of function of CYP72A67 that operates downstream of CYP716A12 in hemolytic saponin biosynthesis were not reported to be dwarf [26].

Comprehensive coexpression analysis of transcripts induced by methyl jasmonate, a key inducer of saponin production in *M. truncatula* cell cultures, identified target uridine diphosphate glycosyltransferases (UGTs) in the saponin biosynthesis pathway [132]. A transposon insertion mutant plant with loss of function of *UGT73F3* showed a severely dwarfed phenotype with defects in reproduction and root development, associated with decreased production of root-specific saponins [132].

Because unconstrained production of saponins leads to severe growth and developmental defects in plants [128], their biosynthesis should be controlled to ensure saponin production at appropriate levels. Plants have adopted a posttranslational negative control triggered by the ectopic accumulation of bioactive saponins. A mutation in the *Makibishi1* gene, which encodes an E3 ubiquitin ligase involved in controlling the activity of HMGR, the rate-limiting enzyme in the mevalonate pathway, led to accumulation of monoglycosylated saponins in malformed *M. truncatula* roots, implying toxic effects of less-tailored saponins on plant cells [133].

Overall, the identification and characterization of the entire set of CYP450s and tailoring enzymes involved in sapo(ge)nin biosynthesis would greatly aid our understanding of the biological roles of sapo(ge)nins in plant development. Furthermore, this would also provide a platform to manipulate synthesis of bioactive saponins in heterologous system, including microorganisms. *M. truncatula* is the organism of choice for such endeavors.

5 Perspectives

Studies of plant metabolic pathways have greatly expanded in the past decade, in part driven by multidisciplinary research using both classical research tools, i.e., molecular genetics, biochemical analysis, and structural biology, and more recently developed "omics" research tools, i.e., genomics, transcriptomics, metabolomics, and bioinformatics. *Arabidopsis* has been used as the prevalent model plant in developmental biology, genetics, genomics, and metabolomics. However, "supermodel" organisms do not represent all species [134]. As described above, loss of function of *M. truncatula* genes can result in different phenotypic, genetic, biochemical, and metabolic responses compared with *Arabidopsis*. Furthermore, results from *M. truncatula* can be more easily translated into its cousin alfalfa (*M. sativa*), which is the most widely used forage legume, than results from *Arabidopsis*, to improve animal performance, soil fertility, biofuel production, and development of eco-friendly nitrogen fertilization through nitrogen fixation. Comparison of the functions of structural and regulatory genes of secondary metabolism in *Arabidopsis*, *M. truncatula*, and other model species will accelerate the development of plants that produce energy-efficient biomass and are beneficial to human health.

References

1. O'Connor SE (2015) Engineering of secondary metabolism. Annu Rev Genet 49:71–94
2. Tsao R (2010) Chemistry and biochemistry of dietary polyphenols. Forum Nutr 2:1231–1246
3. Weng JK, Chapple C (2010) The origin and evolution of lignin biosynthesis. New Phytol 187:273–285
4. Chang W-C, Song H, Liu H-W, Liu P (2013) Current development in isoprenoid precursor biosynthesis and regulation. Curr Opin Chem Biol 17:571–579
5. Bennett RN, Wallsgrove RM (1994) Secondary metabolites in plant defence mechanisms. New Phytol 127:617–633
6. Rose RJ (2008) *Medicago truncatula* as a model for understanding plant interactions with other organisms, plant development and stress biology: past, present and future. Funct Plant Biol 35:253–264
7. Wang Y, Chen R (2013) Regulation of compound leaf development. Plants (Basel) 3:1–17
8. Benlloch R, Navarro C, Beltrán J, Cañas LA (2003) Floral development of the model legume *Medicago truncatula*: ontogeny studies as a tool to better characterize homeotic mutations. Sex Plant Reprod 15:231–241
9. Peng J, Chen R (2011) Auxin efflux transporter MtPIN10 regulates compound leaf and flower development in *Medicago truncatula*. Plant Signal Behav 6:1537–1544
10. Weller JL, Ortega R (2015) Genetic control of flowering time in legumes. Front Plant Sci 6:207
11. Harrison MJ, Dewbre GR, Liu J (2002) A phosphate transporter from *Medicago truncatula* involved in the acquisition of phosphate released by arbuscular mycorrhizal fungi. Plant Cell 14:2413–2429
12. Genre A, Chabaud M, Timmers T, Bonfante P, Barker DG (2005) Arbuscular mycorrhizal fungi elicit a novel intracellular apparatus in *Medicago truncatula* root epidermal cells before infection. Plant Cell 17:3489–3499
13. Guo S, Kamphuis LG, Gao L, Edwards OR, Singh KB (2009) Two independent resistance genes in the *Medicago truncatula* cultivar jester confer resistance to two different aphid species of the genus *Acyrthosiphon*. Plant Signal Behav 4:328–331
14. Chen T, Duan L, Zhou B, Yu H, Zhu H, Cao Y, Zhang Z (2017) Interplay of pathogen-induced defense responses and symbiotic establishment in *Medicago truncatula*. Front Microbiol 8:973
15. Badri M, Chardon F, Huguet T, Aouani ME (2011) Quantitative trait loci associated with drought tolerance in the model legume *Medicago truncatula*. Euphytica 181:415
16. Gil-Quintana E, Lyon D, Staudinger C, Wienkoop S, González EM (2015) *Medicago truncatula* and *Glycine max*: different drought tolerance and similar local response of the root nodule proteome. J Proteome Res 14:5240–5251
17. Liu L, Zhang Z, Dong J, Wang T (2016) Overexpression of MtWRKY76 increases both salt and drought tolerance in *Medicago truncatula*. Environ Exp Bot 123:50–58
18. Dixon RA, Steele CL (1999) Flavonoids and isoflavonoids – a gold mine for metabolic engineering. Trends Plant Sci 4:394–400
19. Tanaka H, Sato M, Fujiwara S, Hirata M, Etoh H, Takeuchi H (2002) Antibacterial activity of isoflavonoids isolated from *Erythrina variegata* against methicillin-resistant *Staphylococcus aureus*. Lett Appl Microbiol 35:494–498
20. Mukne AP, Viswanathan V, Phadatare AG (2011) Structure pre-requisites for isoflavones as effective antibacterial agents. Pharmacogn Rev 5:13–18
21. Kosslak RM, Bookland R, Barkei J, Paaren HE, Appelbaum ER (1987) Induction of *Bradyrhizobium japonicum* common nod genes by isoflavones isolated from *Glycine max*. Proc Natl Acad Sci U S A 84:7428–7432
22. Pueppke SG, Bolaños-Vásquez MC, Werner D, Bec-Ferté M-P, Promé J-C, Krishnan HB (1998) Release of flavonoids by the soybean cultivars McCall and Peking and their perception as signals by the nitrogen-fixing symbiont *Sinorhizobium fredii*. Plant Physiol 117:599–606
23. Dixon RA (2004) Phytoestrogens. Annu Rev Plant Biol 55:225–261
24. Huhman DV, Sumner LW (2002) Metabolic profiling of saponins in *Medicago sativa* and *Medicago truncatula* using HPLC coupled to an electrospray ion-trap mass spectrometer. Phytochemistry 59:347–360
25. Tava A, Scotti C, Avato P (2011) Biosynthesis of saponins in the genus *Medicago*. Phytochem Rev 10:459–469
26. Biazzi E, Carelli M, Tava A, Abbruscato P, Losini I, Avato P, Scotti C, Calderini O (2015) CYP72A67 catalyzes a key oxidative step in *Medicago truncatula* hemolytic saponin biosynthesis. Mol Plant 8:1493–1506
27. Mertens J, Pollier J, Vanden Bossche R, Lopez-Vidriero I, Franco-Zorrilla JM, Goossens A (2016) The bHLH transcription factors TSAR1 and TSAR2 regulate triterpene saponin biosynthesis in *Medicago truncatula*. Plant Physiol 170:194–210

28. Reddy MSS, Chen F, Shadle G, Jackson L, Aljoe H, Dixon RA (2005) Targeted down-regulation of cytochrome P450 enzymes for forage quality improvement in alfalfa (*Medicago sativa* L.). Proc Natl Acad Sci U S A 102:16573–16578
29. McMahon LR, McAllister T, Berg BP, Majak W, Acharya SN, Popp JD, Coulman BE, Wang Y, Cheng KJ (2000) A review of the effect of forage condensed tannins on ruminal fermentation and bloat in grazing cattle. Can J Plant Sci 80:469–485
30. Pang Y, Wenger JP, Saathoff K, Peel GJ, Wen J, Huhman D, Allen SN, Tang Y, Cheng X, Tadege M, Ratet P, Mysore KS, Sumner LW, Marks MD, Dixon RA (2009) A WD40 repeat protein from *Medicago truncatula* is necessary for tissue-specific anthocyanin and proanthocyanidin biosynthesis but not for trichome development. Plant Physiol 151:1114–1129
31. Liu C, Jun JH, Dixon RA (2014) MYB5 and MYB14 play pivotal roles in seed coat polymer biosynthesis in *Medicago truncatula*. Plant Physiol 165:1424–1439
32. Jun JH, Liu C, Xiao X, Dixon RA (2015) The transcriptional repressor MYB2 regulates both spatial and temporal patterns of proanthocyandin and anthocyanin pigmentation in *Medicago truncatula*. Plant Cell 27:2860–2879
33. Wang H, Avci U, Nakashima J, Hahn MG, Chen F, Dixon RA (2010) Mutation of WRKY transcription factors initiates pith secondary wall formation and increases stem biomass in dicotyledonous plants. Proc Natl Acad Sci U S A 107:22338–22343
34. Zhao Q, Wang H, Yin Y, Xu Y, Chen F, Dixon RA (2010) Syringyl lignin biosynthesis is directly regulated by a secondary cell wall master switch. Proc Natl Acad Sci U S A 107:14496–14501
35. Zhou R, Jackson L, Shadle G, Nakashima J, Temple S, Chen F, Dixon RA (2010) Distinct cinnamoyl CoA reductases involved in parallel routes to lignin in *Medicago truncatula*. Proc Natl Acad Sci U S A 107:17803–17808
36. Ha CM, Escamilla-Trevino L, Yarce JC, Kim H, Ralph J, Chen F, Dixon RA (2016) An essential role of caffeoyl shikimate esterase in monolignol biosynthesis in *Medicago truncatula*. Plant J 86:363–375
37. Young ND, Debelle F, Oldroyd GED, Geurts R, Cannon SB, Udvardi MK, Benedito VA, Mayer KFX, Gouzy J, Schoof H, Van de Peer Y, Proost S, Cook DR, Meyers BC, Spannagl M, Cheung F, De Mita S, Krishnakumar V, Gundlach H, Zhou S, Mudge J, Bharti AK, Murray JD, Naoumkina MA, Rosen B, Silverstein KAT, Tang H, Rombauts S, Zhao PX, Zhou P, Barbe V, Bardou P, Bechner M, Bellec A, Berger A, Berges H, Bidwell S, Bisseling T, Choisne N, Couloux A, Denny R, Deshpande S, Dai X, Doyle JJ, Dudez A-M, Farmer AD, Fouteau S, Franken C, Gibelin C, Gish J, Goldstein S, Gonzalez AJ, Green PJ, Hallab A, Hartog M, Hua A, Humphray SJ, Jeong D-H, Jing Y, Jocker A, Kenton SM, Kim D-J, Klee K, Lai H, Lang C, Lin S, Macmil SL, Magdelenat G, Matthews L, McCorrison J, Monaghan EL, Mun J-H, Najar FZ, Nicholson C, Noirot C, O'Bleness M, Paule CR, Poulain J, Prion F, Qin B, Qu C, Retzel EF, Riddle C, Sallet E, Samain S, Samson N, Sanders I, Saurat O, Scarpelli C, Schiex T, Segurens B, Severin AJ, Sherrier DJ, Shi R, Sims S, Singer SR, Sinharoy S, Sterck L, Viollet A, Wang B-B, Wang K, Wang M, Wang X, Warfsmann J, Weissenbach J, White DD, White JD, Wiley GB, Wincker P, Xing Y, Yang L, Yao Z, Ying F, Zhai J, Zhou L, Zuber A, Denarie J, Dixon RA, May GD, Schwartz DC, Rogers J, Quetier F, Town CD, Roe BA (2011) The *Medicago* genome provides insight into the evolution of rhizobial symbioses. Nature 480:520–524
38. Penmetsa RV, Cook DR (2000) Production and characterization of diverse developmental mutants of *Medicago truncatula*. Plant Physiol 123:1387–1398
39. Rogers C, Wen J, Chen R, Oldroyd G (2009) Deletion-based reverse genetics in *Medicago truncatula*. Plant Physiol 151:1077–1086
40. Le Signor C, Savois V, Aubert G, Verdier J, Nicolas M, Pagny G, Moussy F, Sanchez M, Baker D, Clarke J, Thompson R (2009) Optimizing TILLING populations for reverse genetics in *Medicago truncatula*. Plant Biotechnol J 7:430–441
41. Carelli M, Calderini O, Panara F, Porceddu A, Losini I, Piffanelli P, Arcioni S, Scotti C (2013) Reverse genetics in *Medicago truncatula* using a TILLING mutant collection. Methods Mol Biol 1069:101–118
42. Cheng X, Wen J, Tadege M, Ratet P, Mysore KS (2011) Reverse genetics in *Medicago truncatula* using *Tnt1* insertion mutants. Methods Mol Biol 678:179–190
43. Tadege M, Wen J, He J, Tu H, Kwak Y, Eschstruth A, Cayrel A, Endre G, Zhao PX, Chabaud M, Ratet P, Mysore KS (2008) Large-scale insertional mutagenesis using the *Tnt1* retrotransposon in the model legume *Medicago truncatula*. Plant J 54:335–347
44. Veerappan V, Jani M, Kadel K, Troiani T, Gale R, Mayes T, Shulaev E, Wen J, Mysore KS, Azad RK, Dickstein R (2016) Rapid identification of causative insertions underlying *Medicago truncatula Tnt1* mutants defective in symbiotic nitrogen fixation from a forward genetic screen by whole genome sequencing. BMC Genomics 17:141

45. Iwashina T (2000) The structure and distribution of the flavonoids in plants. J Plant Res 113:287–299
46. Kumar S, Pandey AK (2013) Chemistry and biological activities of flavonoids: an overview. Sci World J 2013:162750
47. Wasson AP, Pellerone FI, Mathesius U (2006) Silencing the flavonoid pathway in *Medicago truncatula* inhibits root nodule formation and prevents auxin transport. Regulation by rhizobia. Plant Cell 18:1617–1629
48. Kowalska I, Stochmal A, Kapusta I, Janda B, Pizza C, Piacente S, Oleszek W (2007) Flavonoids from barrel medic (*Medicago truncatula*) aerial parts. J Agric Food Chem 55:2645–2652
49. Pang Y, Peel GJ, Wright E, Wang Z, Dixon RA (2007) Early steps in proanthocyanidin biosynthesis in the model legume *Medicago truncatula*. Plant Physiol 145:601–615
50. Jasiński M, Kachlicki P, Rodziewicz P, Figlerowicz M, Stobiecki M (2009) Changes in the profile of flavonoid accumulation in *Medicago truncatula* leaves during infection with fungal pathogen *Phoma medicaginis*. Plant Physiol Biochem 47:847–853
51. Ng JL, Hassan S, Truong TT, Hocart CH, Laffont C, Frugier F, Mathesius U (2015) Flavonoids and auxin transport inhibitors rescue symbiotic nodulation in the *Medicago truncatula* cytokinin perception mutant *cre1*. Plant Cell 27:2210–2226
52. Le Roy J, Huss B, Creach A, Hawkins S, Neutelings G (2016) Glycosylation is a major regulator of phenylpropanoid availability and biological activity in plants. Front Plant Sci 7:735
53. Tanner GJ, Francki KT, Abrahams S, Watson JM, Larkin PJ, Ashton AR (2003) Proanthocyanidin biosynthesis in plants. Purification of legume leucoanthocyanidin reductase and molecular cloning of its cDNA. J Biol Chem 278:31647–31656
54. Bogs J, Downey MO, Harvey JS, Ashton AR, Tanner GJ, Robinson SP (2005) Proanthocyanidin synthesis and expression of genes encoding leucoanthocyanidin reductase and anthocyanidin reductase in developing grape berries and grapevine leaves. Plant Physiol 139:652–663
55. Liu Y, Shi Z, Maximova S, Payne MJ, Guiltinan MJ (2013) Proanthocyanidin synthesis in *Theobroma cacao*: genes encoding anthocyanidin synthase, anthocyanidin reductase, and leucoanthocyanidin reductase. BMC Plant Biol 13:202–202
56. Ferraro K, Jin AL, Nguyen T-D, Reinecke DM, Ozga JA, Ro D-K (2014) Characterization of proanthocyanidin metabolism in pea (*Pisum sativum*) seeds. BMC Plant Biol 14:238
57. Liu C, Wang X, Shulaev V, Dixon RA (2016) A role for leucoanthocyanidin reductase in the extension of proanthocyanidins. Nat Plants 2:16182
58. Dixon RA, Xie DY, Sharma SB (2005) Proanthocyanidins--a final frontier in flavonoid research? New Phytol 165:9–28
59. Modolo LV, Blount JW, Achnine L, Naoumkina MA, Wang X, Dixon RA (2007) A functional genomics approach to (iso)flavonoid glycosylation in the model legume *Medicago truncatula*. Plant Mol Biol 64:499–518
60. Peel GJ, Pang Y, Modolo LV, Dixon RA (2009) The LAP1 MYB transcription factor orchestrates anthocyanidin biosynthesis and glycosylation in *Medicago*. Plant J 59:136–149
61. Pang Y, Peel GJ, Sharma SB, Tang Y, Dixon RA (2008) A transcript profiling approach reveals an epicatechin-specific glucosyltransferase expressed in the seed coat of *Medicago truncatula*. Proc Natl Acad Sci U S A 105:14210–14215
62. Pang Y, Cheng X, Huhman DV, Ma J, Peel GJ, Yonekura-Sakakibara K, Saito K, Shen G, Sumner LW, Tang Y, Wen J, Yun J, Dixon RA (2013) *Medicago* glucosyltransferase UGT72L1: potential roles in proanthocyanidin biosynthesis. Planta 238:139–154
63. Zhao J, Dixon RA (2009) MATE transporters facilitate vacuolar uptake of epicatechin 3′-O-glucoside for proanthocyanidin biosynthesis in *Medicago truncatula* and Arabidopsis. Plant Cell 21:2323–2340
64. Marinova K, Pourcel L, Weder B, Schwarz M, Barron D, Routaboul JM, Debeaujon I, Klein M (2007) The Arabidopsis MATE transporter TT12 acts as a vacuolar flavonoid/H+ -antiporter active in proanthocyanidin-accumulating cells of the seed coat. Plant Cell 19:2023–2038
65. Zhao J, Huhman D, Shadle G, He X-Z, Sumner LW, Tang Y, Dixon RA (2011) MATE2 mediates vacuolar sequestration of flavonoid glycosides and glycoside malonates in *Medicago truncatula*. Plant Cell 23:1536–1555
66. Stafford HA, Lester HH (1982) Enzymic and nonenzymic reduction of (+)-dihydroquercetin to its 3,4,-diol. Plant Physiol 70:695–698
67. Baudry A, Heim MA, Dubreucq B, Caboche M, Weisshaar B, Lepiniec L (2004) TT2, TT8, and TTG1 synergistically specify the expression of BANYULS and proanthocyanidin biosynthesis in *Arabidopsis thaliana*. Plant J 39:366–380
68. Lloyd A, Brockman A, Aguirre L, Campbell A, Bean A, Cantero A, Gonzalez A (2017) Advances in the MYB-BHLH-WD repeat

(MBW) pigment regulatory model: addition of a WRKY factor and co-option of an anthocyanin MYB for betalain regulation. Plant Cell Physiol 58:1431–1441
69. Walker AR, Davison PA, Bolognesi-Winfield AC, James CM, Srinivasan N, Blundell TL, Esch JJ, Marks MD, Gray JC (1999) The TRANSPARENT TESTA GLABRA1 locus, which regulates trichome differentiation and anthocyanin biosynthesis in Arabidopsis, encodes a WD40 repeat protein. Plant Cell 11:1337–1350
70. Broun P (2005) Transcriptional control of flavonoid biosynthesis: a complex network of conserved regulators involved in multiple aspects of differentiation in Arabidopsis. Curr Opin Plant Biol 8:272–279
71. Nesi N, Debeaujon I, Jond C, Pelletier G, Caboche M, Lepiniec L (2000) The TT8 gene encodes a basic helix-loop-helix domain protein required for expression of DFR and BAN genes in Arabidopsis siliques. Plant Cell 12:1863–1878
72. Payne CT, Zhang F, Lloyd AM (2000) GL3 encodes a bHLH protein that regulates trichome development in arabidopsis through interaction with GL1 and TTG1. Genetics 156:1349–1362
73. Morohashi K, Zhao M, Yang M, Read B, Lloyd A, Lamb R, Grotewold E (2007) Participation of the Arabidopsis bHLH factor GL3 in trichome initiation regulatory events. Plant Physiol 145:736–746
74. Zhang F, Gonzalez A, Zhao M, Payne CT, Lloyd A (2003) A network of redundant bHLH proteins functions in all TTG1-dependent pathways of Arabidopsis. Development 130:4859–4869
75. Li P, Chen B, Zhang G, Chen L, Dong Q, Wen J, Mysore KS, Zhao J (2016) Regulation of anthocyanin and proanthocyanidin biosynthesis by *Medicago truncatula* bHLH transcription factor MtTT8. New Phytol 210:905–921
76. Xu W, Dubos C, Lepiniec L (2015) Transcriptional control of flavonoid biosynthesis by MYB-bHLH-WDR complexes. Trends Plant Sci 20:176–185
77. Gonzalez A, Mendenhall J, Huo Y, Lloyd A (2009) TTG1 complex MYBs, MYB5 and TT2, control outer seed coat differentiation. Dev Biol 325:412–421
78. Nesi N, Jond C, Debeaujon I, Caboche M, Lepiniec L (2001) The arabidopsis TT2 gene encodes an R2R3 MYB domain protein that acts as a key determinant for proanthocyanidin accumulation in developing seed. Plant Cell 13:2099–2114
79. Li SF, Milliken ON, Pham H, Seyit R, Napoli R, Preston J, Koltunow AM, Parish RW (2009) The Arabidopsis MYB5 transcription factor regulates mucilage synthesis, seed coat development, and trichome morphogenesis. Plant Cell 21:72–89
80. Verdier J, Zhao J, Torres-Jerez I, Ge S, Liu C, He X, Mysore KS, Dixon RA, Udvardi MK (2012) MtPAR MYB transcription factor acts as an on switch for proanthocyanidin biosynthesis in *Medicago truncatula*. Proc Natl Acad Sci U S A 109:1766–1771
81. Aharoni A, De Vos CH, Wein M, Sun Z, Greco R, Kroon A, Mol JN, O'Connell AP (2001) The strawberry FaMYB1 transcription factor suppresses anthocyanin and flavonol accumulation in transgenic tobacco. Plant J 28:319–332
82. Paolocci F, Robbins MP, Passeri V, Hauck B, Morris P, Rubini A, Arcioni S, Damiani F (2011) The strawberry transcription factor FaMYB1 inhibits the biosynthesis of proanthocyanidins in *Lotus corniculatus* leaves. J Exp Bot 62:1189–1200
83. Albert NW, Davies KM, Lewis DH, Zhang H, Montefiori M, Brendolise C, Boase MR, Ngo H, Jameson PE, Schwinn KE (2014) A conserved network of transcriptional activators and repressors regulates anthocyanin pigmentation in eudicots. Plant Cell 26:962–980
84. Yoshida K, Ma D, Constabel CP (2015) The MYB182 protein down-regulates proanthocyanidin and anthocyanin biosynthesis in poplar by repressing both structural and regulatory flavonoid genes. Plant Physiol 167:693–710
85. Zhu H-F, Fitzsimmons K, Khandelwal A, Kranz RG (2009) CPC, a single-repeat R3 MYB, is a negative regulator of anthocyanin biosynthesis in Arabidopsis. Mol Plant 2:790–802
86. Bhalla A, Bansal N, Kumar S, Bischoff KM, Sani RK (2013) Improved lignocellulose conversion to biofuels with thermophilic bacteria and thermostable enzymes. Bioresour Technol 128:751–759
87. Boerjan W, Ralph J, Baucher M (2003) Lignin biosynthesis. Annu Rev Plant Biol 54:519–546
88. Sun Y, Cheng J (2002) Hydrolysis of lignocellulosic materials for ethanol production: a review. Bioresour Technol 83:1–11
89. Vanholme R, Demedts B, Morreel K, Ralph J, Boerjan W (2010) Lignin biosynthesis and structure. Plant Physiol 153:895–905
90. Guo D, Chen F, Wheeler J, Winder J, Selman S, Peterson M, Dixon RA (2001) Improvement of in-rumen digestibility of alfalfa forage by genetic manipulation of lignin O-methyltransferases. Transgenic Res 10:457–464
91. Chen F, Dixon RA (2007) Lignin modification improves fermentable sugar yields for biofuel production. Nat Biotechnol 25:759–761

92. Fu C, Mielenz JR, Xiao X, Ge Y, Hamilton CY, Rodriguez M Jr, Chen F, Foston M, Ragauskas A, Bouton J, Dixon RA, Wang ZY (2011) Genetic manipulation of lignin reduces recalcitrance and improves ethanol production from switchgrass. Proc Natl Acad Sci U S A 108:3803–3808
93. Li M, Pu YQ, Ragauskas AJ (2016) Current understanding of the correlation of lignin structure with biomass recalcitrance. Front Chem 4:45
94. Gall DL, Ralph J, Donohue TJ, Noguera DR (2017) Biochemical transformation of lignin for deriving valued commodities from lignocellulose. Curr Opin Biotechnol 45:120–126
95. Fritz JO, Cantrell RP, Lechtenberg VL, Axtell JD, Hertel JM (1981) Brown midrib mutants in sudangrass and grain sorghum. Crop Sci 21:706–709
96. Cherney JH, Moore KJ, Volenec JJ, Axtell JD (1986) Rate and extent of digestion of cell wall components of brown-midrib sorghum species. Crop Sci 26:1055–1059
97. Getachew G, Ibanez AM, Pittroff W, Dandekar AM, McCaslin M, Goyal S, Reisen P, DePeters EJ, Putnam DH (2011) A comparative study between lignin down regulated alfalfa lines and their respective unmodified controls on the nutritional characteristics of hay. Anim Feed Sci Technol 170:192–200
98. Tong ZY, Li H, Zhang RX, Ma L, Dong JL, Wang T (2015) Co-downregulation of the hydroxycinnamoyl-CoA:shikimate hydroxycinnamoyl transferase and coumarate 3-hydroxylase significantly increases cellulose content in transgenic alfalfa (*Medicago sativa* L.). Plant Sci 239:230–237
99. Hisano H, Nandakumar R, Wang ZY (2009) Genetic modification of lignin biosynthesis for improved biofuel production. In Vitro Cell Dev Biol-Plant 45:306–313
100. Vanholme R, Morreel K, Darrah C, Oyarce P, Grabber JH, Ralph J, Boerjan W (2012) Metabolic engineering of novel lignin in biomass crops. New Phytol 196:978–1000
101. Eudes A, Liang Y, Mitra P, Loque D (2014) Lignin bioengineering. Curr Opin Biotechnol 26:189–198
102. Meng XZ, Ragauskas AJ (2014) Recent advances in understanding the role of cellulose accessibility in enzymatic hydrolysis of lignocellulosic substrates. Curr Opin Biotechnol 27:150–158
103. Wang Y, Fan C, Hu H, Li Y, Sun D, Wang Y, Peng L (2016) Genetic modification of plant cell walls to enhance biomass yield and biofuel production in bioenergy crops. Biotechnol Adv 34:997–1017
104. Lei L (2017) Lignin evolution: invasion of land. Nat Plants 3:17042
105. Barros J, Serrani-Yarce JC, Chen F, Baxter D, Venables BJ, Dixon RA (2016) Role of bifunctional ammonia-lyase in grass cell wall biosynthesis. Nat Plants 2:16050
106. Bonawitz ND, Chapple C (2010) The genetics of lignin biosynthesis: connecting genotype to phenotype. Annu Rev Genet 44:337–363
107. Vanholme R, Cesarino I, Rataj K, Xiao Y, Sundin L, Goeminne G, Kim H, Cross J, Morreel K, Araujo P, Welsh L, Haustraete J, McClellan C, Vanholme B, Ralph J, Simpson GG, Halpin C, Boerjan W (2013) Caffeoyl shikimate esterase (CSE) is an enzyme in the lignin biosynthetic pathway in Arabidopsis. Science 341:1103–1106
108. Zhou CE, Han L, Pislariu C, Nakashima J, Fu CX, Jiang QZ, Quan L, Blancaflor EB, Tang YH, Bouton JH, Udvardi M, Xia GM, Wang ZY (2011) From model to crop: functional analysis of a stay-green gene in the model legume *Medicago truncatula* and effective use of the gene for alfalfa improvement. Plant Physiol 157:1483–1496
109. Committee on Genetically Engineered Crops: Experiences and Prospects (2016) Genetically engineered crops: experiences and prospects. National Academies Press, Washington, DC
110. Vallet C, Chabbert B, Czaninski Y, Monties B (1996) Histochemistry of lignin deposition during sclerenchyma differentiation in alfalfa stems. Ann Bot 78:625–632
111. Zhao Q, Tobimatsu Y, Zhou R, Pattathil S, Gallego-Giraldo L, Fu C, Jackson LA, Hahn MG, Kim H, Chen F, Ralph J, Dixon RA (2013) Loss of function of cinnamyl alcohol dehydrogenase 1 leads to unconventional lignin and a temperature-sensitive growth defect in *Medicago truncatula*. Proc Natl Acad Sci U S A 110:13660–13665
112. Zhao Q, Gallego-Giraldo L, Wang H, Zeng Y, Ding SY, Chen F, Dixon RA (2010) An NAC transcription factor orchestrates multiple features of cell wall development in *Medicago truncatula*. Plant J 63:100–114
113. Fagerstedt KV, Saranpaa P, Tapanila T, Immanen J, Serra JA, Nieminen K (2015) Determining the composition of lignins in different tissues of silver birch. Plants (Basel) 4:183–195
114. Harman-Ware AE, Foster C, Happs RM, Doeppke C, Meunier K, Gehan J, Yue FX, Lu FC, Davis MF (2016) A thioacidolysis method tailored for higher-throughput quantitative analysis of lignin monomers. Biotechnol J 11:1268–1273
115. Hatfield R, Fukushima RS (2005) Can lignin be accurately measured? Crop Sci 45:832–839
116. Moreira-Vilar FC, Siqueira-Soares RD, Finger-Teixeira A, de Oliveira DM, Ferro AP, da Rocha GJ, Ferrarese MDL, dos Santos

WD, Ferrarese O (2014) The acetyl bromide method is faster, simpler and presents best recovery of lignin in different herbaceous tissues than klason and thioglycolic acid methods. PLoS One 9(10):e110000. https://doi.org/10.1371/journal.pone.0110000
117. Nakano J, Meshitsuka G (1992) The detection of lignin. In: Lin SY, Dence CW (eds) Methods in lignin chemistry. Springer Berlin Heidelberg, Berlin, Heidelberg, pp 23–32
118. Tobimatsu Y, Chen F, Nakashima J, Escamilla-Trevino LL, Jackson L, Dixon RA, Ralph J (2013) Coexistence but independent biosynthesis of catechyl and guaiacyl/syringyl lignin polymers in seed coats. Plant Cell 25:2587–2600
119. Lauvergeat V, Lacomme C, Lacombe E, Lasserre E, Roby D, Grima-Pettenati J (2001) Two cinnamoyl-CoA reductase (CCR) genes from Arabidopsis thaliana are differentially expressed during development and in response to infection with pathogenic bacteria. Phytochemistry 57:1187–1195
120. Mir Derikvand M, Sierra JB, Ruel K, Pollet B, Do CT, Thevenin J, Buffard D, Jouanin L, Lapierre C (2008) Redirection of the phenylpropanoid pathway to feruloyl malate in Arabidopsis mutants deficient for cinnamoyl-CoA reductase 1. Planta 227:943–956
121. Sibout R, Eudes A, Mouille G, Pollet B, Lapierre C, Jouanin L, Seguin A (2005) CINNAMYL ALCOHOL DEHYDROGENASE-C and -D are the primary genes involved in lignin biosynthesis in the floral stem of Arabidopsis. Plant Cell 17:2059–2076
122. Zhong R, Ye Z-H (2009) Transcriptional regulation of lignin biosynthesis. Plant Signal Behav 4:1028–1034
123. Ko JH, Jeon HW, Kim WC, Kim JY, Han KH (2014) The MYB46/MYB83-mediated transcriptional regulatory programme is a gatekeeper of secondary wall biosynthesis. Ann Bot 114:1099–1107
124. Nakano Y, Yamaguchiz M, Endo H, Rejab NA, Ohtani M (2015) NAC-MYB-based transcriptional regulation of secondary cell wall biosynthesis in land plants. Front Plant Sci 6:288
125. Ohashi-Ito K, Oda Y, Fukuda H (2010) Arabidopsis vascular-related NAC-DOMAIN6 directly regulates the genes that govern programmed cell death and secondary wall formation during xylem differentiation. Plant Cell 22:3461–3473
126. Yamaguchi M, Mitsuda N, Ohtani M, Ohme-Takagi M, Kato K, Demura T (2011) VASCULAR-RELATED NAC-DOMAIN7 directly regulates the expression of a broad range of genes for xylem vessel formation. Plant J 66:579–590
127. Augustin JM, Kuzina V, Andersen SB, Bak S (2011) Molecular activities, biosynthesis and evolution of triterpenoid saponins. Phytochemistry 72:435–457
128. Moses T, Papadopoulou KK, Osbourn A (2014) Metabolic and functional diversity of saponins, biosynthetic intermediates and semi-synthetic derivatives. Crit Rev Biochem Mol Biol 49:439–462
129. Thimmappa R, Geisler K, Louveau T, O'Maille P, Osbourn A (2014) Triterpene biosynthesis in plants. Annu Rev Plant Biol 65:225–257
130. Tava A, Avato P (2006) Chemical and biological activity of triterpene saponins from *Medicago* species. Nat Prod Commun 1:1159–1180
131. Carelli M, Biazzi E, Panara F, Tava A, Scaramelli L, Porceddu A, Graham N, Odoardi M, Piano E, Arcioni S, May S, Scotti C, Calderini O (2011) *Medicago truncatula* CYP716A12 is a multifunctional oxidase involved in the biosynthesis of hemolytic saponins. Plant Cell 23:3070–3081
132. Naoumkina MA, Modolo LV, Huhman DV, Urbanczyk-Wochniak E, Tang YH, Sumner LW, Dixon RA (2010) Genomic and coexpression analyses predict multiple genes involved in triterpene saponin biosynthesis in *Medicago truncatula*. Plant Cell 22:850–866
133. Pollier J, Moses T, Gonzalez-Guzman M, De Geyter N, Lippens S, Vanden Bossche R, Marhavy P, Kremer A, Morreel K, Guerin CJ, Tava A, Oleszek W, Thevelein JM, Campos N, Goormachtig S, Goossens A (2013) The protein quality control system manages plant defence compound synthesis. Nature 504:148–152
134. Alfred J, Baldwin IT (2015) New opportunities at the wild frontier. elife 4:e06956

Index

A

B

C

Luis A. Cañas and José Pío Beltrán (eds.), *Functional genomics in* Medicago truncatula*: Methods and Protocols*, Methods in Molecular Biology, vol. 1822, https://doi.org/10.1007/978-1-4939-8633-0,

M

N

O

P

R

S

T

U

V

W

Z

Printed by Printforce, the Netherlands